GENETICS - RESEARCH AND ISSUES

GENE-ENVIRONMENT INTERACTIONS AND HUMAN DISEASES

GENETICS - RESEARCH AND ISSUES

Additional books in this series can be found on Nova's website under the Series tab.

Additional e-books in this series can be found on Nova's website under the e-book tab.

GENETICS - RESEARCH AND ISSUES

GENE-ENVIRONMENT INTERACTIONS AND HUMAN DISEASES

LU QI
EDITOR

nova publishers
New York

Library of Congress Cataloging-in-Publication Data

ISBN: 978-1-63483-680-7
Library of Congress Control Number: 2015948673

Published by Nova Science Publishers, Inc. † New York

CONTENTS

PREFACE

Non-communicable, chronic diseases such as obesity, diabetes, cardiovascular diseases, cancer, respiratory and inflammatory diseases have been rapidly increasing in the past few decades and are by far the leading causes of mortality in the world. Epidemiological studies have related an array of environmental factors—especially diet habits and lifestyle—with these disorders. Recently, genetic research has also successfully identified thousands of variations in the human genomic sequence underlying susceptibility to these chronic diseases. However, the "nature/nurture" debate has confounded research on chronic diseases for a long time and how environmental factors and genetic factors jointly affect disease risk remains unresolved. This book is written for individuals who seek to understand the interplays between human genome and environmental factors and their relative contributions in determining human health, particularly the risk of chronic diseases.

The concept of risk factors was first introduced in the Framingham Heart Study for prediction of cardiovascular disease (CVD) in 1961. Since then, identifying risk factors for chronic diseases has become one of the primary aims in epidemiological studies. The majority of the established risk factors, which usually explain a large proportion of the events, are related to environmental factors like dietary intakes and lifestyle. A long-standing view is that the environmental factors may play a dominant role in determining disease risk. However, the recent findings in genetics have clearly indicated genetic components in almost all the chronic diseases. In addition, genetic factors may play an important role in affecting the pattern of disease epidemic through interaction with environmental risk factors. Mounting data are now lending support to the notion of gene-environment interaction, the manner which these two types of risk factors play together in affecting disease risk. As Dr. Francis Collins once said, "Genes load the gun; environment pulls the trigger."

The book aims to provide a comprehensive overview on the advances in the field of gene-environment interaction on multifactorial, chronic diseases. Chapter 1 is an introduction to the field. In this chapter, L. Qi provides an outline of the discovery of environmental and genetic risk factors, and discusses the concept of gene-environment interaction. New areas including epigenome, metagenome, and their interactions with environment, as well as personalized medicine and prevention are also touched upon in this chapter. In Chapter 2, G. Gibson and A. Nabb consider the collective contributions of culture, genetics, and genotype-by-environment interactions to human health and fitness in the context of public health; they particularly highlight various scenarios in which genotype-by-environment interactions could be statistically detected. Chapter 3 by J. Biernacka j. Usset, B. Fridley and S. Winham delivers an overview of study designs and analytical approaches for investigating gene-environment interaction and potential sources of biases. The chapter particularly addresses powerful, statistical approaches suitable for genome-wide analyses, including two-step filtering approaches, data-mining methods, and gene-level and pathway-level analyses.

The advances in gene-environment interactions on some common risk factors for chronic diseases are presented in two chapters. In Chapter 4, L. Qi deals with obesity, which is one of the most shared common risk factors for a variety of chronic diseases. He provides a comprehensive summary on research related to the obesogenic environment, particularly diet and lifestyle, genetics, and their interactions on body weight and obesity risk in observational studies and randomized intervention trials. In Chapters 5, D. Corella focuses on another commonly shared risk factor for many chronic diseases – lipids. The chapter goes through in detail about lipid metabolism and research on genetic and environmental factors closely related to lipids in humans. She comprehensively summarizes studies addressing genetic interactions with several lifestyle factors such as smoking, physical activity, and particularly highlights the importance of diet in its interaction with genetic factors on lipids.

The next 6 chapters encapsulate current literature of studies on a wide range of chronic diseases, including type 2 diabetes, cancer, cardiovascular disease, Parkinson's disease (PD), multiple sclerosis (MS), and pancreatitis. In Chapter 6, M. Xu and L. Qi offer comprehensive reviews on gene-environment interactions on type 2 diabetes (T2D). Even though it is widely believed that the epidemic of diabetes in the past few decades is largely the consequence of interplay between transition of lifestyle and genetic factors, reproducible evidence from population-based studies is still sparse. Cancer incidence and clinical outcomes show great variations across nations and between populations. The field of genetic epidemiology and gene-environment interaction in explaining such variations is carefully exampled in Chapter 7 by T. Wallace and S. Ambs. Coronary Artery Disease (CAD) is among the chronic diseases whose genetic basis is most extensively explored; however, collective efforts of genomic studies conducted to date have not revealed the substantial causation or major susceptibilities to CAD, indicating considerable 'missing heritability'. In Chapter 8, P. Zalloua and D. Platt addressed this puzzling issue. While development of novel analytical tools is promising in the discovery of new genetic components that may explain some missing heritability, it is believed that gene-environment interaction also plays an important role in the story. In Chapter 9, J. Wise and J. Cannon illustrate the contribution of genetic and environmental risks to PD, a very complex progressive neurodegenerative disorder affecting millions of people worldwide. They also present studies on gene-environment interactions and experimental tools that are used to study these interactions and mechanisms related to PD. R. Alvarez-Lafuente's Chapter 10 on MS introduces readers to the etiology and risk factors of this inflammatory demyelinating disease. Different from some other metabolic chronic diseases, the genetic determinants for MS primarily involve immune genes, and viruses play an important role among other environmental factors such as sun and vitamin D exposure. Large-scale studies on gene-environment interaction of MS are scarce; however, those that focus on the special risk factors would be promising. In Chapter 11, L. Tan and C. Ooi offer a current review of the topic related to pancreatitis. Genetic factors have only recently been discovered to play an important role in the pathophysiology of pancreatitis, and the majority of research in genetics and gene-environment interactions focus on a limited number of genes and environmental factors such as alcohol consumption.

In addition to various chronic diseases, the book also summarizes the studies of gene-environment interactions from a different angle – on longevity and lifespan without major diseases. In Chapter 12, C. Lai, Y. Ma, and L. Parnell provide a wide range of discussion on genetics, environment, and their interaction on longevity. The genetic findings are limited, though extensive efforts have been made on the whole genome scale. Environment-wide association study (EWAS) is emerging as a novel strategy to identify the relevant dietary and lifestyle factors, and integrating this approach with GWAS appears promising to shed light on gene-environment interactions on longevity. In addition, environmental factors can induce epigenetic changes that in turn affect healthy lifespan.

The next two chapters address emerging research on effects of exposure to prenatal risk factors on human health, and potential mechanisms particularly related to epigenetic changes, which are the heritable changes that affect gene expression through mechanisms not associated with DNA sequence variations. The Developmental Origins of Health and Disease (DOHaD) hypothesis proposes that exposures during early life play a critical role in determining the risk of developing metabolic diseases over the lifespan. In Chapter 13, R. Kelishadi and P Poursafa provide a broad review on epigenetic factors related to the developmental origins of health and disease from early life to adulthood. The roles of dynamic changes in DNA as a consequence of environmental interactions in the development of obesity and related chronic diseases are presented in Chapter 14 by M. Garcia-Lacarte, F. Milagro, and J. Martinez. In this chapter, the common epigenetic events such as DNA methylation, histone modifications, and non-coding RNA are introduced; global DNA methylation and its relation with chronic diseases and environmental factors are comprehensively reviewed.

Human genome is not the only genome that affects human health. The Microbiome, defined as the collection of genomes of the microbiota present in the human body, has recently emerged as an important factor in human physiology and disease. In Chapter 15, N. Cortes-Perez and L. Bermúdez-Humarán highlight the role of host-microbiome interactions in affecting health, and application of new technologies (such as metagenomics) in revealing important aspects of such interactions. In addition, the chapter also summarizes the influence of gastrointestinal, metabolic, and other disorders on microbiome.

Understanding gene-environment interactions in relation to chronic disease will promote more targeted, person-centric treatment and prevention, thus reducing mortality burden by these diseases. As editor of this book, I believe that it is not only relevant, but also appropriate to conclude the book with a chapter on the potential application of the findings in the field of gene-environment interaction in clinical and preventive practice. Therefore, the final part of the book is dedicated to the discipline of public health implication of gene-environment interaction research. In Chapter 16, D. Bauer provides detailed elucidation on personalized health management and use of genome information in diagnosing rare genetic diseases, modulating treatment, and devising risk-stratification to prevent or delay the onset of common chronic diseases. The chapter also lists priority areas that need to be particularly underscored in the future research and health practice.

This book is of interest to a variety of audiences, primarily researchers and practitioners in public health and medicine. All chapters include a list of references, pertinent statistical information, tables, and graphs.

The preface would not be complete without acknowledging all those who contributed to the production of this book. I acknowledge with gratitude the collaboration of the authors for accepting my invitation to contribute to the chapters. I am also indebted to the Nova editorial team, particularly Cara, whose professionalism and cordiality made the entire enterprise a very pleasant and efficient experience. I am also grateful to the publisher.

Finally, this book would not have been possible without the support of our institutes. My sincere gratitude goes to Tulane University School of Public Health and Tropical Medicine, Harvard School of Public Health and Harvard Medical School.

In: Gene-Environment Interactions and Human Diseases
Editor: Lu Qi
ISBN: 978-1-63483-680-7

Chapter 1

GENOME, ENVIRONMENT AND HUMAN DISEASES

Lu Qi[1,2,*]

[1]Department of Epidemiology, School of Public Health and Tropical Medicine, Tulane University, New Orleans, LA, US

[2]Department of Nutrition, Harvard School of Public Health, Boston, MA, US

ABSTRACT

In the last century, non-communicable, chronic diseases have overtaken infectious disease as the leading cause of morbidity and mortality and now account for nearly two-thirds of deaths worldwide. Most chronic diseases are complex in etiology, involving multiple environmental and genomic factors and their interactions. In the past decades, tremendous efforts have been put into clinical and epidemiological studies, which have uncovered a broad array of dietary, lifestyle, and other environmental risk factors for chronic diseases. In addition, genomics has recently developed into a mature science; and genome-wide association studies (GWAS) have successfully identified thousands of genetic variants underlying susceptibility to various chronic diseases. However, how environmental factors and genetic factors synergistically affect vulnerability to these diseases remains unresolved. Recent studies have, therefore, focused on the role of gene-environment interactions in determining disease risk. In addition, emerging evidence has shown that sequence variation is not the only component of genomics that may modify the environmental effects; and research on other genomic factors in human body, such as epigenome and gut microbiome, has been rapidly growing. One of the greatest challenges in the post-GWAS era lies in translating the findings from these aforementioned areas into medicine and public health practice, known as personalized medicine and prevention that focus on individuals' susceptibility to chronic diseases and their response to treatment and intervention, with hope to ultimately endorse person-centric approachesto mitigate these disorders and improve human health.

---

* Correspondence and requests for reprint: Dr. Lu Qi. Department of Epidemiology, School of Public Health and Tropical Medicine, Tulane University, New Orleans, Louisiana 70112. Telephone: 504-988-7259; Fax: 504-988-3548; Email address: lqi1@tulane.edu; luqi@hsph.harvard.edu.

INTRODUCTION

During the past decades the average life expectancy of humans considerably increased, and the major causes of dying have changed from infectious diseases to noncommunicable, chronic diseases such as cancer, cardiovascular disease, diabetes, respiratory and autoimmune diseases, as well as other aging related disorders [1, 2]. In 2010-2011, the World Health Organization (WHO) has estimated that approximately two-thirds of all deaths worldwide were caused by chronic diseases [3]. In the USA, approximately half of adults are affected by at least one chronic condition and 26% have two or more conditions. It is apparent that chronic diseaseshave become the major drivers of disability and health-care costs [4].

The current epidemic of chronic diseases is known to have coincided with profound shift of our living environment, here broadly taken to include dietary patterns, sedentary habits such as television watching and computing, unhealthy lifestyle such as physical inactivity, smoking, bad sleeping manners, as well as demographic and cultural features. In epidemiological studies, many of these environmental factors have been related to risk of chronic diseases. Considering the environmental factors are largely modifiable, these major causes of chronic diseases are essentially preventable [5]. According to a recent WHO report, at least 80% of cardiovascular diseases and type 2 diabetes, and over 40% of cancer would be prevented, if the modifiable risk factors were eliminated [6].

The genetic contribution to chronic diseases has also been acknowledged and studied for long time. In the recent decade, high-throughput genotyping has enabled genome-wide association studies (GWAS) of many chronic diseases. These studies have discovered thousands of disease susceptible loci and unveiled unexpected mechanisms. However, predicting disease risk from genotype is greatly complicated by complex interactions between the genetic variants and environment.

Gene-environment interactions are ubiquitous, and may account for the greater part of disease risk seen across genotypes [7]. Accumulating evidence has suggested a "multiple hit" phenomenon that precipitates chronic diseases, i.e., genomic makeup may act together with environment factors. This hypothesis is partly driven by the observation that the genomic variations themselves only account for a small proportion of disease risk (missing heritability), and has been evidenced by emerging studies on gene-environment interactions [8-10].

In this book we discuss recent progress in deciphering gene-environment interactions on a broad range of chronic diseases including obesity, diabetes, cardiovascular disease, cancer, multiple sclerosis, pancretitis, and Parkinson's disease, as well as other health related traits such as lipids and longevity.

In addition, we highlight insights gleaned from studies on epigenome and gut microbiome, and how this knowledge informs understanding of more complex architecture of gene-environment interactions.

Moreover, we articulate how the confluence of the rapid emerging knowledge from research of gene-environment interaction will facilitate personalized medicine and prevention, mitigate disease risk, and improve human health.

Environmental Risk Factors

Environmental factors including dietary intakes, lifestyle behaviors, socioeconomic status, air pollution, built environment, and demographic and cultural features are the major sources of risk factors associated withchronic diseases. The research conducted over the last decades in the field of epidemiology has been instrumental in gaining a better insight into the roles of environmental components in development of chronic diseases; and it is believed that the recent global burden of chronic diseases is likely attributable to a relatively short list of risk factors, such as smoking, poor diet (high intakes of fast foods, red meat, and sugary drink; low intake of whole grains, fiber, nuts), low physical activity and sedentary lifestyle, as well asexposure to air pollution and toxic chemicals [4-11]. For example, regular exercise and health diet high in intakes of whole grains and fiber have been related to a reduced risk of a range of chronic disorders such as obesity, CVD, diabetes, cancer and their related mortalities [12-15]. At the same time, evidence supporting the roles of many other environmental factors in chronic diseases is highly controversial and their contributions remain poorly understood.

The environment factors responsible for the development of chronic diseases are not easy to tie down. In addition, although epidemiological evidence clearly points to an environmental influence, not all the subjects exposed to the same environment have disease. Emerging data have shown that different response under the umbrella of a single exposure is quite striking, suggesting other inborn susceptibility must thus play a role in determining disease propensity in collision with environmental risk factors [16-18]. Therefore, it may be more appropriate to consider chronic disease as the outcome of combination of inborn susceptibility in conjunction with the environment. Several strong indicators suggest that the marked individual differences in sensitivity to potentially pathogenic exposures are, at least partially, due to genetic factors. Understanding interactions between the intrinsic and extrinsic mechanisms of chronic diseases will be the key to their future treatment and prevention through modifying environmental factors such as diet and lifestyle.

Genetic Risk Factors

More than 150 years after Mendel discovered the laws of heredity, the genetic architecture of most of the chronic diseases remains elusive. However, decipher of genetic code for human health conditions has been recently accelerated. The completion of the human genome sequence in 2003 clearly marked the beginning of a new era for detection of geneticrisk factors [19]. With rapid advances in genotyping and sequencing technologies, genetic discovery has been undergoing a paradigm shift from hypothesis-driven to hypothesis-free manner, and analyses using GWAS represent one of the most important advances in the field. These new technologies present an unprecedented opportunity to move from approaches using only partial genomic information, such as candidate-gene association study and linkage analysis to complete analysis of the relationship between genomic variation and disease risk. In GWAS, allele frequencies of sequence variations such as single nucleotide polymorphisms (SNPs) or copy number variants (CNVs) are compared on the whole genome scale between the affected patients and healthy controls. A genetic variant is determined to be associated with disease risk if the different distribution reaches genome-wide significance level (e.g., $5*10^{-8}$) and shows reproducible evidence in multiple populations

[20]. The strategy is based on the common variant–common disease hypothesis, which states that common genetic variation plays a major role in common diseases [21].

The first wave of large-scale GWAS was published in 2007, starting from identification of genomic loci associated with T2D from analyses of up to nearly 500,000 SNPs [22, 23]. Now GWAS has been applied to hundreds of different diseases and phenotypes. At present, up to several thousand SNPs have been established for associations with chronic diseases such as T2D, CVD, cancers, asthma, osteoporosis, multiple sclerosis, Parkinson's disease, and other diseases [24-29]. These SNPs are spread throughout the human genome (*Figure 1*; see the Catalog of Published Genome-Wide Association Studies at https://www.genome.gov/26525384).

The common disease-associated variants share several common features. First, a large proportion of the SNPs discovered are intergenic or lie within the intronic regions of genes, and the functional relevance of these variants is often not obvious. It is possible that the SNPs discovered through GWAS are in linkage disequilibrium (LD) with the real causal variant that is not captured by the platform. Second, the effect sizes associated with individual variants are generally modest, and the identified variants thus far together only account for a small proportion of disease risk, generally 5-15% [28-31]. This is known as 'missing heritability'. Common explanations for "missingheritability" include massive polygenicity, rare variants, epigenetics, and epistasis [31]. While new techniques are being applied to unravel these more complex architecture embedded in human genome [32, 33], emerging efforts also pay attention on an alternative explanation – gene-environment interaction [34]. The potential exist of gene-environment interaction is partly supported by traditional genetic studies in twins and families [34], as well as observation that people living in different environment are disproportionately affected by chronic diseases.

Gene-Environment Interaction

The epidemic of chronic diseases occurred in the blink of an eye in geological terms. Whereas the alterations in human genomic makeup are negligible, it is clear that dramatic shifts of living environment, particularly lifestyle and dietary habits, act as a catalyst for the genetic susceptibility to manifest.

Interestingly, the past decades coincided with an era of mass emigration and globalization on a scale unparalleled in the history of humanity; this further inflated the effects of collision between genome and environment on human health. A better understanding of the effects of genetic predictors of disease requires that genetic determinants be placed within the context of known environmental risk factors. Genes can trigger the occurrence of diseases when a person with a high-risk genetic profile is exposed to high-risk environmental factors, a phenomenon called gene–environment interaction [16]. Gene–environment interaction may reflect a causal mechanism where the genetic variants and environmental factors contribute to the causation of a condition in the same individual with the genetic factors influencing the sensitivity to environmental exposures.

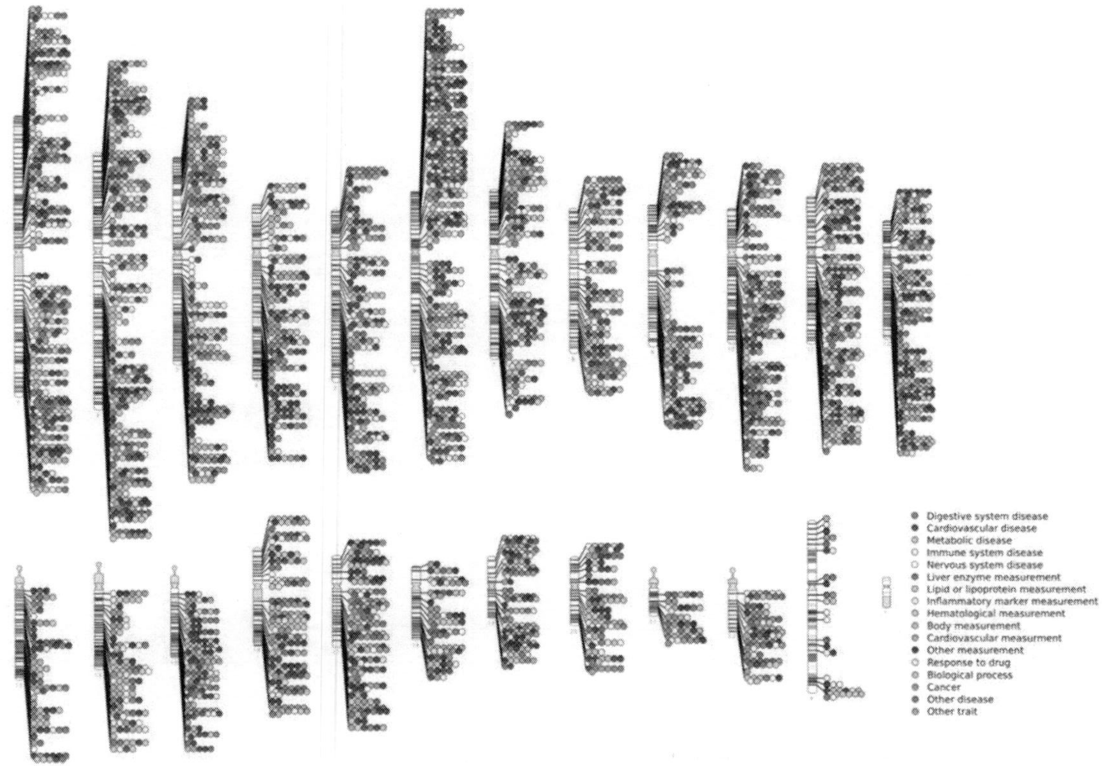

Figure 1. Published Genome-Wide Associations through 12/2013. (Published GWA at p≤5X10-8 for 17 trait categories). Modified from https://www.genome.gov/26525384.

By definition, a gene-environment interaction occurs when there is a mutual dependency between these two types of factors on the trait of interest.The majority of population studies of gene-environment interaction are designed to test whether the additive effect of genetic and environmental factors is greater than the contribution of either risk factor. Evidence supporting gene-environment interactions on various chronic diseases and their risk factors is growing exponentially. Recently emerging data have shown reproducible evidence for gene-environment interactions, such as smoking-*BRCA2*gene interaction on lung cancer, [35] alcohol-*CASP8* gene interaction on breast cancer, [36] sugar sweetened beverage-obesity genes interaction on BMI (Figure 2), [8] fried foods-obesity genes interaction on BMI, [9] dietary patterns-CHD gene interaction on cardiovascular disease, [37] 25-hydroxyvitamin D-vitamin D related genes interaction on composite outcome of incident hip facture, myocardial infarction, cancer, and mortality, [38] physical activity-obesity genes interaction on obesity, [10] smoking-pulmonary function related genes interaction on pulmonary function, [39] and lifestyle-diabetes genes interaction on diabetes [40]. Even though the mechanisms behind most of the observed gene–environment interactions remain unknown, these interactions are biologically plausible [16]. Nevertheless, paramount of challenges exist in the field. A considerable amount of the identified interactions are not reproducible, and validation of these findings is still a demanding and intricate task. In addition, statistical significant gene-environment interactions often do not correspond to actual biological interaction. Therefore, multiple methods are required to establish functional relevance for the interactions observed.

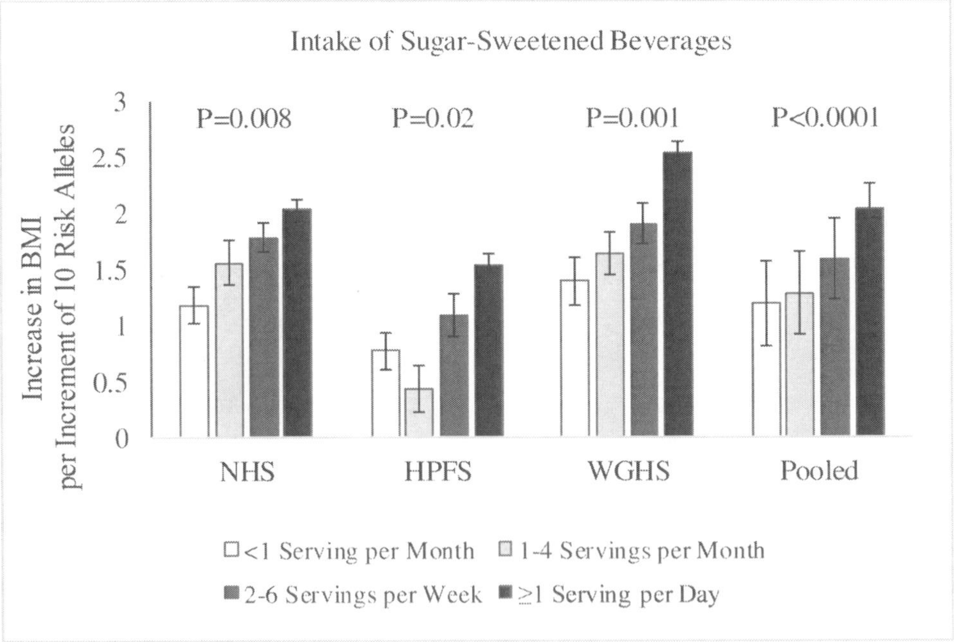

Figure 2. Gene-diet interaction on body mass index (BMI).Increase in BMI per increment of 10 units of genetic risk score which was calculated by summing up of the number of risk alleles of 32 BMI-associated genetic variants, according to intake of sugar-sweetened beverages in the Nurses' Health Study (NHS), Health Professionals Follow-up Study (HPFS) and Women Genome Health Study (WGHS).

Other challenges include imprecise approaches to assess environmental factors (such as dietary intakes, physical activity, and environmental pollution), difficulty in defining the causal variants, and lack of standardized statistical models to detect interactions in different patterns. The extent of exposure to modern environment could be more finely determined to explore threshold-dependent models of gene-environment interaction [41, 42]. Although statistical methods to investigate gene-environment interactions are available, further development and extension of these methods are urgently needed, especially in the context of GWAS [43]. Moreover, individual genetic variant is usually related to moderate effect in gene-environment interaction on disease risk. A recent study indicates that gene-environment interactions may more likely to be detected by considering multiple relevant genetic variants together [44]. This makes sense considering that the combined genetic variants not only present the largest effect sizes, but also have multiplied them in the high-risk environment.

Epigenome-Environment Interactions

Variation in human genome sequence might not be the only determining factor in affecting gene-environment interactions. In the post-GWAS era a major challenge lies in accounting for missing heritability by additional genomic factors to characterize theirinteractions with environmental factors. Compelling evidence has implicated epigenetic mechanism as an important candidate that may directly interact with environment and act at the interface of genomic variation and environment in affecting human health. An epigenetic trait can be defined as a "stably heritable phenotype resulting from changes in a chromosome without alterations in the DNA sequence." [45] Epigenetic events include DNA methylation, histone modifications, chromatin remodeling and noncoding RNAs. A comprehensive picture of epigenetic patterns and the relations of these epigenetic patterns with human diseases are now emerging [46, 47]. Among these epigenetic patterns, DNA methylation has been best studied in the context of various chronic diseases [48, 49].

Epigenetic alterations are critical for establish of a particular gene expression signature for each cell type during early developmental processes. The "Developmental Origins of Health and Disease" theory emphasizes the critical early-life period when environmental stimuli may induce long-lasting changes such as in epigenomethat affect disease risk with long latency period.

The epigenome-environment interaction may occur before and after birth. DNA methylation at cytosine residues in CpG dinucleotides is one of the major epigenetic patterns that affected by prenatal exposures to mal- and over-nutrition, maternal smoking, and other environmental contaminants [50, 51]. Later life exposure to environmental risk factors and aging are also associated with profound epigenetic changes, resulting in alterations of gene expression and subsequent risk of chronic diseases [52]. Importantly, there is already some evidence for an interaction of environment with epigenetics [53]. Many techniques have been developed to study epigenome; and these new approaches may provide keys to unravel environmental factors that impinge on epigenome to affect later life risk of chronic diseases [46].

Metagenome-Environment Interaction

The human gut is colonized by a wide diversity of micro-organisms, which are now known to play a key role in the human host by regulating metabolic functions and immune homeostasis [54]. Many studies have indicated that the genomes of gut microbiota, known as metagenome – the gut microbiome or our "other genome," could also play an important role in interacting with environmental factors in affects far reaching consequences on host health [55-57]. The composition or abundance of metagenomeis significantly influenced by diet such as macronutrients carbohydrates, proteins, and fats, as well as lifestyle. Recent investigations in humans have shown exciting evidence that modifications on dietary intakes of the macronutrients significantly altered richness or compositions of gut microbiome, and subsequently affect metabolic profiles including obesity and diabetes risk factors [57, 58]. The study of metagenome and metagenome-environment interaction on human health would be one of the most exciting advancements in biomedicine in the near future.

Personalized Medicine and Prevention

About 30 years ago, epidemiologist Geoffrey Rose has proposed two approaches to disease prevention: the individual and the population approaches [59]. The population approach focuses on identifying the underlying causes of disease and providing a generalized intervention that shifts the distribution of risk at the population level, whereas the individual approach focuses on identifying individuals at high risk and providing intervention at individual level. Along with the revolutionary advances in genomics, epigenomics, metagenomics, and other omics research, it has become plentifully clear that each individual has personalized biological content that determines widely-ranged diversity in response to prevention and treatment. The gathering of individual-based big data from the omics studies set up a remarkable and unprecedented opportunity to improve medical treatment and develop preventive strategies to preserve health, and facilitate burgeoning personalized medicine and prevention that places the individual at the center of health care [60, 61].

Pharmacogenetics is one example of how genomic informationis already being taken into account in screening response to drug at individual level [62]. Such applications represent one component of personalized medicine. In addition, application of large-scale genotyping methodologies increasingly allows to define the individual predisposition for a given disease in apparently healthy populations, and to predict disease risk when individuals with distinct genotypes exposed to the same environment. The new approach necessitates personalized interventions to lower disease risk through genomic-tailored modifications of diet and lifestyle factors. For example, out of the nutrigenomic discipline that aims to investigate the interactions between dietary factors and genomic variations and, ultimately, improve the treatment and prevention of nutrition-related diseases [63, 64]. Accumulation of knowledge in these new areas would improve the efficacy of disease management and prevention. However, to what extent personalized treatment and prevention would add to existing strategists and recommendations remains uncertain. The new paradigm of developing personalized medicine and prevention is timely, given the burden of reducing cost while improving quality in healthcare system. Consumer genomic tests are becoming increasingly available at a relatively affordable price, and direct-to-consumer (DTC) genetic test has added a new dimension to health screening [65]. It has been assumed that individual differences in disease perceptions may influence psychological outcomes following genetic testing [41]. However, research on the influence of genetic testing on alterations of human behaviors is still in its infancy and proven evidence for understanding personal response in the multitude of available genetic tests is still sparse; whereas such studies are desperately needed [66].

SUMMARY

Etiology of chronic diseases is complex, involving contributions from both environmental and genetic factors. The recent epidemic of a variety of chronic diseases has been ascribed to a collision between the genes and the environment. Genome-wide profiling efforts have given informative insights into biological processes of these diseases; but our understanding about how these genomic factors interact with environment is still extremely little. In the past few years, studies on gene-environment interactions have been rapidly increasing. The findings from these studies have broad appeal and will no doubt lead to a great deal of further analysis by the scientific community. However, considering the wealth of variation, the major challenge still remains in their meaningful interpretation.

The findings from gene-environment interactions hold great promise to accelerate personalized treatment and intervention. Even though genomic information is generally not useful for disease prediction, it could be used along with other risk factors in screening and classifying high-risk populations to tailored intervention or treatment [67]. Nevertheless, important gaps in translation of the new knowledge into clinic and preventive practice pose a major challenge. In addition, implementation challenges include difficulties in integration of genetic testing in the current health-care system, the setting of appropriate risk thresholds, and ethical, legal, and social issues; and empirical data are currently lacking on these issues [68]. To accelerate this momentum and move our understanding into this new frontier, solutions need to be applied to efficiently integrate phenotypic and genomic data to better interpret. More collaborative and multi-disciplinary research is needed to integrate the fruits from scientific research into public health and clinical practice.

REFERENCES

[1] Yach, D., Hawkes, C., Gould, C. L. & Hofman, K. J. The global burden of chronic diseases: overcoming impediments to prevention and control. *JAMA* 291, 2616-2622, (2004).

[2] Murray, C. J. et al. Disability-adjusted life years (DALYs) for 291 diseases and injuries in 21 regions, 1990-2010: a systematic analysis for the Global Burden of Disease Study 2010. *Lancet* 380, 2197-2223, (2012).

[3] World Health Organization. Hypertension et diabète sont en forte hausse selon les dernières statistiques de l'OMS [Hypertension and diabetes are rising sharply according to the latest WHO statistics] (2012)

[4] Bauer, U. E., Briss, P. A., Goodman, R. A. & Bowman, B. A. Prevention of chronic disease in the 21st century: elimination of the leading preventable causes of premature

death and disability in the USA. *Lancet* 384, 45-52, (2014).

[5] Fontana, L. Modulating human aging and age-associated diseases. *Biochim Biophys Acta* 1790, 1133-1138, (2009).

[6] Preventing chronic diseases: a vital investment — WHO global report. Geneva: World Health Organization.

[7] Grishkevich, V. & Yanai, I. The genomic determinants of genotype x environment interactions in gene expression. *Trends Genet* 29, 479-487, (2013).

[8] Qi, Q. et al. Sugar-sweetened beverages and genetic risk of obesity. *N Engl J Med* 367, 1387-1396, (2012).

[9] Qi, Q. et al. Fried food consumption, genetic risk, and body mass index: gene-diet interaction analysis in three US cohort studies. *BMJ* 348, g1610, (2014).

[10] Ahmad, S. et al. Gene x physical activity interactions in obesity: combined analysis of 111,421 individuals of European ancestry. *PLoS Genet* 9, e1003607, (2013).

[11] Mokdad, A. H., Marks, J. S., Stroup, D. F. & Gerberding, J. L. Actual causes of death in the United States, 2000. *JAMA* 291, 1238-1245, (2004).

[12] Park, Y., Subar, A. F., Hollenbeck, A. & Schatzkin, A. Dietary fiber intake and mortality in the NIH-AARP diet and health study. *Arch Intern Med* 171, 1061-1068, (2011).

[13] Cho, S. S., Qi, L., Fahey, G. C., Jr. & Klurfeld, D. M. Consumption of cereal fiber, mixtures of whole grains and bran, and whole grains and risk reduction in type 2 diabetes, obesity, and cardiovascular disease. *Am J Clin Nutr* 98, 594-619, (2013).

[14] Vogel, T. et al. Health benefits of physical activity in older patients: a review. *Int J Clin Pract* 63, 303-320, (2009).

[15] Reiner, M., Niermann, C., Jekauc, D. & Woll, A. Long-term health benefits of physical activity--a systematic review of longitudinal studies. *BMC Public Health* 13, 813, (2013).

[16] Hunter, D. J. Gene-environment interactions in human diseases. *Nat Rev Genet* 6, 287-298, (2005).

[17] Qi, L. & Cho, Y. A. Gene-environment interaction and obesity. *Nutr Rev* 66, 684-694, (2008).

[18] Qi, L. & Liang, J. Interactions between genetic factors that predict diabetes and dietary factors that ultimately impact on risk of diabetes. *Curr Opin Lipidol* 21, 31-37, (2010).

[19] Altshuler, D., Daly, M. J. & Lander, E. S. Genetic mapping in human disease. *Science* 322, 881-888, (2008).

[20] Manolio, T. A. Genomewide association studies and assessment of the risk of disease. *N Engl J Med* 363, 166-176, (2010).

[21] Schork, N. J., Murray, S. S., Frazer, K. A. & Topol, E. J. Common vs. rare allele hypotheses for complex diseases. *Curr Opin Genet Dev* 19, 212-219, (2009).

[22] Sladek, R. et al. A genome-wide association study identifies novel risk loci for type 2 diabetes. *Nature* 445, 881-885, (2007).

[23] Zeggini, E. et al. Replication of genome-wide association signals in UK samples reveals risk loci for type 2 diabetes. *Science* 316, 1336-1341, (2007).

[24] Lin, M. K. & Farrer, M. J. Genetics and genomics of Parkinson's disease. *Genome Med* 6, 48, (2014).

[25] Urano, T. & Inoue, S. Genetics of osteoporosis. *Biochem Biophys Res Commun* 452, 287-293, (2014).

[26] Sayols-Baixeras, S., Lluis-Ganella, C., Lucas, G. & Elosua, R. Pathogenesis of

coronary artery disease: focus on genetic risk factors and identification of genetic variants. *Appl Clin Genet* 7, 15-32, (2014).

[27] Cree, B. A. Multiple sclerosis genetics. *Handb Clin Neurol* 122, 193-209, (2014).

[28] O'Donnell, C. J. & Nabel, E. G. Genomics of cardiovascular disease. *N Engl J Med* 365, 2098-2109, (2011).

[29] McCarthy, M. I. Genomics, type 2 diabetes, and obesity. *N Engl J Med* 363, 2339-2350, (2010).

[30] Voight, B. F. et al. Twelve type 2 diabetes susceptibility loci identified through large-scale association analysis. *Nat Genet* 42, 579-589, (2010).

[31] Manolio, T. A. et al. Finding the missing heritability of complex diseases. *Nature* 461, 747-753, (2009).

[32] Saint Pierre, A. & Genin, E. How important are rare variants in common disease? *Brief Funct Genomics* 13, 353-361, (2014).

[33] Hou, L. & Zhao, H. A review of post-GWAS prioritization approaches. *Front Genet* 4, 280, (2013).

[34] Kaprio, J. Twins and the mystery of missing heritability: the contribution of gene-environment interactions. *J Intern Med* 272, 440-448, (2012).

[35] Wang, Y. et al. Rare variants of large effect in BRCA2 and CHEK2 affect risk of lung cancer. *Nat Genet* 46, 736-741, (2014).

[36] Nickels, S. et al. Evidence of gene-environment interactions between common breast cancer susceptibility loci and established environmental risk factors. *PLoS Genet* 9, e1003284, (2013).

[37] Do, R. et al. The effect of chromosome 9p21 variants on cardiovascular disease may be modified by dietary intake: evidence from a case/control and a prospective study. *PLoS Med* 8, e1001106, (2011).

[38] Levin, G. P. et al. Genetic variants and associations of 25-hydroxyvitamin D concentrations with major clinical outcomes. *JAMA* 308, 1898-1905, (2012).

[39] Hancock, D. B. et al. Genome-wide joint meta-analysis of SNP and SNP-by-smoking interaction identifies novel loci for pulmonary function. *PLoS Genet* 8, e1003098, (2012).

[40] Langenberg, C. et al. Gene-lifestyle interaction and type 2 diabetes: the EPIC interact case-cohort study. *PLoS Med* 11, e1001647, (2014).

[41] Boeldt, D. L., Schork, N. J., Topol, E. J. & Bloss, C. S. Influence of individual differences in disease perception on consumer response to direct-to-consumer genomic testing. *Clin Genet* 87, 225-232, (2015).

[42] Tucker, K. L., Smith, C. E., Lai, C. Q. & Ordovas, J. M. Quantifying diet for nutrigenomic studies. *Annu Rev Nutr* 33, 349-371, (2013).

[43] Winham, S. J. & Biernacka, J. M. Gene-environment interactions in genome-wide association studies: current approaches and new directions. *J Child Psychol Psychiatry* 54, 1120-1134, (2013).

[44] Marigorta, U. M. & Gibson, G. A simulation study of gene-by-environment interactions in GWAS implies ample hidden effects. *Front Genet* 5, 225, (2014).

[45] Berger, S. L., Kouzarides, T., Shiekhattar, R. & Shilatifard, A. An operational definition of epigenetics. *Genes Dev* 23, 781-783, (2009).

[46] Weichenhan, D. & Plass, C. The evolving epigenome. *Hum Mol Genet* 22, R1-6, (2013).

[47] Rivera, C. M. & Ren, B. Mapping human epigenomes. *Cell* 155, 39-55, (2013).
[48] Kaur, P., Shorey, L. E., Ho, E., Dashwood, R. H. & Williams, D. E. The epigenome as a potential mediator of cancer and disease prevention in prenatal development. *Nutr Rev* 71, 441-457, (2013).
[49] LaSalle, J. M. Epigenomic strategies at the interface of genetic and environmental risk factors for autism. *J Hum Genet* 58, 396-401, (2013).
[50] Drake, A. J. et al. An unbalanced maternal diet in pregnancy associates with offspring epigenetic changes in genes controlling glucocorticoid action and foetal growth. *Clin Endocrinol (Oxf)* 77, 808-815, (2012).
[51] DelCurto, H., Wu, G. & Satterfield, M. C. Nutrition and reproduction: links to epigenetics and metabolic syndrome in offspring. *Curr Opin Clin Nutr Metab Care* 16, 385-391, (2013).
[52] Brunet, A. & Berger, S. L. Epigenetics of aging and aging-related disease. *J Gerontol A Biol Sci Med Sci* 69 Suppl 1, S17-20, (2014).
[53] Martin-Nunez, G. M. et al. Methylation levels of the SCD1 gene promoter and LINE-1 repeat region are associated with weight change: an intervention study. *Mol Nutr Food Res* 58, 1528-1536, (2014).
[54] Pluznick, J. L. Gut microbes and host physiology: what happens when you host billions of guests? *Front Endocrinol (Lausanne)* 5, 91, (2014).
[55] Cox, A. J., West, N. P. & Cripps, A. W. Obesity, inflammation, and the gut microbiota. *Lancet Diabetes Endocrinol* 3, 207-215, (2015).
[56] Qin, N. et al. Alterations of the human gut microbiome in liver cirrhosis. *Nature* 513, 59-64, (2014).
[57] David, L. A. et al. Diet rapidly and reproducibly alters the human gut microbiome. *Nature* 505, 559-563, (2014).
[58] Cotillard, A. et al. Dietary intervention impact on gut microbial gene richness. *Nature* 500, 585-588, (2013).
[59] Rose, G. Sick individuals and sick populations. *Int J Epidemiol* 14, 32-38, (1985).
[60] Chan, I. S. & Ginsburg, G. S. Personalized medicine: progress and promise. *Annu Rev Genomics Hum Genet* 12, 217-244, (2011).
[61] Agboola, S. O., Ball, M., Kvedar, J. C. & Jethwani, K. The future of Connected Health in preventive medicine. *QJM* 106, 791-794, (2013).
[62] Chung, W. H. et al. Genetic variants associated with phenytoin-related severe cutaneous adverse reactions. *JAMA* 312, 525-534, (2014).
[63] Ordovas, J. M. & Mooser, V. Nutrigenomics and nutrigenetics. *Curr Opin Lipidol* 15, 101-108, (2004).
[64] Sales, N. M., Pelegrini, P. B. & Goersch, M. C. Nutrigenomics: definitions and advances of this new science. *J Nutr Metab* 2014, 202759, (2014).
[65] Caulfield, T. & McGuire, A. L. Direct-to-consumer genetic testing: perceptions, problems, and policy responses. *Annu Rev Med* 63, 23-33, (2012).
[66] Bloss, C. S., Darst, B. F., Topol, E. J. & Schork, N. J. Direct-to-consumer personalized genomic testing. *Hum Mol Genet* 20, R132-141, (2011).
[67] Burton, H. et al. Public health implications from COGS and potential for risk stratification and screening. *Nat Genet* 45, 349-351, (2013).
[68] Bloss, C. S., Jeste, D. V. & Schork, N. J. Genomics for disease treatment and prevention. *Psychiatr Clin North Am* 34, 147-166, (2011).

In: Gene-Environment Interactions and Human Diseases
Editor: Lu Qi ISBN: 978-1-63483-680-7

Chapter 2

ARE HUMANS GETTING FITTER, SICKER, NEITHER, OR BOTH?

Greg Gibson and Alec Nabb
School of Biology, Georgia Institute of Technology, Atlanta, GA, US

ABSTRACT

This chapter considers the contributions of culture, genetics, and genotype-by-environment interactions to human health and fitness. Contradictory trends are observed at many levels. From a physical fitness perspective, both chronic and rare diseases are increasing in prevalence, but longevity and quality of life are also increasing. From a Darwinian fitness perspective, developed countries are seeing a dramatic turnaround in the long-term trend of population growth, as fecundity has dropped well below replacement values – but this is a result of behavioral choice, not genetic constraint. Mutation accumulation is likely to contribute significantly to morbidity as rare deleterious variants increase in frequency, but globally admixture is shuffling the gene pool and thereby altering genetic risk profiles for all common diseases that have diverged, for unknown reasons, among the major population groups. Although individual genotype-by-environment interactions are rare in general, interactions could in theory be pervasive at the level of composite genetic risk scores, and would have been missed by current analytical procedures. A particular type of genotype-by-environment interaction known as decanalization is the increase in variance resulting from genetic and cultural perturbation of buffered genetic systems, and could be found to be a major source of shifting epidemiological patterns.

INTRODUCTION

Despite the astonishing gains in our understanding of the genetics of human disease since the implementation of genome-wide association studies and whole exome sequencing [1], two central questions have barely been addressed: why do populations differ in susceptibility to specific diseases, and why is the incidence of so many chronic diseases increasing globally? The simple answers, because of the impacts of selection and cultural change, while to some extent true, belie complexity and likely obscure more relevant conclusions. For example, it is highly debatable whether selection against disease susceptibility has been the primary driver of population divergence in risk profiles [2], and the role of cultural change may lie more in decreasing buffering than directionally shifting susceptibility thresholds [3]. This chapter argues that genotype-by-environment interactions are in fact a major contributor to the shifting burden of disease within and among populations.

We start with some epidemiological considerations regarding whether human fitness is increasing or decreasing, concluding that any trends are largely driven by culture and behavior rather than genetics. However, this does not mean that genetics is irrelevant, so the second section discusses whether and how distributions of disease risk may be evolving. The concluding section attempts to reconcile these viewpoints within the framework of the decanalization of disease risk, a phenomenon that is explicitly molded by genotype-by-genotype and genotype-by-environment (G×E) interactions [4].

MAIN TEXT

The Epidemiology of Fitness

An immediate inclination when asked the question "are we getting fitter?" is to point to morbidity maps, such as those provided by the CDC, which generally paint a grim picture. Figure 1 shows how over a 20 year period, obesity (defined as a Body Mass Index greater than 30) in the United States has more than doubled – we are unquestionably getting fatter [5]. Spreading out from the stroke belt, which stretches from southern Texas through Arkansas, Mississippi, and Georgia to South Carolina, prevalence has increased from 15% in the states with the highest burden, to over 25% nationwide.

Metabolic disease, most notably type 2 diabetes but many inflammatory conditions as well, follow and are the major public health concern not just for the US, but for most Western nations [6].

The increasing prevalence of disease is by no means restricted to obesity-related illness. Signs of coronary artery disease are found in over half the adult US population [7], despite being completely absent from some human populations such as the Masai people of east Africa [8]. On a recent trip to Fiji, physicians informed the first author that hypertension was unknown two generations ago among Melanesians, but is now one of the major health concerns even for highland villagers. Childhood asthma prevalence rose from 3.6% to 7.5% in the US in the fifteen years from 1980 to 1995 [9], and is an even larger concern in Australia where severe cases debilitate a small percentage of children. The incidence of hip fractures in the elderly, exacerbated by osteoporosis, is expected to quadruple by 2030 [10]. Meanwhile, the World Health Organization estimates that mental health issues, particularly monopolar depression, will represent the greatest source of lost quality of life globally in the next century [11]. Some of the increased burden of these and many other diseases may be attributed to ascertainment bias in more surveillance-oriented healthcare systems, and some simply to aging, but few would doubt that there is not also a true increase in the age-adjusted incidence of many diseases.

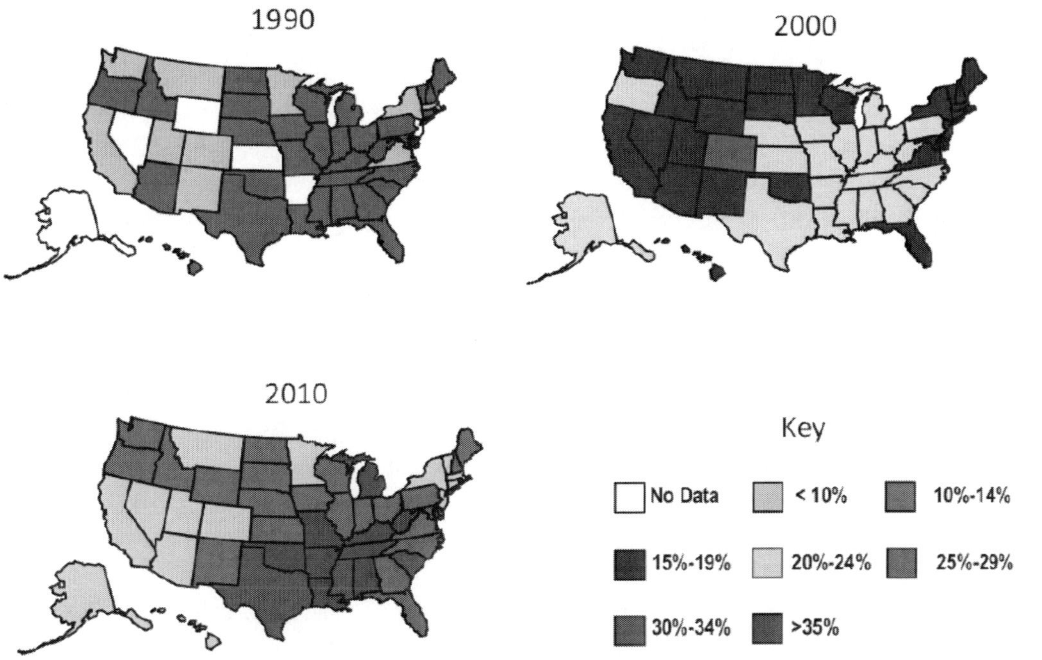

Figure 1. Increase in prevalence of obesity in the United States. Each map shows the prevalence of obesity, defined as BMI greater than 30, by state in the indicated year. Data from the CDC maps available at http://www.cdc.gov/obesity/data/adult.html#History.

On the other hand, there are also compelling signs of improved human health and well-being. Most obviously, longevity continues to rise. From the Paleolithic through the medieval period, human life expectancy in Europe is thought to have been consistently below 40, though these numbers are likely to be skewed by very high infant mortality [12, 13]. Those who survive into their teens seem to have enjoyed a lifespan of 60 or more years in many human populations, including contemporary nomadic and hunter-gatherer people. Infectious disease has of course been a major source of mortality, most recently with HIV reducing life expectancy back into the 40s in several southern African countries [14]. The industrial revolution was not a happy time for many, and its legacy persists in some communities (people in East Glasgow for example until very recently could have expected to live 20 years less than counterparts in more affluent communities across the River Clyde [15]). Health disparities among ethnic groups are similarly a major source of concern in the US, but overall life expectancy is nudging 80 years and only expected to rise. Another astonishing measure of improved human well-being, known as the Flynn effect, is a gain in mean IQ of around 20 points over the past two generations. Steven Pinker has discussed this phenomenon as well as a persistent millennial global decline in violence in his book, *The Better Angels of Our Nature* [16, 17], attributing greater empathy, self-control, morality and reason to gains in education and rationality since the Humanitarian revolution of the 18th century.

It is debatable whether any of these observations are at all relevant with respect to the issue of whether human fitness is evolving in a Darwinian sense. Fecundity is the more relevant parameter than disease or lifespan as a measure of the likelihood that an individual transmits his or her genes to the next generation. Humanity as a whole remains in growth mode, in just fifty years having risen from around 4 billion people to over 6 billion. This is up from an effective population size of 10,000 at the origin of the species no more than 10,000 generations ago, a fact that immediately calls into question standard assumptions about the role of selection shaping population genetic variation at equilibrium. Just in the past two decades, the situation has dramatically shifted once more since it is now believed that in most developed nations, the fertility rate is significantly below the replacement level of two children per couple [18]. At this most basic level, human fitness is unquestionably declining, and it is doing so for cultural and social, not genetic, reasons.

The Big Picture of Evolving Rare-Variant Genetic Risk

Perhaps the most important consequence of several thousand years of expanding population size has been an estimated doubling of the total burden of rare deleterious variants in the human gene pool over the past 5,000 years [19]. The average number of deleterious variants per person does not vary greatly among populations [20], because most such variants per genome actually have a minor allele frequency greater than 1%. However, there is a clear increase in the prevalence of very rare disruptive variants (non-sense, splice-site disrupting, and indel mutations) relative to neutral expectations, consistent with strong purifying selection against them as a class. Following an argument first highlighted by Mike Lynch [21], if we assume that improved health care reduces the strength of selection against strongly deleterious variants, then a decline in fitness of around one percent per generation compounded over the next 10 generations may begin to have meaningful consequences for humanity. This argument has been critiqued from the perspective of fecundity [22], but it may remain valid in relation to health, particularly in an aging population since selection against variants that act late in life (post-reproduction) will be particularly weak.

The most likely domain of health where the impact may be first seen is mental health, simply because such a large proportion of the genome influences it, presenting the largest mutational target. Indeed, empirical estimates from whole exome sequencing studies are that of the order of 5% of cases of schizophrenia can be attributed largely to de novo mutations in a core set of several hundred genes with functions in brain growth and development and synaptogenesis [23]. (For other neurological disorders such as autism [26], similar burdens can be attributed to rare copy number variants [24] and very rare segregating deleterious polymorphisms [25]). Debate continues surrounding the notion of causality for such variants, since focused family studies that would establish the penetrance and expressivity of the alleles have not been reported, but reasonable estimates place the genotype relative risks in the range of 5 to 10. Given the low incidence of the diseases in the general population, a reasonable interpretation is that no more than 10% of people with a particular variant will have a severe manifestation of psychiatric disease. Whether or not such variants contribute to an increased burden of sub-clinical mental health issues is very hard to tell.

Counteracting the effects of mutation accumulation are two demographic processes that are particularly relevant with respect to recessive mutations. The predominant trend for contemporary humans is admixture, namely the mixing of gene pools that inevitably follow within a couple of generations of migration. Rare recessive variants can reach measurable frequencies as a consequence of genetic drift in individual populations, but their frequencies are reduced upon admixture. It would be a stretch to claim hybrid vigor in humans, since the divergence has not been great to begin with, but rare homozygote frequencies should decline in general. On the other hand, compound heterozygosity may be more common than homozygous identity by descent, in which case admixture may not have much of an effect since the total burden of different rare variants is similar across populations. The other demographic shift offsetting the effects of inbreeding is the declining incidence of consanguinity. One in eight humans worldwide still marry first or second cousins, the rate being particularly high in Pakistan and Iran for example, but quantitative estimates of its contribution to congenital and/or chronic disease are surprisingly difficult to find [27].

Rare variants are receiving the bulk of the attention of geneticists currently as whole exome and genome sequencing have supplanted genotype-driven genome-wide association studies. Some of this reflects disappointment over the magnitude of the missing heritability, which many assumed must be attributable to rare variants. However, it is now clear both from SNP-based heritability estimation with mixed models, and from extrapolation of the spectrum of observed effects, that in fact much, if not most, of the heritability is more hidden than missing [28]. That is to say, the effect sizes of alleles are so small on average that sample sizes of several hundreds of thousands will be required to detect individual SNPs at stringent GWAS levels. Even for rare psychiatric conditions that have well-documented rare variant contributions, common polymorphisms seem to explain much more of the genetic variance [29]. At a minimum they are the main genetic modifiers of penetrance and expressivity of rare variants.

The Big Picture of Evolving Common-Variant Genetic Risk

This raises the question: is the distribution of common variant effects similar across human populations? There are essentially two non-exclusive explanations for differential genetic contributions to disease susceptibility between populations: different allelic effect sizes, or different allele frequencies. The most careful analyses to date suggest that the majority of allelic effects are relatively similar, at least between Asian and Caucasian populations [30, 31], and probably in Africans as well (although without knowing the actual causal variant responsible for each association signal, this is less clear). There are certainly exceptions, but it seems that the bigger source of divergence in risk is divergence in allele frequencies. Disease-associated alleles do have elevated Fst values on the whole [32], in some cases large enough for the minor-and major-allele status to flip between populations, which also influences the relative risk of heterozygotes.

Since individual genotypes rarely explain more than a few percent of the variance in liability, it is natural to generate genetic risk scores that summarize the joint risk across multiple alleles. The simplest risk scores are allelic sums, which are unbiased by effect size estimates. Summation of the number of risk alleles each individual carries classifies them relative to the remainder of a population, and comparison of the distribution of risk scores among populations provides a crude measure of average divergence in risk.

Figure 2 shows four such compilations from the 1000 Genomes Project [33] (AN, J Zhao and GG, unpublished), for Parkinson's Disease, Coronary Artery Disease, Type 2 Diabetes, and Prostate Cancer. There is no significant divergence in risk for Parkinson's Disease, but the other three diseases each show elevated risk in East Asians or Africans, broadly concordant with known relatively high risk of CAD, or of Type 2 Diabetes and Prostate Cancer, respectively.

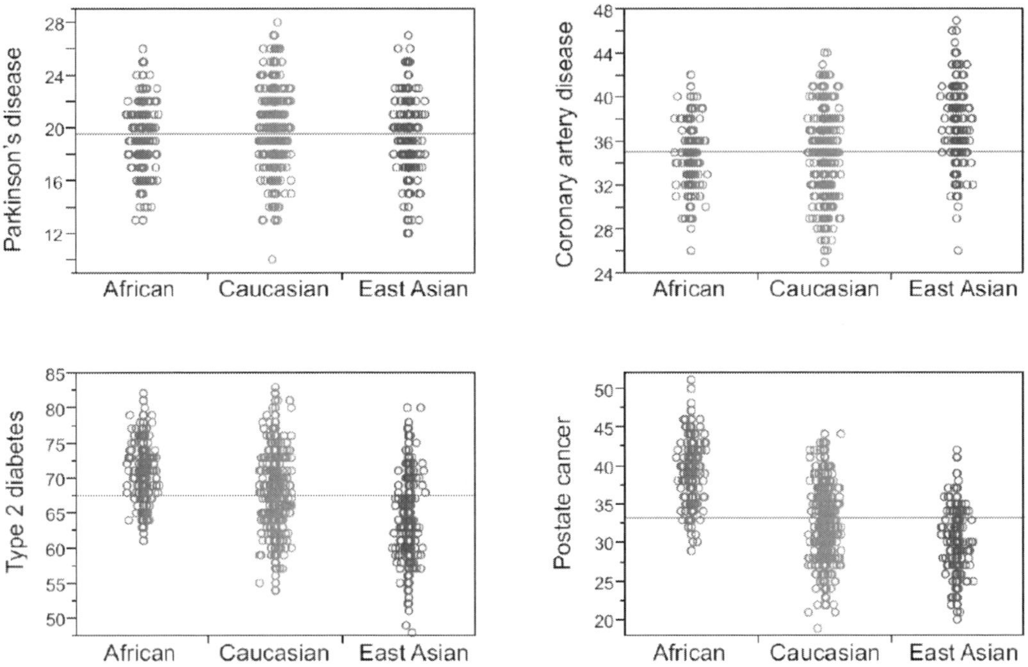

Figure 2. Genetic risk scores for four common dseases (diseases) in the 1000 Genomes dataset. GRS were calculated as the simple sum of the number of risk alleles at 29 loci for Parkinson's disease, 38 loci for coronary artery disease, 77 loci for type 2 diabetes, and 49 loci for prostate cancer. Genotype data downloaded from the 1000 Genomes website, and risk alleles obtained from the Genome-wide association study browser. Details available upon request.

It is important to emphasize two caveats to this analysis, namely that the variants were predominantly discovered in European ancestry individuals, and they only explain of the order of 10% of the genetic risk for each disease. Consequently, the analysis does not allow us to attribute a proportion of the divergence in risk among populations to these genetic risk scores. However, we can say that there has been substantial divergence between the three major human population groups in the distribution of frequencies of the discovered variants that make relatively large contributions to the diseases. Why this is the case is anyone's guess. Evolutionary medicine provides speculation in some cases, and very often conjures up scenarios that have popular appeal, but essentially no grounding in fact [34]. We do know that the signatures of selection from sophisticated tests of neutrality applied to the analysis of nucleotide diversity provide little hint of recent strong directional or balancing selection for most disease-associated variants [35].

There is no good evidence that selection against disease has been a major driver of shifts in the frequencies of the alleles that associate with those diseases. A more parsimonious explanation, at least in our view, is that some combination of founder effects, drift, and weak multivariate selection related to diverse and complex contributions of the alleles to a range of traits, have contributed to divergence [36].

Genotype-by-Environment Influences on Human Health and Fitness

The second potential source of heritability, in addition to rare variants, that many assumed might be discovered in the wake of GWAS is G×E. At the level of individual genotypes, though, there has been little evidence that interaction effects are a major source of variance. There are good technical explanations for this situation, including the difficulty of defining and measuring the appropriate environmental covariate(s), and the multiple comparison burden, particularly under the expectation that average G×E interaction effects may be small relative to main genotype effects that are themselves small [37]. Journals have been conservative in filtering reports, perhaps recognizing that interactions need to stand the test of time: given 5 discovered variants and 1 environment, a p-value for the interaction of 0.001 may be experiment-wide significant, but should the number of discovered variants increase to 100 at some point in the future, the inferred interaction effect would need to be re-evaluated. Only a quite small number of replicated specific G×E interactions have been documented in humans, many discussed elsewhere in this book. One prominent instance is the FTO locus, at which physical activity attenuates the influence of the high-weight gain allele on obesity [38].

A likely more pervasive source of G×E is at the level of genetic risk scores. For example, GRS×E for the interaction between a weighted allelic sum and sugary beverage consumption in the Nurses Health Study that replicates in other cohorts [39]. That is to say, individuals with a high GRS for obesity gain weight in a 10 year period more than those with a low score, as do individuals who drink a Coca-cola or equivalent, or more, per day. Those with both a high GRS and sugary beverage consumption gain weight even more quickly, to the tune of a couple of extra pounds over a decade. With this in mind, we simulated GRS×E under a variety of models in the context of GWAS and arrived at two important conclusions [40]. First, there is a wide range of realistic conditions under which GRS×E interactions can be highly significant (10^{-8} or less) in a sample of several thousand individuals, even in the complete absence of individually significant G×E. The simulations assumed two environments (ancestral and contemporary), no divergence in allele frequencies, and a shift in the phenotype distribution similar to that observed for BMI in the past two decades. The core result was obtained assuming up to 2-fold increased effect size in the contemporary population for a random sample of 1% of 500 significant SNPs, or just a 1.3-fold increase for up to 20% of the common polymorphisms.

Second, we demonstrated that under most of these conditions, the power to detect any SNPs in the GWAS setting in a mixed population is dramatically reduced relative to the contributing populations alone. The reason is that the variance associated with each individual genotype is inflated by the greater variability due to the genetic contributions at all of the other loci. In other words, assuming precisely the circumstance under which G×E is likely to occur, namely interaction with those at highest genetic risk across multiple loci, the likelihood of finding the contributing SNPs is greatly reduced; when they are found they are unlikely to individually show G×E, yet GRS×E would be readily detected. To the extent that contemporary populations are a mixture of people living different lifestyles, we propose that this mechanism could be a considerable source of undetected heritability.

It follows that the importance of G×E is likely to be under-appreciated [41]. Assuming it is important, there are three major models for how the interactions may be shaping the evolving risk for multiple diseases, as illustrated in Figure 3.

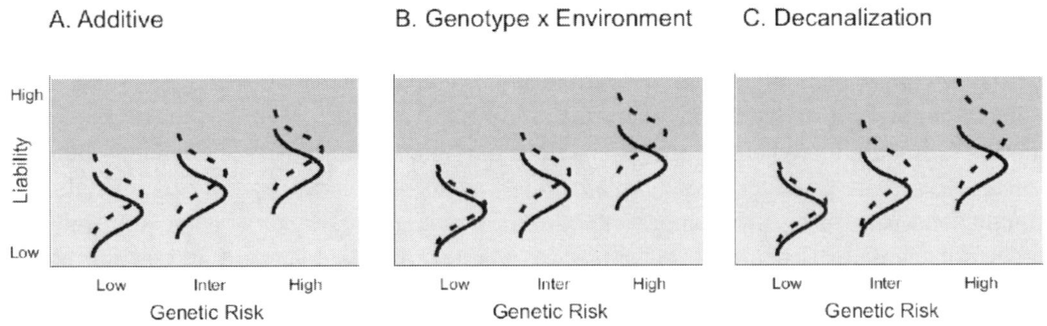

Figure 3. Three models for the impact of environment on genetic risk for disease. Solid dark curves approximate the liability distribution for individuals with low, intermediate, and high genetic risk in an ancestral environment. Dashed curves show liability distributions in a contemporary environment where more individuals exceed the threshold for high risk of disease (indicated by the purple region). (A) The standard additive model assumes that the environment simply shifts the distribution of liability for each genotype class toward higher risk independent of genotypic risk. However, most individuals with disease have high genotypic risk in the modern environment, so GxE is observed at the phenotype level. (B) The GxE model assumes that the shift in liability in the modern environment increases as genetic risk increases. (C) The decanalization model assumes that an additional effect of the modern environment is to increase the variability of liability for individuals in higher risk categories, with the result that more individuals cross into the zone of high disease likelihood.

Common wisdom is that there is no interaction at the level of underlying liability, but given a liability threshold there can well be G×E for disease (Figure 3A). Thus, there is no change in the distribution of genetic effect sizes on BMI in a novel environment, but if only 1% of the population is beyond the BMI>30 threshold in the ancestral environment, then 10% would be beyond it in the presence of an environment that adds 2 BMI units to everyone. This would induce a GRS×E for categorically define obesity, since people with the combination of high risk score and modern lifestyle are much more likely to be obese than those with just one of the factors. Note that the figure exaggerates the effect for illustrative purposes. The second model (Figure 3B) is the one outlined above, where there is a true GRS×E at the level of the liability itself, for example of BMI. This would also exacerbate the interaction effect at the level of obesity or obesity-related disease.

The third scenario is called decanalization of disease [3]. Canalization is evolved homeostasis [4]. Under persistent stabilizing selection, genetic systems are expected to evolve not just to the optimum phenotype, but also to ensure that as many individuals in the population are close to the optimum [42]. Mechanistically, this is attributed to the evolution of buffering mechanisms such as Hsp70 or miRNA mediation of regulatory homeostasis [43]; but it can also arise more generally as a property of the network of interactions among alleles. One conceptualization of canalization supposes a complex trait, such as blood sugar levels, which is a non-linear function of two contributing traits, such as insulin sensitivity and insulin production. Evolution favors homeostasis where small offsets in either response to or production of insulin is offset by a small compensatory change in the other biochemical parameter (production or response). Individuals who chronically produce insulin in an effort to overcome the resistance they have built up after many years of high calorie consumption, are no longer able to buffer small changes, and fall into impaired glucose tolerance or eventually type 2 diabetes. Similar models can be conceived for many other medical conditions.

The notion of the adolescent human genome is that, like adolescents themselves, the genome has not yet reached a state of stable equilibrium, and finds itself in environments that it is not quite comfortable with [44]. The twin perturbations of rapid genome evolution, and contemporary cultural environments that are quite different from those encountered throughout human evolution, push populations away from the optimal zone of canalization, increasing overall genetic variability (Figure 3C). Rather than there necessarily being a directional shift in liability, this idea supposes that there is an increase in genetic variance that results in more people being beyond the disease threshold.

Given that we are only just beginning to be able to generate genetic risk scores, and measure their interaction effects, it must be said that there is as yet little empirical support for the decanalization of disease model. A meta-analysis of type 2 diabetes did find that Africans, Europeans, and Asians occupy different portions of the curve relating insulin sensitivity to insulin production [45]: Africans tend to have greater insulin resistance and production, so are away from what may be expected to be the canalized optimum. It is though not yet clear how much of African elevated genetic risk is for either production or release of insulin, and certainly not established that stabilizing selection has had any role in shaping the optimum (which may not be the same under all dietary conditions).

Summary and Perspective

Returning to the issue of whether human fitness is evolving, my argument is that (i) health risks per se are evolving, (ii) their relationship to classical fitness is complex and likely overwhelmed by cultural trends, (iii) genotypic disease risks have certainly diverged among human populations but for reasons we will not understand for a long time, and (iv) the interaction between environment and composite risk score needs to be reconsidered now that we have a large number of identified alleles for many diseases.

Cultural and behavioral practices, in conjunction with a shift in the age distribution, at least in developed societies, have had a greater impact on the epidemiological transition of disease than genotypic change. Failure to detect interaction effects at the level of individual genotypes does not mean that they would not be detected if genetic risk scores could be routinely evaluated with relevant environmental factors included in the model. We expect that once hundreds of loci, and eventually causal SNPs, are associated with most common diseases, it will be possible to evaluate which biochemical and cellular mechanisms they regulate. To the extent that those mechanisms interact non-linearly, it will become possible to test the hypothesis that decanalization is a particular class of genotype-by-environment interaction that in a general sense helps explain the rising incidence of disease. Perhaps too it may help highlight which cultural practices are best subject to preventative interventions.

Key Facts

1. Many common complex diseases, including diabetes, coronary heart disease, asthma, and depression, not to mention dementia, are globally increasing in prevalence.
2. Human fecundity has declined rapidly in the past two generations resulting in a drop in fitness in developed societies that has a behavioral, not genetic, basis.
3. Rare deleterious variants are strongly selected against, but likely increasing in frequency in human populations as a result of relaxed purifying selection and population expansion.
4. Common variant genetic risk profiles have diverged among human populations but are being homogenized by admixture as a result of migration, which is the major influence on evolving genetic risk.
5. Genetic risk score-by-environment interactions would have been undetected by genome-wide association studies, but could easily be a pervasive contributor to increasing disease incidence.
6. The human genome is adolescent in the sense that it we are still relatively young as a species and find ourselves in environments with which we are uncomfortable, and as a consequence decanalization may also contribute to the increasing burden of disease.

Acknowledgment

We are grateful to Jing Zhao for her help in analysis of the genetic risk scores. GG is supported in part by 1-P01-GM GM0996568.

Glossary

genome-wide association studies - Usually referred to as GWAS, the approach to genetic mapping that relies on testing for the association between genotypes and a phenotype, often disease status, at hundreds of thousands of variable sites throughout the genome.

genotype-by-environment interactions - Abbreviated G×E, situations where the effect of a genotype on a trait is modified by the environment, for example where risk of obesity due to a specific locus is present in individuals who consume sugary beverages, but absent in those who do not.

Fitness - In evolutionary biology, fitness refers to the differential production of viable and fertile offspring: fitter individuals have a higher likelihood of their genes propagating into the next generation.

rare deleterious variants - Genetic polymorphisms that have a minor allele frequency less than 5% and are associated with reduced fitness, either decreasing the health or fertility of the individual.

mutational target - That portion of the genome that will affect a trait of interest when mutations affect the expression or function of genes is known as the mutational target of the trait.

admixture - The process whereby two previously isolated populations come together, resulting in the mixture of alleles from one population into the gene pool of the second one.

compound heterozygosity - Analagous to a recessive homozygote, except that the two mutations are different alleles, derived from parents who are heterozygous carriers for two different mutations.

missing heritability - The term coined to described the observation that genetic variation explained by genetic polymorphisms identified by genome-wide association studies explains much less than the variation that is known to be due to genotypic differences. Much of the missing heritability is now known to be due to rare or very small effect polymorphisms whose effects are hidden by low statistical power.

Fst values - Fst is a statistic that describes the reduction in heterozygosity in a total population due to slight divergence in frequencies between sub-populations; it is thus a measure of population structure.

genetic risk scores - Abbreviated GRS, genetic risk scores summarize the contributions of multiple loci to a trait, for example adding the number of risk alleles across loci, or generating a cumulative log likelihood.

balancing selection - Natural selection that tends to maintain a balance of two alleles at a locus, for example because the alternate alleles have higher fitness in males and females, or in two environments.

founder effects - The influence on the gene pool due to change in allele frequency when a new population is founded by emigration of a small number of individuals from a source population.

canalization - The process by which evolution favors genotypes that favor the generation of individuals with a near-optimal phenotype. Under persistent stabilizing selection, the genome evolves not just to toward the optimum, but also to be more robust to new mutations or environmental perturbations.

homeostasis - Robustness to perturbation by mutations or environmental stress

stabilizing selection - Natural selection that favors individuals that have an intermediate phenotype, as opposed to directional selection that favors higher or lower values. It does not however generally result in maintenance of genetic variation at individual loci.

REFERENCES

[1] Visscher, P.M. et al. Five years of GWAS discovery. *Am. J. Hum. Genet.* 90, 7-24 (2012).

[2] Barreiro, L.B. et al. Natural selection has driven population differentiation in modern humans. *Nat. Genet.* 40, 340-345 (2008).

[3] Gibson, G. Decanalization and the origin of complex disease. *Nat. Rev. Genet.* 10, 134-140 (2009).

[4] Gibson, G., Wagner, G. Canalization in evolutionary genetics: a stabilizing theory? *Bioessays* 22, 372-380 (2000).

[5] Wang, Y., Beydoun, M.A. The obesity epidemic in the United States—gender, age, socioeconomic, racial/ethnic, and geographic characteristics: a systematic review and meta-regression analysis. *Epidemiol. Rev.* 29, 6-28 (2007).

[6] Ford, E.S., et al. Prevalence of the metabolic syndrome among US adults. Findings from the Third National Health and Nutrition Examination Survey. *JAMA*. 287, 356-359 (2002).

[7] Keys, A. Coronary heart disease – the global picture. *Atherosclerosis* 22, 149-192 (1975).

[8] Mann, G.V., et al. Cardiovascular disease in the masai. *J Atheroscl. Res.* 4, 289-312 (1964).

[9] Akinbami, L.J., Schoendorf, K.C. Trends in childhood asthma: prevalence, health care utilization, and mortality. *Pediatrics* 110, 315-322 (2002).

[10] Dubey, A., et al. Hip fracture prevention: a review. *Am. J. Orthoped.* 27, 407-412 (1998).

[11] Lépine, J-P, Briley, M. The increasing burden of depression. *Neuropsychiatr. Dis. Treat.* 7 (Suppl.1), 3-7 (2011).

[12] Sagan, L.A. The health of nations: true causes of sickness and well-being. *Basic Books*. New York, NY. (1987).

[13] Cutler, D.M., et al. The determinants of mortality. *NBER Working Paper* No. 11963. http://www.nber.org/papers/w11963 (2006).

[14] Fortson, J.G. Mortality risk and human capital investment: the impact of HIV/AIDS in Sub-Saharan Africa. Rev. *Econ. Stat.* 93, 1-15 (2011).

[15] Walsh, D., et al. The aftershock of deindustrialization—trends in mortality in Scotland and other parts of post-industrial Europe. *Eur. J. Pub. Health* 1-7 doi:10.1093/eurpub/ckp063 (2009).

[16] Flynn, J.R. Massive IQ gains in 14 nations: what IQ tests really measure. *Psychol. Bull.* 101, 171-191 (1987).

[17] Pinker, S. *The Better Angels of Our Nature: Why Violence Has Declined.* Penguin, NY. (2012).

[18] Skakkaebeck, N.E., et al. Is human fecundity declining? *Int. J. Androl.* 29, 2-11 (2006).

[19] Fu, W., et al. Analysis of 6,515 exomes reveals the recent origin of most human protein-coding variants. *Nature* 493, 216-220 (2012).

[20] Simons, Y.B., et al. The deleterious mutation load is insensitive to recent population history. *Nat Genet*. 46, 220-224 (2014).

[21] Lynch, M. Rate, molecular spectrum, and consequences of human mutation. *Proc. Natl Acad. Sci. USA.* 107, 961-968 (2010).

[22] Lesecque, Y., et al. A resolution of the mutation load paradox in humans. *Genetics*, 191, 1321-1330 (2012).

[23] Fromer M., et al. De novo mutations in schizophrenia implicate synaptic networks. *Nature.* 506, 179-184 (2014).

[24] Gratten, J., et al. Large-scale genomics unveils the genetic architecture of psychiatric disorders. *Nat. Neurosci.* 17, 782-790 (2014).

[25] Kirov G, et al. De novo CNV analysis implicates specific abnormalities of postsynaptic signalling complexes in the pathogenesis of schizophrenia. *Mol Psychiatry*. 17, 142-153 (2012).

[26] Purcell, S.M., et al. A polygenic burden of rare disruptive mutations in schizophrenia. *Nature* 506, 185-190 (2014).

[27] Bittles, A.H., Black, M.L. Consanguinity, human evolution, and complex diseases. *Proc. Natl Acad. Sci. USA.* 107 Suppl 1, 1779-1786 (2009).

[28] Gibson, G. Hints of hidden heritability in GWAS. *Nat. Genet.* 42, 558-560 (2010).
[29] Schizophrenia Working Group of the Psychiatric Genomics Consortium. Biological insights from 108 schizophrenia-associated genetic loci. *Nature* 511, 421-427 (2014).
[30] Marigorta, U.M., Navarro, A. High trans-ethnic replicability of GWAS results implies common causal variants. *PLoS Genet.* 9, e1003566 (2013).
[31] Carlson, C.S., et al. Generalization and dilution of association results from European GWAS in populations of non-European ancestry: the PAGE study. *PLoS Biology* 11, e1001661 (2013).
[32] Myles S, et al. Worldwide population differentiation at disease-associated SNPs. *BMC Med. Genom.* 1,22 (2008).
[33] 1000 Genomes Project Consortium. An integrated map of genetic variation from 1,092 human genomes. *Nature* 491, 56-65 (2012).
[34] Wade, N. A. Troublesome inheritance: genes, race and human history. *Penguin*, NY. (2014).
[35] Fu, W., Akey, J.M. Selection and adaptation in the human genome. *Annu. Rev. Genomics Hum. Genet.* 14, 467-489 (2013).
[36] Agarwala, V., et al. Evaluating empirical bounds on complex disease genetic architecture. *Nat. Genet.* 45, 1418-1427 (2013).
[37] Kauffmann, F., Demenais, F. Gene-environment interactions in asthma and allergic diseases: challenges and perspectives. *J Allergy Clin Immunol.* 130, 1229-1240 (2012).
[38] Kilpeläinen, T.O., et al. Physical activity attenuates the influence of FTO variants on obesity risk: a meta-analysis of 218,166 adults and 19,268 children. *PLoS Med.* 8, e1001116 (2011).
[39] Qi, Q, et al. Sugar-sweetened beverages and genetic risk of obesity. *N. Engl. J. Med.* 367, 1387-1396 (2012).
[40] Marigorta, U.M., Gibson, G. A simulation study of gene-by-environment interactions in GWAS implies ample hidden effects. *Front. Genet.* 5, 225 (2014).
[41] Moffitt, T.E., et al. Strategy for investigating interactions between measured genes and measured environments. *Arch. Gen. Psychiatry* 62, 473-481 (2005).
[42] Hermisson, J., Wagner, G.P. The population genetic theory of hidden variation and genetic robustness. *Genetics* 168, 2271-2284 (2004).
[43] Rutherford, S.L. Between genotype and phenotype: protein chaperones and evolvability. *Nat. Rev. Genet.* 4, 263-274 (2003).
[44] Gibson, G. *It Takes a Genome*. FT Press, NY. (2010).
[45] Kodama, K., et al. Ethnic differences in the relationship between insulin sensitivity and insulin response: a systematic review and meta-analysis. *Diabetes Care* 36, 1789-1796 (2013).

In: Gene-Environment Interactions and Human Diseases ISBN: 978-1-63483-680-7
Editor: Lu Qi

Chapter 3

STUDY DESIGN AND ANALYSIS CONSIDERATIONS IN GENOME-WIDE STUDIES OF GENE-ENVIRONMENT INTERACTIONS

Joanna M. Biernacka[1,2], Joseph Usset[3], Brooke L. Fridley[3] and Stacey J. Winham[1]

[1]Division of Biomedical Statistics and Informatics, Department of Health Sciences Research, Mayo Clinic, Rochester, MN, US

[2]Department of Psychiatry and Psychology, Mayo Clinic, Rochester, MN, US

[3]Department of Biostatistics, University of Kansas Medical Center, Kansas City, KS, US

ABSTRACT

Most complex human diseases are believed to be influenced by a combination of genetic and environmental factors, and gene-environment interactions are thought to play a crucial role in the development of complex diseases and disease prognosis. Numerous candidate gene studies to investigate specific gene-environment interactions have been performed, and hypothesis-generating genome-wide gene-environment interaction studies are beginning to emerge. This chapter provides an overview of study designs, various analytical approaches for studying gene-environment interactions, and potential sources of biases. Powerful statistical approaches suitable for genome-wide analyses, including two-step filtering approaches, data-mining methods, and gene-level and pathway-level analyses are also discussed.

Keywords: Gene-environment interaction, bias, genome-wide gene-environment studies, multi-marker, data-mining

INTRODUCTION

Genome-wide association studies of complex human traits have now identified hundreds of associated genetic variants [1]. Despite this progress, replicated associations explain a small proportion of the heritability of complex traits [2], and ongoing research is largely focused on discovering additional risk factors [3]. Because both genetic and environmental factors are known to contribute to most complex traits, with gene-environment (G-E) interactions playing a role (meaning that for some genetic variants, the effect of the variant depends on certain environmental exposures, or equivalently, that the effect of an environmental factor depends on an individual's genetic makeup), investigation of G-E interactions may enable the identification of novel genes or environmental factors that interact to influence the trait. Quantifying G-E interactions also has the potential to improve predictive models of disease onset or treatment outcomes. Finally, studying G-E interactions may guide further studies of disease-causing mechanisms, by demonstrating which factors work together to alter risk.

Despite the widespread belief that G-E interactions play an important role in human health, relatively few comprehensive investigations of G-E effects have been conducted. Most G-E studies performed to date have used relatively small samples and investigated few candidate genetic variations and specific environmental factors, leading to numerous un-replicated findings [4]. Duncan and Keller discussed reasons behind the lack of consistent G-E interaction findings, focusing specifically on studies of psychiatric traits [4]. Factors contributing to the non-replicability problem include low power (inadequate sample sizes), publication bias, high extent of multiple testing that is not adequately corrected leading to reports of false positive findings, and inconsistent definitions of phenotypes and environmental variables across studies.

The study of complex disease genetics has shifted from investigating candidate genes to agnostic genome-wide searches for trait-associated single nucleotide polymorphisms (SNPs). Furthermore, more sophisticated analyses that make use of prior knowledge, such as gene-level tests [5] and pathway-based gene-set analyses [6], are now routinely used. Similar to studies of main genetic effects, studies of G-E interactions are also beginning to shift from candidate gene to genome-wide approaches, with an emergence of the concepts of the "GWEIS" (genome-wide environmental interaction study) [7] or the "GEWIS" (gene-environment wide interaction study) [8, 9]. However, the idea of applying advanced multi-SNP methods to investigate G-E interactions and uncover novel mechanisms contributing to complex trait etiology is not well developed or widely utilized.

This chapter introduces common study designs and analysis techniques used in G-E studies, pointing out some potential sources of bias and proposed solutions. Novel types of analyses that warrant further development and wider application in the context of G-E GWAS are then discussed.

STUDY DESIGN

Many study designs commonly used for investigating genetic or environmental effects separately, including standard case-control and cohort samples, are readily adapted to study

G-E interactions, while other designs, such as the case-only design, are specifically designed for testing interaction effects [10]. Thomas [9] provided a detailed discussion of various types of study designs for investigating G-E interactions, along with their advantages and disadvantages. The following overview of methods used in G-E interaction analysis focuses on studies of unrelated individuals with available genetic and environmental data (e.g., population-based case-control or cohort studies). G-E interactions can also be studied in other contexts, including family-based studies of heritability and linkage, which will not be discussed here.

Cohort studies with longitudinal follow-up can be particularly advantageous for investigating G-E effects. Such designs can allow for investigation of the effects of early-life environmental exposures and changes in genetic and environmental effects over time. While researchers recognize the need for longitudinal studies, and several well-known large cohorts with longitudinal data have been studied (e.g., Framingham Health Study [11]), only a few methods have been developed for analyzing G-E effects using repeated-measures data [12, 13].

For rare diseases, case-control or family-based designs with case ascertainment are more powerful than cohort designs. Because case-control studies are most common due to their efficiency and cost-effectiveness, the discussion of statistical methods in this chapter focuses primarily on this study design. The case-only design is a notable alternative, as it is a powerful approach for testing for G-E (or G-G) interactions when the environmental factor is not under genetic influence (or when the two genes of interest are independent in the general population). However, this design has important limitations – it cannot be used to test for genetic or environmental main effects, and its validity depends on the assumption of G-E independence.

Although G-E interaction studies often use previously collected samples of convenience, this practice can lead to substantial loss of power if the available samples are not well suited for addressing the G-E interaction hypothesis. For example, use of previously-collected case and control samples, for which the environmental data was collected in a different way, may either preclude a case-control analysis of G-E interactions, or may bias the analysis. Furthermore, differences between cases and controls in the distribution of relevant covariates (e.g., age and sex) as a result of subject ascertainment may impact the distribution of the environmental exposure of interest and lead to loss of power for detecting G-E effects. For example, in a recent study of BMI-SNP interaction effects, we found that the power of the analysis was substantially affected by the choice of controls (data not shown).

Another important consideration for studies of G-E interaction is the sample size, as it impacts the power of the study to detect G-E effects. While many studies are underpowered to detect genetic effects on complex traits, which are usually modest, studies of G-E interactions are especially vulnerable to this problem as larger samples are needed to detect interactions [14, 15]. Several studies have investigated power to detect G-E interactions with particular analysis methods in the context of genome-wide studies, providing power and sample size estimates for a range of G-E effect sizes [15, 16]. Although such estimates are useful for comparing power across methods, their utility in estimating absolute power to detect G-E effects is limited by the fact that numerous parameters in addition to sample size contribute to the power to detect G-E interactions, including the genetic variant frequency, the distribution/frequency of the environmental variable, the genetic and environmental marginal effects, and the interaction effect size. Nevertheless, power estimates demonstrate that under realistic models, sample sizes exceeding those of typical G-E interaction studies will be needed to detect most interaction effects [15].

TRADITIONAL STATISTICAL ANALYSIS OF G-E INTERACTIONS

G-E interaction arises when the effect of a genetic variant (e.g., SNP) on the phenotype is modified by an environmental factor, which in statistical terms is defined as a departure from an additive model with only marginal genetic and environmental effects on the phenotype. In this definition of interaction, it is important to note that a departure from additivity is scale-dependent, which has been discussed extensively by others [17]. Here we define interaction in terms of two models (one for a quantitative trait, and one for a binary trait) that are typically used in statistical genetic analyses.

Let G be the genotype (typically coded as 0, 1, 2 representing the number of copies of the minor allele) and E be the environmental factor (e.g., E=0, 1 for a binary exposure). For a quantitative trait Y, the expected value of the trait can be modeled as:

$$E(Y) = \beta_0 + \beta_G{*}G + \beta_E{*}E + \beta_{GE}{*}G{*}E \qquad \text{Model (1)}$$

The parameter β_{GE} represents the G-E interaction effect, and $\beta_{GE}=0$ implies no interaction, while $\beta_{GE}\neq 0$ indicates that the effect of E modifies the effect of G on the trait Y. A similar model can be defined for a binary trait D, where the logit of the probability of D (log odds of the disease) is modeled as:

$$\log(P(D)/1\text{-}P(D)) = \beta_0 + \beta_G{*}G + \beta_E{*}E + \beta_{GE}{*}G{*}E \qquad \text{Model (2)}$$

In this model, β_{GE} again represents the G-E interaction effect, such that $\beta_{GE}=0$ implies no interaction, while $\beta_{GE}\neq 0$ is interpreted as indicating a G-E interaction effect. The parameters in Models (1) and (2) are typically estimated using regression techniques, and statistical evidence against the null hypothesis of $\beta_{GE}=0$ is interpreted as evidence of a G-E interaction effect on the trait.

Suppose a logistic regression model including a G-E interaction effect on the risk of a disease D (Model (2)) is fit to the data, with G=0, 1, 2 being the genotype at a particular locus (e.g., AA, AT, TT) and E=0,1 being a binary environmental exposure (such as smoking no/yes). If no G-E interaction is present (i.e., $\beta_{GE} = 0$), $\exp(\beta_G)$ is the odds ratio for the genetic effect, which describes the odds of disease for an individual with genotype G=1 compared to an individual with genotype G=0, regardless of environmental exposure status E. Similarly, $\exp(\beta_E)$ is the odds ratio for the environmental effect, which gives the ratio of the odds of disease for an exposed individual (E=1) compared to an unexposed individual (E=0), regardless of the value of genotype G. However, if a G-E interaction exists then $\beta_{GE} \neq 0$ and the effect of G modifies the effect of E (and vice versa). In this case the OR of G=1 vs. G=0 is given by $OR=\exp(\beta_G{*}1+\beta_{GE}{*}E)$, which is dependent on the exposure status E. If E=0, then the OR for the genetic effect is $OR=\exp(\beta_G)$; whereas if E=1, then the OR for the genetic effect is $OR= \exp(\beta_G+\beta_{GE})$. A numerical example is shown in Table 1 and Figure 1, where the odds ratios of any G*E combination is given relative to the baseline G=0, E=0. For the particular models in this example, the genetic effect becomes higher in the presence of the environmental exposure under the model with a G-E interaction (Figure 1b). For the example shown in Table 1, when no interaction is present (left side of Table 1), the odds ratio of disease due to the genetic effect is the same regardless of exposure status, while under the model with G-E interaction (right side of Table 1), the odds ratio of disease due to the genetic effect is higher for an exposed individual than for an unexposed one.

Typically G-E interaction is statistically evaluated by testing the null hypothesis that β_{GE}=0 in Models (1) or (2) using a one degree of freedom (1df) test. If a researcher is not interested in testing for G-E interaction explicitly, but rather detecting relevant genetic variants while allowing for the possibility of a G-E interaction, a joint two degree of freedom (2df) test of H_0: β_G=β_{GE}=0 can be used instead, and has been shown to be a statistically powerful approach [18]. Depending on the research goals, either the 1df or 2df tests may be appropriate.

The methods above apply to the commonly used case-control design; however, many G-E studies utilize a case-only design (i.e., analysis is subset to cases), which necessitates a different analysis approach. Because unaffected individuals are excluded, marginal genetic or environmental effects cannot be tested. Under the assumption that the genetic and environmental factors are independent, case-only methods directly evaluate hypotheses of G-E interaction by testing the association between the genetic and the environmental factors; a G-E interaction is inferred when G and E are associated in cases despite being independent in the general population.

Under the assumption of G-E independence, the case-only design is more powerful than the case-control design, but leads to biased results if the assumption is violated [19]. This assumption may be false if the environmental variable of interest is under genetic influence (e.g., body mass index). Hybrid methods that combine the case-only and case-control estimates of G-E interaction [20] have been proposed to avoid potential bias and preserve power, including an empirical Bayes approach that calculates a weighted average of both estimates [21].

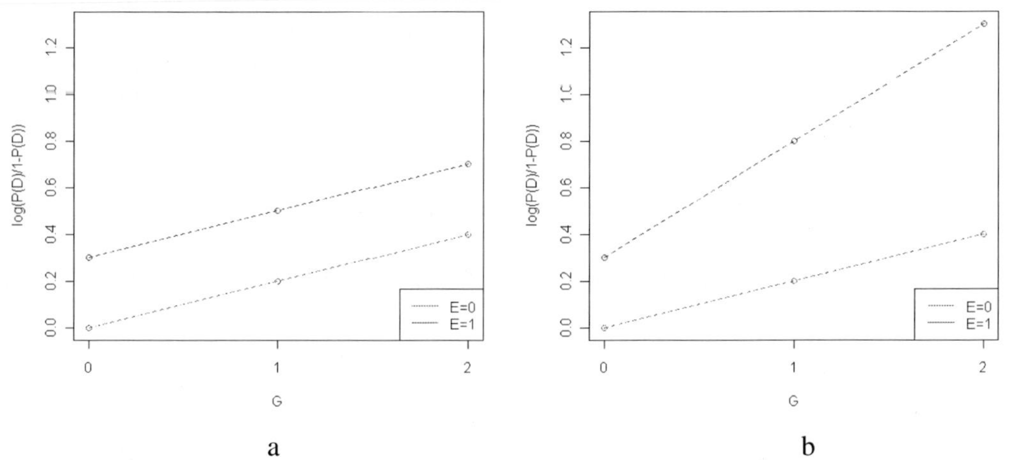

Figure 1. Logistic regression model of a gene-environment interaction effect on a binary trait, modeled as log(P(D)/1-P(D)) = 0 + 0.2*G + 0.2*E + β_{GE}*G*E with (A) β_{GE}=0, i.e., no G-E interaction effect and (B) $\beta_{GE}\neq$0 demonstrating a G-E interaction effect on risk of the disease.

Table 1. Odds ratios for the model log(P(D)/1-P(D)) = 0 + 0.2*G + 0.3*E + β_{GE}*G*E under a situation with no G-E interaction and a situation with a G-E interaction effect on the disease, D. The E=0 (no environmental exposure) and G=0 (0 copies of minor allele, i.e., common homozygote) is the baseline category that other G/E combinations are compared to. When no interaction is present (left side of Table), the odds ratio of disease attributed to the genetic effect is 1.22 regardless of exposure status; for an individual with G=1 as compared to G=0, the OR for an unexposed individual (E=0) is 1.22/1=1.22--the same as for an exposed individual with E=1 (1.65/1.35=1.22). However, under the model with G-E interaction (right side of Table), the odds ratio of disease attributed to the genetic effect is higher for an exposed individual; when E=0, the OR for an individual with G=1 compared to G=0 is again 1.22/1=1.22, but when E=1, the respective OR is 2.23/1.35=1.65

	No G-E interaction (β_{GE}=0)		G-E interaction (β_{GE}= 0.3)	
	E=0	E=1	E=0	E=1
G=0	1	1.35	1	1.35
G=1	1.22	1.65	1.22	2.23
G=2	1.49	2.01	1.49	3.67

Potential Sources of Bias in G-E Interaction Analyses and Proposed Solutions

Bias in G-E interaction analyses can lead to invalid and non-replicated conclusions; thus reducing the potential for bias is a critical issue in G-E interaction analyses, especially due to the non-experimental nature of most G-E studies [22]. Bias in G-E studies can arise both from study design and the chosen method for data analysis. A comprehensive review of potential biases attributed to different epidemiological designs (see, for example, Rothman et al. [23]), and various complex statistical methods, is not provided here. Instead, this section summarizes several specific sources of bias that arise in population-based G-E studies of unrelated individuals (e.g., case-control studies) analyzed using generalized linear models with the environmental factor, individual genetic marker, and their interaction (as in Model (1) and (2)), and highlights recently proposed solutions to minimize the potential bias.

Measurement and collection of environmental data is a key challenge in studies of G-E interactions [24], and measurement error or misclassification may lead to bias or reduced power [25, 26]. Definition of the environmental factor is critical to the success of G-E investigations, and careful thought must be given to ensure that the environmental factor is measured reliably and consistently and is appropriate for the research question. Often the environmental variable is dichotomized or categorized for analysis even if it was measured quantitatively; however, this typically decreases power. In replication studies, the environmental exposure should be measured as similarly as possible to the original study. Unfortunately, the environmental variable is usually measured retrospectively, which can be prone to recall bias. The timing of measurement and the need for multiple measurements per subject are also important aspects to consider when designing a G-E study. Cohort studies that incorporate repeated measurements of environmental variables over time can improve the quality of exposure assessment [24]. If the environmental exposure is difficult to assess, there is a trade-off between sample size and accuracy of measurement; in some cases, a two-stage design with a more accurate measurement of the environment in a subsample may maximize power [24].

Analogously to epidemiological studies in general, confounding factors can bias G-E interaction analyses. In many epidemiological studies, known confounding factors can be controlled for by including them as covariates in a regression model, or through stratified analyses [23, 27]. Population stratification is a particular concern for genetic studies [28], and approaches have been developed to control for this potential source of bias [29, 30]. While bias is less likely in studies of G-E interaction [9], certain situations can give rise to bias due to confounding. In particular, when sub-populations (e.g., ancestral group) are associated with both the environmental exposure and phenotype, differences in SNP allele frequencies across sub-populations could lead to spurious claims about G-E interactions at individual SNP loci [7, 22]. In such cases, covariates traditionally used to control for bias in analyses of genetic main effects (e.g., principal components to account for population substructure) may fail to prevent bias in interaction estimates [7, 22]. Keller [22] discusses the fact that to avoid confounding by population structure, the interaction between the environmental variable and the principal components (used to control for population substructure) should be included as a covariate in the model. More generally, Keller proposes that models used to test for G-E interaction effects should include interactions among all relevant covariates and the environmental and genetic variables of interest [22]. A drawback of this approach is that the inclusion of all possible interactions among covariates and variables of interest could lead to model over-fitting, and reduced power; thus a thoughtful approach to selection of terms to include in the model would be recommended.

Measurement error in SNP markers is another potential source of bias in tests of G-E interaction effects. Often SNPs are not directly associated with phenotype status, but rather are in linkage disequilibrium (LD) with a causative variant [1]. Hein et al. show through simulation that this artifact of study design can bias the G-E interaction effect of the causative marker towards the null, leading to larger sample size requirements to identify statistically significant interaction effects [31]. Dudbridge and Fletcher show that in the presence of both G-E dependence and marginal association between a phenotype and causal marker, imperfect LD between a SNP marker and the causal variant can also lead to spurious association between the phenotype and G-E interaction [32]. The unknown true relationship between the causal marker and phenotype presents a major obstacle to correcting the observed associations; one proposed option is to conduct sensitivity analyses across a range of posited true relationships between the causal marker and phenotype [32].

Model mis-specification has been discussed as another potential source of bias in G-E interaction studies [33, 34]. Voorman et al. illustrated that a non-linear marginal effect of the environmental variable can lead to underestimated model-based errors [34], while Almli et al. described a similar downward bias in model-based errors caused by heteroskedasticity in a quantitative phenotype at different levels of the environmental variable [33]. These papers were motivated by genome-wide analyses in which mis-specified models led to inflated null test statistics, unrelated to problems of population substructure. Both Voorman et al. [34] and Almli et al. [33] showed that alternative robust regression methods [35, 36], which are available in standard statistical software [37, 38], can effectively minimize bias caused by such model mis-specification. A drawback of the proposed robust regression methods is reduced power when the model is specified correctly, although the reduction is usually modest.

Another important source of bias for all research, including G-E studies, is publication bias, which arises because statistically significant findings have a higher likelihood of being published. Publication bias is a known problem in many lines of research and has been discussed by others [39]. However, underpowered G-E interaction studies may be particularly prone to publication bias, which has led to the suggestion that statistically significant results from G-E studies are more likely to be false positives [4, 40]. Root causes of publication bias include the fact that "significant" results are deemed more interesting by readers, editors, and publishers, the fact that negative results often require more work for researchers to disseminate, and the fact that negative studies are often deemed underpowered, while similarly sized positive studies are not judged as such. To reduce publication bias, editors should be encouraged to accept null reports from well-designed studies, and author incentives to submit negative findings need to improve. In 2012, Hewitt outlined an editorial policy on candidate G-E interaction studies for the journal *Behavior Genetics* [41], to address some of the concerns of failed replication. Establishing similar criteria for the broader scientific community for performing and publishing adequately powered genome-wide G-E studies, could guide researchers and help reduce the problem of publication bias.

Analysis Methods for Genome-Wide Gene-Environment Interaction Studies

Currently, most genome-wide evaluations of G-E interactions are conducted by testing for G-E interaction at each genotyped SNP across the genome. However, this approach leads to a large number of tests of interaction. With so many tests, strict significance thresholds are necessary to correct for multiple testing, leading to a requirement for large samples sizes to have adequate power to detect even moderate effect sizes [15]. In order to improve power of G-E studies, two-stage screening approaches have been developed to reduce the number of G-E effects tested. In the first stage, a screening procedure is used to select a subset of SNPs to be evaluated for G-E interaction in the second stage. For example, only variants with evidence of a marginal effect on the phenotype (based on a relaxed level of significance) in this first stage may be considered in the second stage test of G-E interaction [42]. Alternatively, rather than screening based on marginal effects, two different interaction analyses can be employed at each stage: a screening test based on a gene-environment association analysis (similar to a case-only analysis but using all subjects), followed by a traditional case-control analysis of the G-E interaction effect in the second stage [43]. Similarly, other multistage extensions have been proposed to combine the case-control, case-only, and Empirical Bayes estimators of G-E interaction with marginal screening into "hybrid" [15] and "cocktail" methods [16]. Such multistage filtering approaches can substantially reduce the multiple testing burden and improve the power of genome-wide G-E interaction studies over traditional approaches.

Recent Developments and New Directions

Although filtering strategies can improve power for G-E analyses of GWAS data, such approaches still tend to evaluate interactions between the environmental factor and each SNP individually and are under-powered to detect small G-E effects or moderate effects obscured by noisy data. Similarly, GWAS of marginal genetic effects have traditionally taken this one-SNP-at-a-time approach. However, secondary analyses of marginal effects using multi-marker methods aimed at increasing power of genetic association studies are now frequently applied. In particular, approaches such as gene-level tests [5] and pathway-based gene-set analyses [6, 44] incorporate prior biological or functional knowledge and can increase power to detect small genetic effects. Because gene-level and pathway-level approaches were developed for GWAS to investigate marginal genetic effects, and have not been extended to include G-E interactions, most G-E GWAS analyses to date have only considered interactions between the environmental factor and individual SNPs. Nevertheless, G-E studies that incorporate multi-marker approaches are beginning to emerge. For example, a recent study of interactions between genetic variants and asbestos exposure on lung cancer evaluated G-E effects at the SNP, gene, and pathway levels [45]. Other studies may benefit from similar approaches to leverage such biological knowledge.

In addition to gene-based and gene-set methods, researchers have begun applying data-mining techniques to analyze high dimensional genetic data, because these methods allow simultaneous modeling of effects of multiple genetic or environmental risk factors. However, data mining methods are more often suggested in the context of studying G-G effects and constructing prediction models based on a large number of genetic variables, and have rarely been utilized for G-E interaction studies. In the following sections, we discuss these emerging approaches.

Data Mining Methods

Data-mining methods that search for patterns in high dimensional data have recently been proposed for detection of G-G and G-E interactions due to their ability to jointly consider a large number of variables and model non-linear effects [46]. This capacity to handle high throughput data is advantageous for both candidate gene studies and GWAS given the issues of multiple testing and computational efficiency. However, data-mining approaches are exploratory with a goal of prediction and hypothesis generation, and therefore validation of results in independent data is an important step in such analyses. Furthermore, as hypothesis testing is not a primary goal, data-mining methods are generally designed to construct joint models that allow for an interaction, rather than explicitly testing the interaction effect. While data-mining methods have not been specifically developed to identify G-E interactions, they are sometimes applied to search for G-E interactions, where the environmental risk factor is treated in the same manner as the genetic variables. In particular, the non-parametric data-mining approaches of Random Forests (RF) and Multifactor Dimensionality Reduction (MDR) have become popular tools to analyze high-dimensional genetic data, and may effective for discovery of potential G-E interaction effects.

MDR models disease associations with multiple SNPs by aggregating combinations of genotypes at multiple loci into a single binary high/low risk variable, and examines the association between a binary phenotype and this new collapsed variable [47]. Although MDR is actually evaluating combinations of genetic factors jointly, it can be used to detect SNP-SNP interactions post-hoc. Often MDR is used for screening by ranking SNPs and combinations of SNPs based on their predictive ability, where specific high-ranking combinations can be explicitly tested for interaction in a second stage or replication study using traditional approaches. A number of freely available software packages exist to conduct MDR analysis and examine post-hoc relationships [48, 49].

Most MDR applications have targeted G-G interactions, although MDR has also been utilized for G-E analysis by considering the environmental factor as any other categorical variable/SNP. For example, MDR has been used in candidate gene studies to investigate the effect of negative life events in major depressive disorder [50, 51]. In this approach, the environmental factor is one of many factors considered in the analysis and is not included in every combination evaluated, and therefore G-E interaction effects are not directly assessed. Furthermore, this also requires that the environmental factor be modeled as a categorical variable. Model-based MDR (MBMDR) has extended the traditional non-parametric version to include a regression framework to accommodate quantitative traits and discrete and continuous covariates [52], although this extension was not designed to address a search for G-E interaction.

Originally designed for prediction with high-dimensional data, Random Forests (RF) is another popular data-mining method that has been utilized to study G-G interactions [53]. RF is an ensemble method, where a 'forest' of many classification or regression trees is aggregated to achieve a final prediction. Each tree is grown using a bootstrap sample of subjects from the original data set, and a random subset of predictors (e.g., SNPs) is evaluated at each node. Non-linear interactions can be captured in the prediction model via the hierarchical splits of the trees, although hypothesis tests are not typically conducted. Unlike MDR, RF builds predictive models using all predictors, and does not perform variable selection. However, variable importance measures are computed that assess the predictive ability of each variable, which can be used for SNP ranking, and therefore selection. Variable importance rankings can be used to identify interacting variables in low-dimensional datasets, although only interacting factors with strong marginal effects are highly ranked in high dimensional genetic data [54]. Such variable importance measures are computed separately for each variable, and therefore do not explicitly determine which factors are interacting; this requires measuring the importance of the joint contribution [55], which is computationally inefficient in high-dimensional GWAS data.

Investigation of G-E interaction effects with RF is complicated by the fact that it is not obvious how to best incorporate environmental covariates. As with MDR, the environmental factor is usually included in the standard model-building process along with the other predictor variables, which allows for possible G-E interactions but does not guarantee the inclusion of the environmental factor into every tree of the forest. This strategy is easy to implement, but it has been shown that building an RF with predictor variables on vastly different scales (e.g., SNPs and a quantitative environmental factor) could bias results, although this can be avoided by using alternative strategies for sampling during tree construction and assessing variable importance [56, 57]. Because RF does not search for interactions explicitly, it is often used as a filter that allows for interactions, where SNPs and environmental factors ranked highly with RF in the first stage can be carried forward to the second stage where their interactions are directly tested with traditional statistical methods. Applications of RF to study G-E interactions have used this screening approach. In a study of age of onset of coronary heart disease, genetic and environmental factors were first screened with RF variable importance measures, and then interactions between genetic variants and smoking were modeled with generalized estimating equations [58].

Similarly, in a study of esophageal cancer RF was used to filter out SNPs most associated with cancer risk, and G-E interactions with the top SNPs were modeled using a case-only analysis [59]. Because these two-stage approaches can lead to over-fitting if the top SNPs are selected from the same dataset that is further analyzed, cross-validation (or similar approaches) should be used to evaluate significance, or each stage should be applied to an independent data set.

Gene and Pathway-Level G-E Analysis

Gene-level tests can improve power over analyses of single variants, particularly when there are several causal variants in a gene [5]. While gene-level approaches generally do not identify specific SNP-trait associations, they can identify relevant genes, leading to improved understanding of disease mechanisms and enabling focused investigation of candidate polymorphisms within the identified genes. Many gene-level analysis approaches have been proposed for studying genetic main effects [6], including the minimum SNP p-value [60], various SNP p-value summary statistics [61], and joint modeling of the effects of all SNPs in a gene [62]. In addition to gene-level analyses of un-phased genotype data, haplotype analyses provide another multi-marker alternative to SNP-level tests (e.g., [63]). Rare variant analyses are also often performed at the gene level; early methods typically assumed rare variants in a gene affect the trait in a similar direction and magnitude [64, 65], while more recent methods avoid this assumption [66, 67].

While numerous multi-marker approaches have been used to study genetic main effects, few have been applied in the context of G-E interaction analysis. Chatterjee et al. [68] proposed an approach that can simultaneously test for main effects of multiple SNP markers and their interaction with an environmental exposure. This omnibus testing method improves power to detect genetic associations in the presence of G-E interaction by using a single interaction parameter; however, it assumes each SNP's interaction effect with the environmental factor is directly proportional to its marginal effect on the trait. Wang et al. [69] suggested an alternative dimension-reduction approach to multi-marker G-E interaction testing. Using partial least squares, their procedure initially finds a linear combination of multiple SNPs within a gene that explain a maximum amount of variation in trait status. The cross-product of this linear combination of SNPs and the environmental variable is then incorporated into a regression model, enabling various tests of interaction that give higher weight to SNPs with larger marginal effects.

Some approaches recently proposed for studying G-G interactions, such as dimension reduction techniques, could be applied to perform gene-level G-E interaction tests. For example, Li et al. derived principal components (PCs) representing variation in each of the genes of interest, and then evaluated G-G interactions by testing the PC-PC interaction effects [70]. He et al. used a similar approach, but derived LD-based SNP weights prior to performing PC analysis on genes [71], and Peng et al. [72] used canonical correlation analysis to evaluate G-G interactions. Similar approaches could be applied in the context of G-E interaction studies.

A few gene-level G-E analysis methods that include both common and rare variants have also been proposed. In particular, for analysis of quantitative traits, Tzeng et al. suggested a method based on gene-trait similarity regression to assess gene-level G-E interactions [73]. When aggregating multiple markers in the gene, this method can implement weights based on allele frequency. The proposed approach is equivalent to a haplotype random-effects model, and allows for testing the joint genetic effects as well as testing G-E interactions specifically. Traditional haplotype analysis methods can also be extended to test for interactions with environmental factors. For example, Li et al. recently used Bayesian hierarchical modeling to more efficiently detect haplotype-haplotype and haplotype-environment interactions [74].

In the data-mining framework, a gene-level MDR method has been suggested [75]. The approach of Oh et al. uses a two-step procedure: in the first step, MDR jointly models multiple SNPs within a gene, while in the second step MDR is applied again to the estimated gene-level effects. Although this method reduces computation time and provides an interpretation at the gene-level, the two-stage approach may create unwanted bias, and hence the statistical properties of this method need to be further evaluated.

Beyond testing for effects at the gene-level, gene set analysis (GSA) approaches evaluate the overall evidence of association of a trait with genetic variation in a set of functionally related genes (e.g., representing a molecular pathway). These approaches incorporate prior biological knowledge and have become widely used in complex trait research [6]. Just as some gene-level tests of genetic main effects have been extended to study G-E interactions, some GSA approaches could also be extended to test for G-E interactions at the pathway (gene-set) level to provide insights regarding which sets of genes interact with particular environmental factors to influence the trait of interest. Thomas [9, 76] discussed the concept of studying G-E interactions in the context of biological pathways, pointing out the potential for Gene Set Enrichment Analysis (GSEA) and hierarchical models to empirically evaluate effects in pre-defined pathways. He also discussed the potential application of Bayesian network analysis for identifying novel pathways in G-E studies. However, as Thomas mentioned, such analyses have not yet been applied to studies of G-E interaction effects.

To date, there are few examples of published studies that investigated G-E interactions at the gene or pathway level. A notable exception is a GWAS of lung cancer reported by Wei et al. [45], which included both gene-level and pathway-level analyses of G-E effects. In this study, G-E analyses at the gene level were performed using a modification of the VErsatile Gene-based Association Study (VEGAS) approach of Liu et al. [77]. While the original VEGAS method involved combining the main-effect p-values from multiple SNPs within a gene, Wei et al. used it to combine SNP-environment interaction p-values within each gene. They then performed pathway-level analyses of G-E interaction effects using an extension of the GSEA approach [78]. While Wei et al.'s investigation of G-E interactions at the gene and pathway level must be commended, the performance of the modified VEGAS approach, and the modified GSEA approach, in the context of G-E interaction analysis has not been adequately investigated. The statistical properties of such approaches need to be evaluated, particularly given the complexities of permutation-based testing in the analysis of interaction effects [79].

In addition to gene and pathway-level approaches, the application of multi-marker genetic risk scores (GRS) is increasingly used in genetic studies of complex traits [80, 81]. The basic idea of the GRS approaches is to combine multiple variants that exhibit association with the phenotype (but have small effect sizes) into a single score typically calculated as a weighted sum of the risk alleles, and use the score for risk prediction [82]. Recent studies have emerged that examine G-E interactions in the context of the GRS. For example, in a G-E study of obesity, a GRS including 32 BMI-associated risk variants was examined for its potential interaction with intake of sugar-sweetened beverages [83]. While a simulation study demonstrated a range of scenarios with adequate power to detect GRS-E effects where G-E effects at individual loci would go undetected [84], further investigation of the potential utility of such approaches is needed.

Conclusion

This chapter provided a brief overview of G-E interaction studies. The main focus was on population-based G-E studies of non-related individuals and standard data analysis methods that linearly model the environmental variable, genetic marker, and gene x environment interaction. We also discussed other important topics such as potential sources of bias, alternative study designs, and promising novel data analysis methods, including two-step filtering procedures, data-mining methods, and gene-level and pathway level analyses.

Key Facts

1. Gene-environment (G-E) interactions can be investigated using several study designs, including traditional epidemiological study designs (e.g., cohort and case-control studies), as well as study designs specific to the study of interactions (e.g., case-only studies).
2. Numerous statistical analysis methods have been proposed for investigating G-E interaction effects. Several approaches have been introduced to increase power of genome-wide G-E interaction analyses (e.g., two-stage filtering strategies that reduce the multiple testing burden).
3. Several potential sources of bias in G-E interaction analyses have been discussed, and solutions have been proposed.
4. While genome-wide searches for G-E interactions are now being undertaken, current studies primarily apply tests of interactions between individual SNPs and the environmental exposure.
5. Another possibility is to use novel multi-marker approaches to reduce the number of tests performed.
6. Novel statistical methods for G-E interaction analysis are under development. These include data-mining methods and multi-marker approaches that can be used to perform gene-level and pathway-level analyses.

The authors declare no conflicts of interest.

REFERENCES

[1] Cantor, R. M., Lange, K. & Sinsheimer, J. S. Prioritizing GWAS results: A review of statistical methods and recommendations for their application. *American journal of human genetics* 86, 6-22 (2010).

[2] Manolio, T. A. et al. Finding the missing heritability of complex diseases. *Nature* 461, 747-753 (2009).

[3] Eichler, E. E. et al. Missing heritability and strategies for finding the underlying causes of complex disease. *Nat Rev Genet* 11, 446-450 (2010).

[4] Duncan, L. E. & Keller, M. C. A critical review of the first 10 years of candidate gene-by-environment interaction research in psychiatry. *Am J Psychiatry* 168, 1041-1049, doi:10.1176/appi.ajp.2011.11020191 (2011).

[5] Neale, B. M. & Sham, P. C. The future of association studies: gene-based analysis and replication. *American journal of human genetics* 75, 353-362, doi:10.1086/423901 (2004).

[6] Fridley, B. L. & Biernacka, J. M. Gene set analysis of SNP data: benefits, challenges, and future directions. *Eur J Hum Genet* 19, 837-843, doi:10.1038/ejhg.2011.57 (2011).

[7] Aschard, H. et al. Challenges and opportunities in genome-wide environmental interaction (GWEI) studies. *Hum Genet* 131, 1591-1613, doi:10.1007/s00439-012-1192-0 (2012).

[8] Khoury, M. J. & Wacholder, S. Invited commentary: from genome-wide association studies to gene-environment-wide interaction studies--challenges and opportunities. *Am J Epidemiol* 169, 227-230; discussion 234-225, doi:10.1093/aje/kwn351 (2009).

[9] Thomas, D. Gene--environment-wide association studies: emerging approaches. *Nature reviews. Genetics* 11, 259-272, doi:10.1038/nrg2764 (2010).

[10] Umbach, D. M. & Weinberg, C. R. Designing and analysing case-control studies to exploit independence of genotype and exposure. *Stat Med* 16, 1731-1743 (1997).

[11] Jaquish, C. E. The Framingham Heart Study, on its way to becoming the gold standard for Cardiovascular Genetic Epidemiology? *BMC Medical Genetics* 8, 63 (2007).

[12] Mukherjee, B. et al. Principal interactions analysis for repeated measures data: application to gene–gene and gene–environment interactions. *Statistics in Medicine* 31, 2531-2551 (2012).

[13] Gu, C. C., Yang, W. W., Kraja, A. T., de las Fuentes, L. & Dávila-Román, V. G. in *BMC Proceedings.* S86 (BioMed Central Ltd).

[14] McClelland, G. H. & Judd, C. M. Statistical difficulties of detecting interactions and moderator effects. *Psychological bulletin* 114, 376-390 (1993).

[15] Murcray, C. E., Lewinger, J. P., Conti, D. V., Thomas, D. C. & Gauderman, W. J. Sample size requirements to detect gene-environment interactions in genome-wide association studies. *Genet Epidemiol* 35, 201-210, doi:10.1002/gepi.20569 (2011).

[16] Hsu, L. et al. Powerful cocktail methods for detecting genome-wide gene-environment interaction. *Genet Epidemiol* 36, 183-194, doi:10.1002/gepi.21610 (2012).

[17] Cordell, H. J. Epistasis: what it means, what it doesn't mean, and statistical methods to detect it in humans. *Hum Mol Genet* 11, 2463-2468 (2002).

[18] Kraft, P., Yen, Y. C., Stram, D. O., Morrison, J. & Gauderman, W. J. Exploiting gene-environment interaction to detect genetic associations. *Hum Hered* 63, 111-119, doi:10.1159/000099183 (2007).

[19] Albert, P. S., Ratnasinghe, D., Tangrea, J. & Wacholder, S. Limitations of the case-only design for identifying gene-environment interactions. *Am J Epidemiol* 154, 687-693 (2001).

[20] Dai, J. Y. et al. Simultaneously testing for marginal genetic association and gene-environment interaction. *Am J Epidemiol* 176, 164-173, doi:10.1093/aje/kwr521 (2012).

[21] Mukherjee, B. & Chatterjee, N. Exploiting gene-environment independence for analysis of case-control studies: an empirical Bayes-type shrinkage estimator to trade-off between bias and efficiency. *Biometrics* 64, 685-694, doi:10.1111/j.1541-0420.2007.00953.x (2008).

[22] Keller, M. C. Gene x environment interaction studies have not properly controlled for potential confounders: the problem and the (simple) solution. *Biol Psychiatry* 75, 18-24, doi:10.1016/j.biopsych.2013.09.006 (2014).

[23] Rothman, K. J., Greenland, S. & Lash, T. L. *Modern epidemiology*. (Lippincott Williams & Wilkins, 2008).

[24] Mechanic, L. E. et al. Next generation analytic tools for large scale genetic epidemiology studies of complex diseases. *Genet Epidemiol* 36, 22-35, doi:10.1002/gepi.20652 (2012).

[25] Greenland, S. The effect of misclassification in the presence of covariates. *Am J Epidemiol* 112, 564-569 (1980).

[26] Wong, M. Y., Day, N. E., Luan, J. A. & Wareham, N. J. Estimation of magnitude in gene-environment interactions in the presence of measurement error. *Stat Med* 23, 987-998, doi:10.1002/sim.1662 (2004).

[27] Demissie, S. & Cupples, L. A. Bias due to two-stage residual-outcome regression analysis in genetic association studies. *Genet Epidemiol* 35, 592-596 (2011).

[28] Tiwari, H. K. et al. Review and evaluation of methods correcting for population stratification with a focus on underlying statistical principles. *Hum Hered* 66, 67-86, doi:10.1159/000119107 (2008).

[29] Price, A. L., Zaitlen, N. A., Reich, D. & Patterson, N. New approaches to population stratification in genome-wide association studies. *Nature reviews. Genetics* 11, 459-463, doi:10.1038/nrg2813 (2010).

[30] Tian, C., Gregersen, P. K. & Seldin, M. F. Accounting for ancestry: population substructure and genome-wide association studies. *Human molecular genetics* 17, R143-150, doi:10.1093/hmg/ddn268 (2008).

[31] Hein, R., Beckmann, L. & Chang-Claude, J. Sample size requirements for indirect association studies of gene-environment interactions (G x E). *Genet Epidemiol* 32, 235-245, doi:10.1002/gepi.20298 (2008).

[32] Dudbridge, F. & Fletcher, O. Gene-environment dependence creates spurious gene-environment interaction. *American journal of human genetics* 95, 301-307, doi:10.1016/j.ajhg.2014.07.014 (2014).

[33] Almli, L. M. et al. Correcting Systematic Inflation in Genetic Association Tests That Consider Interaction Effects: Application to a Genome-wide Association Study of Posttraumatic Stress Disorder. *JAMA psychiatry*, doi:10.1001/jamapsychiatry.2014.1339 (2014).

[34] Voorman, A., Lumley, T., McKnight, B. & Rice, K. Behavior of QQ-Plots and Genomic Control in Studies of Gene-Environment Interaction. *Plos One* 6, e19416, doi:10.1371/journal.pone.0019416 (2011).

[35] Huber, P. J. in *Proceedings of the fifth Berkeley symposium on mathematical statistics and probability.* 221-233.

[36] White, H. A heteroskedasticity-consistent covariance matrix estimator and a direct test for heteroskedasticity. *Econometrica: Journal of the Econometric Society*, 817-838 (1980).

[37] Zeileis, A. Object-oriented computation of sandwich estimators. (2006).

[38] Zeileis, A. Econometric computing with HC and HAC covariance matrix estimators. (2004).

[39] Ioannidis, J. P. Why most published research findings are false. *PLoS medicine* 2, e124 (2005).

[40] Burton, P. R. et al. Size matters: just how big is BIG? Quantifying realistic sample size requirements for human genome epidemiology. *International Journal of Epidemiology* 38, 263-273 (2009).

[41] Hewitt, J. K. Editorial policy on candidate gene association and candidate gene-by-environment interaction studies of complex traits. *Behav Genet* 42, 1-2, doi:10.1007/s10519-011-9504-z (2012).

[42] Kooperberg, C. & Leblanc, M. Increasing the power of identifying gene x gene interactions in genome-wide association studies. *Genet Epidemiol* 32, 255-263, doi:10.1002/gepi.20300 (2008).

[43] Murcray, C. E., Lewinger, J. P. & Gauderman, W. J. Gene-environment interaction in genome-wide association studies. *Am J Epidemiol* 169, 219-226, doi:10.1093/aje/kwn353 (2009).

[44] Biernacka, J. M., Jenkins, G. D., Wang, L., Moyer, A. M. & Fridley, B. L. Use of the gamma method for self-contained gene-set analysis of SNP data. *Eur J Hum Genet* 20, 565-571, doi:10.1038/ejhg.2011.236 (2012).

[45] Wei, S. et al. Genome-wide gene-environment interaction analysis for asbestos exposure in lung cancer susceptibility. *Carcinogenesis* 33, 1531-1537, doi:10.1093/carcin/bgs188 (2012).

[46] Szymczak, S. et al. Machine learning in genome-wide association studies. *Genet Epidemiol* 33 Suppl 1, S51-57, doi:10.1002/gepi.20473 (2009).

[47] Ritchie, M. D. et al. Multifactor-dimensionality reduction reveals high-order interactions among estrogen-metabolism genes in sporadic breast cancer. *American journal of human genetics* 69, 138-147, doi:10.1086/321276 (2001).

[48] Winham, S. J. & Motsinger-Reif, A. A. An R package implementation of multifactor dimensionality reduction. *BioData Min* 4, 24, doi:10.1186/1756-0381-4-24 (2011).

[49] Bush, W. S., Dudek, S. M. & Ritchie, M. D. Parallel multifactor dimensionality reduction: a tool for the large-scale analysis of gene-gene interactions. *Bioinformatics* 22, 2173-2174, doi:10.1093/bioinformatics/btl347 (2006).

[50] Yang, C. X. et al. The combined effects of the BDNF and GSK3B genes modulate the relationship between negative life events and major depressive disorder. *Brain Research* 1355, 1-6, doi:10.1016/j.brainres.2010.07.079 (2010).

[51] Zhang, K. R. et al. The combined effects of the 5-HTTLPR and 5-HTR1A genes modulates the relationship between negative life events and major depressive disorder

in a Chinese population. *J Affect Disord* 114, 224-231, doi:10.1016/j.jad.2008.07.012 (2009).

[52] Calle, M. L., Urrea, V., Malats, N. & Van Steen, K. mbmdr: an R package for exploring gene-gene interactions associated with binary or quantitative traits. *Bioinformatics* 26, 2198-2199, doi:10.1093/bioinformatics/btq352 (2010).

[53] Breiman, L. Random forests. *Machine Learning* 45, 5-32, doi:10.1023/a:1010933404324 (2001).

[54] Winham, S. J. et al. SNP interaction detection with Random Forests in high-dimensional genetic data. *BMC Bioinformatics* 13, 164, doi:10.1186/1471-2105-13-164 (2012).

[55] Bureau, A. et al. Identifying SNPs predictive of phenotype using random forests. *Genet Epidemiol* 28, 171-182, doi:10.1002/gepi.20041 (2005).

[56] Strobl, C., Boulesteix, A. L., Zeileis, A. & Hothorn, T. Bias in random forest variable importance measures: Illustrations, sources and a solution. *BMC Bioinformatics* 8, doi:2510.1186/1471-2105-8-25 (2007).

[57] Strobl, C., Boulesteix, A. L., Kneib, T., Augustin, T. & Zeileis, A. Conditional variable importance for random forests. *BMC Bioinformatics* 9, 307, doi:10.1186/1471-2105-9-307 (2008).

[58] Maenner, M. J. et al. Detecting gene-by-smoking interactions in a genome-wide association study of early-onset coronary heart disease using random forests. *BMC Proc* 3 Suppl 7, S88 (2009).

[59] Zhai, R. H. et al. Interactions between environmental factors and polymorphisms in angiogenesis pathway genes in esophageal adenocarcinoma risk: A case-only study. *Cancer* 118, 804-811, doi:10.1002/cncr.26325 (2012).

[60] Wang, K., Li, M. & Bucan, M. Pathway-based approaches for analysis of genomewide association studies. *The American Journal of Human Genetics* 81, 1278-1283 (2007).

[61] De la Cruz, O., Wen, X., Ke, B., Song, M. & Nicolae, D. L. Gene, region and pathway level analyses in whole-genome studies. *Genet Epidemiol* 34, 222-231, doi:10.1002/gepi.20452 (2010).

[62] Ballard, D. H., Cho, J. & Zhao, H. Comparisons of multi-marker association methods to detect association between a candidate region and disease. *Genet Epidemiol* 34, 201-212, doi:10.1002/gepi.20448 (2010).

[63] Liu, N., Zhang, K. & Zhao, H. Haplotype-association analysis. *Adv Genet* 60, 335-405, doi:10.1016/S0065-2660(07)00414-2 (2008).

[64] Li, B. & Leal, S. M. Methods for detecting associations with rare variants for common diseases: application to analysis of sequence data. *The American Journal of Human Genetics* 83, 311-321 (2008).

[65] Madsen, B. E. & Browning, S. R. A groupwise association test for rare mutations using a weighted sum statistic. *PLoS Genet* 5, e1000384 (2009).

[66] Wu, M. C. et al. Rare-variant association testing for sequencing data with the sequence kernel association test. *The American Journal of Human Genetics* 89, 82-93 (2011).

[67] Lee, S., Wu, M. C. & Lin, X. Optimal tests for rare variant effects in sequencing association studies. *Biostatistics* 13, 762-775 (2012).

[68] Chatterjee, N., Kalaylioglu, Z., Moslehi, R., Peters, U. & Wacholder, S. Powerful multilocus tests of genetic association in the presence of gene-gene and gene-

environment interactions. *American journal of human genetics* 79, 1002-1016, doi:10.1086/509704 (2006).

[69] Wang, T., Ho, G., Ye, K., Strickler, H. & Elston, R. C. A partial least-square approach for modeling gene-gene and gene-environment interactions when multiple markers are genotyped. *Genet Epidemiol* 33, 6-15 (2009).

[70] Li, J., Tang, R., Biernacka, J. M. & de Andrade, M. Identification of gene-gene interaction using principal components. *BMC proceedings* 3 Suppl 7, S78 (2009).

[71] He, J. et al. Gene-based interaction analysis by incorporating external linkage disequilibrium information. *Eur J Hum Genet* 19, 164-172, doi:10.1038/ejhg.2010.164 (2011).

[72] Peng, Q., Zhao, J. & Xue, F. A gene-based method for detecting gene-gene co-association in a case-control association study. *Eur J Hum Genet* 18, 582-587, doi:10.1038/ejhg.2009.223 (2010).

[73] Tzeng, J. Y. et al. Studying Gene and Gene-Environment Effects of Uncommon and Common Variants on Continuous Traits: A Marker-Set Approach Using Gene-Trait Similarity Regression. *American journal of human genetics* 89, 277-288, doi:10.1016/j.ajhg.2011.07.007 (2011).

[74] Li, J., Zhang, K. & Yi, N. A Bayesian hierarchical model for detecting haplotype-haplotype and haplotype-environment interactions in genetic association studies. *Hum Hered* 71, 148-160, doi:10.1159/000324841 (2011).

[75] Oh, S. et al. A novel method to identify high order gene-gene interactions in genome-wide association studies: gene-based MDR. *BMC Bioinformatics* 13 Suppl 9, S5, doi:10.1186/1471-2105-13-S9-S5 (2012).

[76] Thomas, D. Methods for investigating gene-environment interactions in candidate pathway and genome-wide association studies. *Annu Rev Public Health* 31, 21-36, doi:10.1146/annurev.publhealth.012809.103619 (2010).

[77] Liu, J. Z. et al. A Versatile Gene-Based Test for Genome-wide Association Studies. *American journal of human genetics* 87, 139-145, doi:10.1016/j.ajhg.2010.06.009 (2010).

[78] Zhang, K., Cui, S., Chang, S., Zhang, L. & Wang, J. i-GSEA4GWAS: a web server for identification of pathways/gene sets associated with traits by applying an improved gene set enrichment analysis to genome-wide association study. *Nucleic Acids Res* 38, W90-95, doi:10.1093/nar/gkq324 (2010).

[79] Buzkova, P., Lumley, T. & Rice, K. Permutation and parametric bootstrap tests for gene-gene and gene-environment interactions. *Ann Hum Genet* 75, 36-45, doi:10.1111/j.1469-1809.2010.00572.x (2011).

[80] Pharoah, P. D., Antoniou, A. C., Easton, D. F. & Ponder, B. A. Polygenes, risk prediction, and targeted prevention of breast cancer. *N Engl J Med* 358, 2796-2803, doi:10.1056/NEJMsa0708739 (2008).

[81] van Hoek, M. et al. Predicting type 2 diabetes based on polymorphisms from genome-wide association studies: a population-based study. *Diabetes* 57, 3122-3128, doi:10.2337/db08-0425 (2008).

[82] Park, J. H. et al. Estimation of effect size distribution from genome-wide association studies and implications for future discoveries. *Nature* genetics 42, 570-575, doi:10.1038/ng.610 (2010).

[83] Qi, Q. et al. Sugar-sweetened beverages and genetic risk of obesity. *N Engl J Med* 367, 1387-1396, doi:10.1056/NEJMoa1203039 (2012).

[84] Marigorta, U. M. & Gibson, G. A simulation study of gene-by-environment interactions in GWAS implies ample hidden effects. *Front Genet* 5, 225, doi:10.3389/fgene.2014.00225 (2014).

In: Gene-Environment Interactions and Human Diseases
Editor: Lu Qi ISBN: 978-1-63483-680-7

Chapter 4

GENE-ENVIRONMENT INTERACTION AND OBESITY

Lu Qi[1,2,*]

[1]Department of Epidemiology, School of Public Health and Tropical Medicine, Tulane University, New Orleans, LA, US
[2]Department of Nutrition, Harvard School of Public Health, Boston, MA, US

ABSTRACT

The past few decades witnessed rapid rise of obesity and its complications such as diabetes in industrialized countries and also in developing countries. It is becoming increasingly clear that contemporary changes of environmental factors, particularly lifestyle and dietary habits, have contributed to the obesity pandemic. Recent advances in genomics have led to identification of the genetic variants determining body weight and obesity risk. These discoveries highlight the importance of genes involved in the control of food intake and energy balance by the central nervous system in obesity development. In addition, emerging data have lent support to interactions between genetic variations and various environmental factors such as physical activity, sedentary lifestyle, and intakes of sugary drink and fried foods in relation to obesity; and studies in randomized intervention trials have also shown that individuals' genotypes might modify their response in weight loss and maintenance to diet and lifestyle interventions. Data collected from investigations on gene-environment interaction would ultimately endorse person-centric diet and lifestyle intervention to mitigate obesity and related disorders.

INTRODUCTION

Obesity is symbolized by excessive fat accumulation, and is strongly associated with a variety of disorders such as type 2 diabetes, cardiovascular disease, asthma, sleep apnea,

* Correspondence and requests for reprint: Dr. Lu Qi. Department of Epidemiology, School of Public Health and Tropical Medicine, Tulane University, New Orleans, Louisiana 70112. Telephone: 504-988-7259; Fax: 504-988-3548; Email address: lqi1@tulane.edu; luqi@hsph.harvard.edu.

hypertension, and certain cancers [1]. In the United States, the prevalence of obesity has been growing for approximate a century; and its increase was obviously accelerated in the past 30 years. Obesity is also an epidemic in most of other developed countries and is becoming a major health hazard in many developing nations as well. A global estimate showed that in 2005 there were at least 400 million adults were obese; and it is predicted that by 2015 more than 700 million will be obese and approximately 2–3 billion adults will be overweight [2]. Obesity is predominantly viewed to result from increased calorie intake and lack of physical activity. This view has a solid ground as the soaring of obesity prevalence has paralleled with histrionic shift of lifestyle from traditional, healthy patterns toward obesogenic, unhealthy patterns featured by increased intakes of energy and reduced consumption of high-fiber foods, as well as sedentary lifestyle.

The contribution of genetics to obesity is also becoming increasingly evident. The past ten years witnessed speedy advances in research of genomics, which has made great strides in detection of genetic variants associated with body weight regulation and obesity [3]. Notably, many of the identified obesity genes are involved in the control of balance between energy intake and energy expenditure by the central nervous system [4]. In addition, emerging data have shown that the genetic variants may interact with environmental factors, especially diet and lifestyle, in relation to obesity and weight change [5-7]

This chapter summarizes recent advances in studying dietary and lifestyle, as well as genetic risk factors for obesity. The interactions between the genetic factors and these environmental factors on adiposity and obesity risk are particularly addressed. In addition, the chapter also reviews the studies that assess the modification effects of genetic variants on diet and lifestyle interventions targeting weight loss or weight loss maintenance; and introduces the potential use of the findings from this new field in prevention and management of obesity, with hope to ultimately endorse person-centric strategies to alleviate the substantial burden of this disorder.

1. Obesogenic Environment and Epidemic of Obesity

The root of obesity etiology is imbalance between dietary energy intake and energy expenditure. In the search for causes of the recent obesity epidemic, emphasis has been placed on the change in obesogenic environment. Numerous studies have unanimously pointed to the same set of environmental factors – sedentary lifestyle and unhealthy diet habits, which have resulted in increased access to calories and decreased energy expenditure – as the culprit leading to the current obesity epidemic. Human evolution has favored a preference for energy-dense and fatty foods, as consequence of exposure to ancestral famine [8]. This leaves humans susceptible to modern obesogenic environment regarding rise of energy intake and reduction of energy expenditure. Obesogenic lifestyle is featured by reduced physical activity at work due to mechanization, improved motorized transport and preferences for viewing television, computers, and video games to outdoor games during leisure time. Physical activity is among the most important factors determining energy balance and body weight regulation [9]. Low levels of physical activity and sedentary lifestyle have been consistently associated with an elevated risk of obesity in epidemiological studies [10]. For example, in a study of 10-year weight change in 3,515 men and 5,810 women, Williamson et al. found that the relative risks for gaining > 13kg in individuals whose activity level was low compared with those in the high-activity level were 2.3 in men and 7.1 in women. Schoeller reviewed cross-sectional data from doubly-labeled water studies and longitudinal studies [11]. The data similarly indicate that the prevalence of overweight, BMI, or body fat increases along with decreasing physical activity. Television watching is among the most common sedentary behaviors in leisure time in the United States and many European countries. Epidemiological studies have shown that prolonged TV watching was associated with an increased risk of obesity and related diseases [12, 13] Of interest, the associations of physical activity and sedentary lifestyle with obesity risk appear to be independent of each other. In addition, an increasing number of epidemiological studies have shown associations of disturbance of sleep habits especially sleep loss with weight gain and an increased obesity risk, even though the evidence is not entirely consistent [14, 15]. Sleep, different from other sedentary behaviors, facilitates appetite control and may thus be considered a 'satiating' activity. Deprivation of sleep leads to disturbed energy metabolism in favor of energy storage and subsequently causes obesity [16].

Obesogenic diet is characterized by increased consumption of energy-dense, processed foods and reduced consumption of nutrient rich foods, such as fruits and vegetables. Data from the National Health and Nutrition Examination Survey (NHANES) have shown marked upward shift of energy intake, increasing by 7% in men and 22% in women from 1971–1974 to 1999–2000 [17]. A variety of dietary factors have been related to obesity. For example, consumption of fast foods, typically high in fat and energy density, and low in fiber, increased from 20% of total calories in 1970 to 40% of total calories in 1995 in the United States. Positive associations between fast foods consumption and obesity risk have been reported in several epidemiology studies, though the data are not entirely consistent [18, 19]. In the United States, there has been a sudden upsurge in consumption of sugar-sweetened beverages (SSB) since mid-1990s, paralleled the rise in obesity prevalence [20] Compelling evidence supports a positive association of high SSB consumption with weight gain and obesity risk [21]. Reduced intakes of vegetable, fruits, nuts, and yogurt have been also related to weight gain and obesity risk [21-23]. However, the data are largely collected from observational studies, and it remains debatable whether these factors were driving the obesity, or vice versa.

Besides diet and lifestyle, obesogens have recently emerged as a third major category of environmental risk for obesity [24]. Obesogens are xenobiotic chemicals in the environment that affect fat accumulation through disturbing adipogenesis and energy balance. Various obesogens such as dichlorodiphenyltrichloroethane (DDT), tributyltin (TBT), and bisphenol A (BPA) have been related to obesity [25].

According to the NHANES, after nearly 30 years of increase [26] the surging trend in the prevalence of obesity in the United States may have crested. The rate of obesity calculated from BMI did not change significantly from 2003-2004 to 2011-2012. This phenomenon suggests that only a proportion of the population becomes obesity when they are exposed to the obesogenic environment; while the others appear to be resistant to such unhealthy environment. The considerable diversity in response to the obesogenic environment within a population may be at least partly explained by modification effect of the innate susceptibility that is largely determined by genomic makeup.

2. Genetics of Obesity

The inherited propensity toward obesity has been supported by numerous family and twin studies. Twin study is useful in decomposing genetic and environmental influences on traits. Classic twin studies in the 1980s and 1990s have estimated that genes contribute anywhere from 10-90% of the variance in body mass index [27]. Although estimates have varied, a general consensus is that about 65% of variation in obesity is familial or genetic [28]. Similar percentage of the variation in other obesity-related phenotypes such as skinfold thickness, fat mass, and leptin levels has been estimated to be heritable [29]. The first evidence directly supporting the effect of genetic variation on obesity in humans came in 1997, when mutations in the leptin gene were found to cause severely obesity in two cousins from Pakistan [30]. Mutations in several other genes involved in the central control of energy balance have been also found to cause extreme obesity [31].

Mapping for obesity genes historically started with linkage analysis and candidate gene association studies [32, 33]. Linkage analysis is designed to identify the disease loci through examining the co-segregation of genetic markers evenly distributed throughout the genome with the disease in families; whereas candidate gene association studies identify candidate genomic loci based on our knowledge of physiological weight regulation mechanisms gleaned from observations in extreme human obesity or animal models. However, these approachese could identify loci, but failed to pinpoint causal genetic variants. In addition, these approaches have major limitations and are underpowered in discovering moderate genetic effects on common-form obesity.

The arrival of the genome-wide association studies (GWAS) brought revolutionary changes in the field. This new approach uses genome-scale high-density scans, large sample size, and multiple-staged study design (usually including discovery and replication stages). In Caucasians, at least 32 loci have been identified through GWAS to be associated with adulthood BMI before 2015 (Figure 1); and addition loci have been detected to be associated with other obesity-related traits such as waist and waist-to-hip ratio [4, 34, 35]. Despite the tremendous success of GWAS, these successfully identified loci only accounted for a fraction of the phenotyping variance. Collectively, the 32 loci associated with BMI explained only 1.45% of the inter-individual BMI variation. The best known locus, FTO, has the largest effect, with each risk allele increasing BMI by 0.39 kg/m^2 and obesity risk by 1.2 times. Despite the high frequency of obesity risk allele in the population, the FTO genotype only explains 0.34% of the inter-individual variation in BMI.

3. Rationales for Gene-Environment Interaction on Obesity

The genetic susceptibility to obesity has its roots in our evolutionary history; however, the impact of genetic factors on the epidemic of obesity in modern societies is likely due their recent collision with dramatic changes in environment. Several theories have been proposed to shed light on obesity pandemic in modern society. These theories attempt to reconcile gene-environment interactions with an understanding of human evolution. The 'thrifty genotype' hypothesis, which was first proposed by James Neel in 1962, [8] asserts that obesity in modern societies is the result of thrifty genes passed down from these ancestors of present day humans who were subject to strong selection pressures that enriched the population with genes that promoted energy storage. As such, feast-or-famine conditions during human evolutionary development naturally selected for people who could store excess energy as body fat for later use. Thus, the ability to conserve more energy as fat offers a genetic advantage for selection of this genotype during periods of food scarcity. However, when individuals are faced with higher caloric loads in a modern context featured by overabundance of foods and a sedentary lifestyle, carrying the thrifty genotype becomes a risk factor for obesity and related metabolic disorders. As such, genetic variations caused by adaptation to famine could have important health consequences in modern society; and obesity may particularly affect those who are still adapted to former famine conditions: i.e., who carry the thrifty genotypes. Although the 'thrifty genotype' hypothesis was met with great skepticism [36], it has been supported by many data. For example, the highest prevalence of obesity is found in certain ethnic groups such as Pima Indians [37]. These populations have undergone a strong genetic selection for survival during unreliable food supply in the past and experienced sudden shift from traditional lifestyle into Western lifestyle.

Several other hypotheses have been also put forward to counter the long-held acceptance of the thrifty genotype hypothesis [38]. Nevertheless, more solid evidence is needed to further demonstrate the validity of these theories. There was a strong polarization of views that the changed environment is the principal cause for recent epidemic of obesity and related complex diseases. However, it is notable that, despite decades of plentiful food supplies and wide spread of sedentary lifestyle, the rate of obesity appear to be reaching a plateau in the United States, with roughly 70% of the population remaining not obese, and ~30% of the population remaining lean [39]. Apparently, people vary dramatically in their response, which is mainly determined by the inbuilt genetic architecture, to the adverse effects of obesogenic environmental factors. This concept is continually reinforced by studies in the burgeoning area of gene-environment interaction.

4. Gene-Diet InteractionS ON Obesity

The FTO gene is among the main culprits in determining genetic susceptibility to obesity; SNPs in introns of the FTO gene are thus far the most strongly associated variants for BMI [40]. Yang et al. noted that although carriers of the FTO risk allele had an increased mean BMI, there was still considerable variability of effect size around the mean, highlighting the importance of environmental factors in influencing the 'genetic burden' of the FTO risk alleles [41]. Experimental studies in animals suggest FTO may play an important role in regulating food intake. Recently, Qi et al. performed large-scale analyses based on data from 177,330 adults (154,439 whites, 5,776 African Americans, and 17,115 Asians) from 40 studies (Table 1) [42]. It was found that the BMI-increasing allele of the FTO variant rs9939609 showed a significant association with higher dietary protein intake, and relative weak associations with lower intake of total energy and carbohydrate. However, the variant did not interact with dietary intakes on BMI.

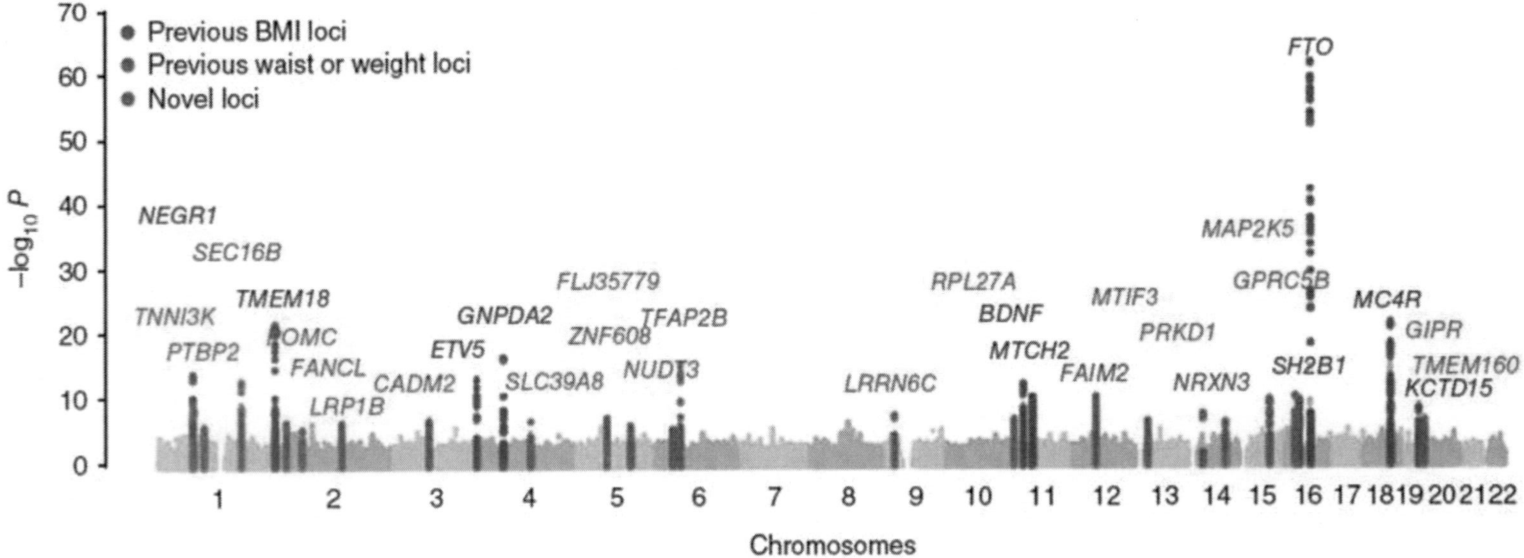

Figure 1. Manhattan plot showing the thirty two body mass associated chromosome loci identified by genome-wide association studies. Modified from Speliotes EK, et al. Nat Genet. 2010.

Table 1. Selected studies on gene-diet/lifestyle interactions in observational studies and randomized clinical trials

Studies	Study design	Diet/lifestyle factors	Genetic factors	Major findings
Cohort studies				
Kilpeläinen T, et al. [52]	218,166 adults and 19,268 children and adolescents	Physical activity	BMI-associated FTO rs9939609	The association of the FTO risk allele with obesity is attenuated by 27% in physically active adults.
Qi Q, et al. 2012 [5]	Three cohorts; N = 6,934, 4,423, and 21,740	Sugar sweetened beverages	32 BMI- and obesity related variants	The genetic association with adiposity were modified by intake of sugar-sweetened beverages
Qi Q, et al. 2012 [55]	Two cohorts; n = 7740 and 4564	Physical activity and television watching	32 BMI- and obesity related variants	The genetic effects on BMI were stronger in those with low physical activity or longer hours of television watching
Ahmad S, et al. 2013[7]	N = 111,421	Physical activity	12 BMI- and obesity related variants	The genetic effects on BMI and obesity risk were stronger in those with low physical activity
Qi Q, et al. 2014 [6]	Three cohorts; N = 9,623, 6,379, and 21,421	Fried foods	32 BMI- and obesity related variants	High fried food consumption amplified the genetic association with BMI
Qi Q, et al. 2014 [42]	N = 177,330	Macronutrients (fat, protein, carbohydrates)	BMI-associated FTO rs9939609	FTO variant did not interact with dietary intake of total energy, protein, carbohydrate or fat on BMI.
Randomized clinical trials				
McCaffery J et al. [64]	N = 3899; diabetic patients	4-y lifestyle intervention in	Obesity variants	The FTO and BDNF genotypes were related to weight regain after weight loss
Zhang X, et al. 2012 [61]	N = 742; overweight or obese subjects	2-y weight-loss diet intervention	Obesity related *FTO* SNP rs1558902	High-protein diet interacted with FTO genotype in relation to weight loss and improvement of body composition and fat distribution
Larson L et al. [63]	N = 742	6-m diet intervention on weight loss maintenance	768 tagSNPs for nutrient-sensitive genes	Genetic variants interacted with diet glycemic index and protein on regain of waist and fat mass
Xu M, et al. 2013 [60]	N = 734; overweight or obese subjects	2-y weight-loss diet intervention	*BCAA associated PPM1K* SNP rs1440581	Dietary fat significantly modified genetic effects on changes in body weight.
Pan Q, et al. 2013 [62]	N = 3,819	2-y intervention; lifestyle modification and metformin	MC4R SNPs	SNP rs17066866 was associated with less weight loss in the lifestyle intervention group, but not in placebo group

Abbreviations: BCAA, branched chain amino acid; BMI, body mass index; SNP, single nucleotide polymorphism.

In another study, Corella et al. reported that the effects of saturated fat acids (SFA) intake significantly interacted with the FTO genotype on BMI in two populations of 1069 and 1094 participants (P-interaction = 0.007 and 0.014; respectively) [43]. It was found that homozygous participants for the FTO risk allele had a higher mean BMI than the other genotypes in the individuals with a high-SFA intake, while no associations with BMI were found at lower SFA intakes. The FTO genotype did not interact with carbohydrate intake. In another study, Dedoussis et al. [44] observed a potential interaction between breastfeeding and the FTO genotype on BMI and other indices of adiposity in two pediatric cohorts; however the finding was not replicated in the third cohort of children.

Several studies have also assessed interactions between genetic factors and dietary patterns, which are more closely parallel with real-world scenarios of food and nutrient combinations and clustering. For example, Roswall et al. examined whether adherence to the Mediterranean Diet measured by using the Mediterranean diet score (MDS) interacted with the FTO and TCF7L2 genotypes in relation to changes in weight, BMI, and waist circumference in 11,048 participants from 5 European countries [45]. A borderline significant interaction was observed between the MDS and TCF7L2 rs7903146 on weight gain (P = 0.05), which showed a beneficial effect of the MDS only in subjects carrying risk allele. The FTO genotype did not modify the observed associations.

In addition to individual SNPs, more studies focused on testing interactions with the genetic risk score (GRS), which is calculated by summing up the number of the risk alleles of multiple BMI- or obesity-associated SNPs and therefore plays a better role in representing the overall genetic susceptibility (Table 1). In a study of US men and women from prospective cohorts the Nurses' Health Study (NHS) and Health Professionals' Follow-up Study (HPFS), [5] Qi et al. assessed interactions between habitual intake of SSB and obesity genetic susceptibility, evaluated by a GRS derived from 32 obesity-associated loci, in relation to BMI and obesity risk. Significant interactions between genetic susceptibility and SSB consumption were consistently observed in the NHS and HPFS. In the combined samples of these cohorts, the increases in BMI (kg/m^2) per 10 risk alleles were 1.00 for sugar sweetened beverage intake of <1 serving/month, 1.03 for 1-4 servings/month, 1.39 for 2-6 servings/week, and 1.77 for ≥1 servings/day (P for interaction<0.001). The findings were successfully replicated in the Women Genome Health Study (WGHS) (*P* for interaction = 0.001). This study for the first time provides reproducible evidence for the interactions between genetic factors and dietary factors in relation to obesity risk. In another study in these cohorts, [6] consistent interactions were also found between fried food consumption and the obesity GRS. Among participants in the highest third of the GRS, the differences in BMI between individuals who consumed fried foods ≥ 4 times a week and those who consumed fried foods < 1 a week were 1.0 in women and 0.7 in men, whereas the corresponding differences were 0.5 and 0.4 in the lowest third of the score. Viewed differently, the genetic association appeared stronger in those with higher consumption of fried foods. The odds ratios (95% CI) for obesity per 10 risk alleles were 1.61 (1.40 to 1.87), 2.12 (1.73 to 2.59), and 2.72 (2.12 to 3.48) across the three categories of consumption (P = 0.002 for interaction). Interestingly, it was found that the variants in or near genes highly expressed in the central nervous system, such as FTO and MC4R, showed the strongest interactions.

Thus far, the published data on gene-diet interactions are all from studies on the GWAS-identified variants or polymorphisms in the candidate genes. Even though genome-wide approaches have been successfully applied in detection of gene-environment interactions on other disorders such as cancers, [46] genome-wide analysis on gene-environment interaction in relation to obesity is still lacking.

5. Gene-Lifestyle Interactions on Obesity

Physical activity is among the most important lifestyle factors that critically affect adiposity and risk of obesity. After identification of the FTO gene, several studies reported that the increased obesity risk associated with the FTO rs9939609 risk allele was attenuated by high level of physical activity, [47, 48] which is generally associated with beneficial changes in obesity related phenotypes. However, the following efforts to replicate these findings generated highly conflicting results [49-51]. To clarify this incongruence, a large meta-analysis comprising 218,166 individuals from 48 different studies was performed (Table 1) [52]. It was found that the effect of the *FTO* genotype on BMI and obesity risk was approximately 30% smaller in physically active than in sedentary individuals in adults. In addition, it was found that the minor allele of the SNP increased the odds of obesity less in the physically active group (odds ratio = 1.22/allele) than in the inactive group (odds ratio = 1.30/allclc). In children and adolescents, the interaction was not observed, probably due to smaller sample size. Of note, most of the studies involved in this meta-analysis are cross-sectional, raising concern on the existence of reverse causation and confounding. In addition, measurements of physical activity were highly inconsistent across studies; this may add more noise to detect the interactions, if any.

Ahmad et al. recently extended the analysis on gene-physical activity interaction by including 12 obesity-susceptibility loci, and a GRS was calculated by summing the BMI-associated alleles of each variant [7]. The study assessed interaction between the GRS and physical activity in 111,421 participants, and found a statistically significant interaction on BMI (P for interaction = 0.015). Interestingly, populations from North American (P for interaction = 0.014) showed stronger interaction than populations from Europe (P for interaction = 0.275). The discrepancy in gene-physical activity interactions might be partly explained by the difference in physical activity levels between populations living in North American and Europe; and the observations highlight the complexity in detecting gene-environment interactions. At individual variant level, variants in the FTO and SEC16B genes showed significant interactions with physical activity.

Television watching is one of the most common sedentary lifestyle in the United States and other western countries, with the average daily TV watching time ranging from 3.5 to 5.0 hours [53]. In epidemiological studies, TV watching has been associated with an increased risk of obesity and related diseases, independent of physical activity levels [53, 54]. Qi et al. analyzed interactions between TV watching, leisure time physical activity, and genetic predisposition in relation to BMI in 7,740 women and 4,564 men from the NHS and HPFS [55]. A weighted GRS was calculated on the basis of 32 established BMI-associated variants. In both women and men, the genetic associations with BMI strengthened with increased hours of TV watching. An increment of 10 points in the weighted genetic risk score was associated with 0.8, 0.8, 1.4, 1.5, and 3.4 kg/m^2 higher BMI across the 5 categories of TV watching (0-1, 2-5, 6-20, 21-40, and >40 hours/week; P for interaction = 0.001). In contrast, the genetic association with BMI weakened with increased levels of physical activity. An increment of 10 points in the weighted genetic risk score was associated with 1.5, 1.3, 1.2, 1.2, and 0.8 kg/m^2 higher BMI across the quintiles of physical activity. The interactions of TV watching and physical activity with genetic predisposition in relation to BMI were independent of each other.

Other lifestyle behaviors may also interact with genetic factors in affecting obesity risk. For example, in a population-based twin study (1,088 pairs, 604 monozygotic, 484 dizygotic), it was found that the heritability of BMI was more than twice as large in individuals with sleep duration < 7 hr (h^2 = 70%) as the heritability in those with sleep duration was ≥ 9 hr (h^2 = 32%). Test for interaction was significant [56]. However, the gene-sleep interaction needs to be replicated in large prospective studies. In addition, previous studies have shown that the relation between sleep duration and obesity risk is probably not linear, [14, 15] and specific statistical modeling would be applied in detecting such interactions.

6. Genetic Modifications on Diet and Lifestyle Interventions

Interventions on diet habits and physical activity, as well as behavior therapy and pharmacotherapy have been continuously implemented to lower obesity and its complications, but still have relatively poor long-term success and mainly scarce adherence. Diet and lifestyle modifications to induce a negative energy balance hold particularly larger appeals in management of body weight in overweight or obesity people due to their wider reach and relative safety. In the past ten years, there are many diet and/or lifestyle intervention trials have been advocated and performed to test whether such modifications are effective in improvement of weight loss and long-term weight maintenance [57, 58]. No single diet has proven to be superior to others overall in terms of weight loss outcomes, whereas considerable interindividual heterogeneity has been noted in participants' response to various interventions [59]. These trials provide excellent models to test interactions of genetic factors with diet and lifestyle changes. As compared with studies on gene-environment interaction in observational settings, analysis of interactions in randomized clinical trials has a unique strength to minimize bias caused by confounding and reverse causation; and therefore provides robust evidence. In addition, evidence-based prevention and treatment rely mainly on the statistical interpretation of data from clinical trials. Therefore, to detect gene-diet interaction in diet intervention trials on weight loss would be essential.

In thus far one of the largest and longest diet intervention trials – the Pounds Lost trial, [57] a series analyses on gene-diet interactions on weight loss have been performed (Table 1). This trial includes 811 overweight or obese adults assigned to 1 of 4 weight-loss diets varying in macronutrient contents for 2 years. Xu et al. recently genotyped the branched-chain amino acid/aromatic amino acid ratio-associated variant rs1440581 near the PPM1K gene in 734 overweight or obese adults from the Pounds Lost trial. At 6 months, dietary fat significantly modified genetic effects on changes in weight, fasting insulin, and homeostasis model assessment of insulin resistance (HOMA-IR) after adjustment for the confounders (all P for interaction ≤0.006) [60]. Zhang et al. found SNP rs1558902 in the FTO gene interacted with dietary protein on 2-year changes in fat-free mass, total percentage of fat mass, and total-, visceral-, and superficial adipose tissue mass [61]. It appeared that a high-protein diet were more beneficial in individuals with the risk allele A. These data highlight considerable genetic heterogeneity in weight loss in response to diet interventions.

In Diabetes Prevention Program (DPP), Pan et al. genotyped 20 tagging SNPs for MC4R in 3,819 participants, whom were randomized into intensive lifestyle modification (eating less fat and calories, and exercising for a total of 150 minutes a week), metformin or placebo control [62]. The minor allele of rs17066866 was associated with less short-term (baseline to 6 months; $P = 0.006$) and long-term (baseline to 2 years, $P = 0.004$) weight loss in the lifestyle intervention group, but not in placebo group. Several studies have also examined the gene-diet interactions in relation to weight loss maintenance. The DIOGENES is a randomized, controlled 6-month dietary intervention study that examines the effects of dietary protein and glycemic index (GI) on weight regain and metabolic risk factors in overweight and obese families, after an 8-week weight loss period on a low-calorie diet. In 742 participants from the DIOGENES, Larsen et al. examined 768 tagging SNPs for nutrient-sensitive candidate genes for obesity and obesity-related diseases [63]. The SNPs in genes *GHRL*, *CCK*, *MLXIPL*, and *LEPR* showed interactions with dietary protein *on weight regain; and t*he SNPs in genes *PPARD*, *FABP1*, *LPIN1*, and *PLAUR showed interactions with dietary protein on* fat mass regain. The Look AHEAD is a randomized trial to determine the effects of diabetes support and education (DSE) and intensive lifestyle intervention (ILI), which combined diet modification and increased physical activity designed to produce an average of 7% weight loss and maintenance, on cardiovascular morbidity and mortality in 5,145 ethnically diverse overweight and obese subjects with type 2 diabetes.McCaffery et al. examined the interaction between 13 obesity-predisposing polymorphisms in 8 genes and randomized treatment arm in predicting weight change at year 1, and weight regain at year 4 among individuals who lost 3% or more of their baseline weight by year 1 [64]. It was found that FTO SNP rs3751812 showed significant prediction in DSE group ($P = 0.005$), but not within ILI group.

The gene-environment interactions on weight loss have been also observed in Children. In a recent study of 168 overweight and obese adolescents participating in a lifestyle intervention and nutrition education weight loss program, 9 obesity-related genetic variants were analyzed and a genetic prediction score was derived for each participant [65]. The study showed that participants with lower prediction score had greater weight loss and improvement in metabolic profile compared with participants with higher scores.

Thus far there are two genome-wide analyses on weight loss after gastric bypass surgery. Hatoum et al. performed GWAS in 693 individuals undergoing Roux-en-Y gastric bypass (RYGB) surgery and replicated in a group of 327 individuals undergoing RYGB. They found a locus on chromosome 15 associated with weight loss after RYGB [66].In another study of 258 individuals undergoing RYGB, several additional loci were identified [67]. The data suggest that genetic variants may also affect weight change by surgery. However, the reported genetic effect did not reach genome-wide significant level (p of 5×10^{-8}), raising concerns about potentially false positive findings. No study has examined gene-diet interaction in randomized clinical trials on the whole genome scale. Because of the dearth of experimental data, the potential mechanisms underlying the observed interactions between the genetic factors and diet/lifestyle interventions are largely unknown.

7. Future Direction and Conclusion Remarks

Obesity is a major risk factor for a variety of chronic diseases that severely affect our modern society; and the recent obesity epidemic has imposed significant health and economic concerns. In epidemiological studies, considerable progress has been made in our knowledge of the roles of environmental factors such as diet habits and lifestyle in development of obesity. In recent years, high-throughput genotyping and sequencing based efforts have provided highly evolving data that significantly expand our understanding of the genetic basis of obesity. In the face of the modern obesity epidemic the previously prevailing view that the condition is purely environment-related has been revised; however, it remains poorly understood how the genetic factors and environmental factors act in concert.

Data from randomized clinical trials advocate that obesity could be prevented or mitigated through modifications on diet and lifestyle [57, 68]. Currently, a one-size-fits-all strategy is widely adopted in prevention and management of obesity. Such approach requires substantial simplification and strong assumption that the response of the whole population to the interventions is homogeneous. However, accumulating evidence from studies of gene-environment interactions in observational settings and in randomized clinical trials has suggested such assumption is not likely to be true. There are examples reflecting some of the triumphs of the studies in deciphering gene-diet and gene-lifestyle interactions in relation to obesity risk and promotion of weight loss and maintenance [5, 55, 69]. Some very general conclusions that can be drawn from these initial data are that genomic information would be taken into consideration in designing public health strategies to alleviate obesity burden. The potential aim of studying gene-environment interactions is to improve our understanding of inter-individual variation in responses to diet and lifestyle modifications, and to translate such knowledge into public health benefit. The next generation of personalized obesity prevention and management would integrate genomic information into the overall regime [70, 71].

Yet the map of the gene-environment interface is far from complete, and many other challenges may exist. In addition to enduring collection of reproducible evidence on gene-environment interaction from population based studies, other tractable lines of investigation would include detailed characterization of the potential mechanisms in experimental settings, and assessment of the impact of genetic testing on change of human behaviors. Although the current evidence level of applying gene-environment interaction to tailor diet and lifestyle modifications is at its early stages, the prospect of widespread incorporation of the novel information to public health practice to prevent obesity and mitigate its complications is highly encouraging.

Key Facts

1. Obesity is a major risk factors for various metabolic disorders;
2. Prevalence of obesity has been rapidly increasing in most of the developed countries, and many developing countries;
3. It is widely believed that obesity is due to environmental factors, genetic factors, and their interactions;

4. In the past few years, studies in large, prospective cohorts have provided reproducible evidence to support gene-diet and gene-lifestyle interactions on obesity;
5. Data from randomized clinical trials also support that the genetic variants may modulate response in body weight to diet or lifestyle interventions;
6. Many challenges exist in detection of moderation gene-environment interaction, in interpretation of the results, and in translation of the findings into personalized interventions to lower risk of obesity and complicated metabolic disorders.

ACKNOWLEDGMENT

Dr Qi is supported by NIH grants from the National Heart, Lung, and Blood Institute (HL071981, HL034594, HL126024), the National Institute of Diabetes and Digestive and Kidney Diseases (DK091718, DK100383, DK078616), the Boston Obesity Nutrition Research Center (DK46200), and United States – Israel Binational Science Foundation Grant 2011036. Dr Qi was a recipient of the American Heart Association Scientist Development Award (0730094N).

No conflict of interest

REFERENCES

[1] Ogden, C. L., Yanovski, S. Z., Carroll, M. D. & Flegal, K. M. The epidemiology of obesity. *Gastroenterology* 132, 2087-2102, (2007).
[2] Malik, V. S., Willett, W. C. & Hu, F. B. Global obesity: trends, risk factors and policy implications. *Nat Rev Endocrinol* 9, 13-27, (2013).
[3] Waalen, J. The genetics of human obesity. *Transl Res* 164, 293-301, (2014).
[4] Speliotes, E. K. et al. Association analyses of 249,796 individuals reveal 18 new loci associated with body mass index. *Nat Genet* 42, 937-948, (2010).
[5] Qi, Q. et al. Sugar-sweetened beverages and genetic risk of obesity. *N Engl J Med* 367, 1387-1396, (2012).
[6] Qi, Q. et al. Fried food consumption, genetic risk, and body mass index: gene-diet interaction analysis in three US cohort studies. *BMJ* 348, g1610, (2014).
[7] Ahmad, S. et al. Gene x physical activity interactions in obesity: combined analysis of 111,421 individuals of European ancestry. *PLoS Genet* 9, e1003607, (2013).
[8] Neel, J. V. Diabetes mellitus: a "thrifty" genotype rendered detrimental by "progress"? *Am J Hum Genet* 14, 353-362, (1962).
[9] Chaput, J. P. et al. Physical activity plays an important role in body weight regulation. *J Obes* 2011, (2011).
[10] Booth, F. W., Roberts, C. K. & Laye, M. J. Lack of exercise is a major cause of chronic diseases. *Compr Physiol* 2, 1143-1211, (2012).
[11] Schoeller, D. A. Balancing energy expenditure and body weight. *Am J Clin Nutr* 68, 956S-961S, (1998).
[12] Hu, F. B., Li, T. Y., Colditz, G. A., Willett, W. C. & Manson, J. E. Television watching

and other sedentary behaviors in relation to risk of obesity and type 2 diabetes mellitus in women. *JAMA* 289, 1785-1791, (2003).

[13] Dunstan, D. W. et al. Television viewing time and mortality: the Australian Diabetes, Obesity and Lifestyle Study (AusDiab). *Circulation* 121, 384-391, (2010).

[14] Schmid, S. M., Hallschmid, M. & Schultes, B. The metabolic burden of sleep loss. *Lancet Diabetes Endocrinol* 3, 52-62, (2015).

[15] Zimberg, I. Z. et al. Short sleep duration and obesity: mechanisms and future perspectives. *Cell Biochem Funct* 30, 524-529, (2012).

[16] Klingenberg, L., Sjodin, A., Holmback, U., Astrup, A. & Chaput, J. P. Short sleep duration and its association with energy metabolism. *Obes Rev* 13, 565-577, (2012).

[17] Centers for Disease Control and Prevention. Trends in intake of energy and macronutrients--United States, 1971-2000. *MMWR Morb Mortal Wkly Rep* 53, 80-82, (2004).

[18] Bahadoran, Z., Mirmiran, P., Hosseini-Esfahani, F. & Azizi, F. Fast food consumption and the risk of metabolic syndrome after 3-years of follow-up: Tehran Lipid and Glucose Study. *Eur J Clin Nutr* 67, 1303-1309, (2013).

[19] Duffey, K. J., Gordon-Larsen, P., Jacobs, D. R., Jr., Williams, O. D. & Popkin, B. M. Differential associations of fast food and restaurant food consumption with 3-y change in body mass index: the Coronary Artery Risk Development in Young Adults Study. *Am J Clin Nutr* 85, 201-208, (2007).

[20] Malik, V. S., Popkin, B. M., Bray, G. A., Despres, J. P. & Hu, F. B. Sugar-sweetened beverages, obesity, type 2 diabetes mellitus, and cardiovascular disease risk. *Circulation* 121, 1356-1364, (2010).

[21] Mozaffarian, D., Hao, T., Rimm, E. B., Willett, W. C. & Hu, F. B. Changes in diet and lifestyle and long-term weight gain in women and men. *N Engl J Med* 364, 2392-2404, (2011).

[22] Cho, S. S., Qi, L., Fahey, G. C., Jr. & Klurfeld, D. M. Consumption of cereal fiber, mixtures of whole grains and bran, and whole grains and risk reduction in type 2 diabetes, obesity, and cardiovascular disease. *Am J Clin Nutr* 98, 594-619, (2013).

[23] Bayon, V., Leger, D., Gomez-Merino, D., Vecchierini, M. F. & Chennaoui, M. Sleep debt and obesity. *Ann Med* 46, 264-272, (2014).

[24] Sharpe, R. M. & Drake, A. J. Obesogens and obesity--an alternative view? *Obesity (Silver Spring)* 21, 1081-1083, (2013).

[25] Kelishadi, R., Poursafa, P. & Jamshidi, F. Role of environmental chemicals in obesity: a systematic review on the current evidence. *J Environ Public Health* 2013, 896789, (2013).

[26] Centers for Disease Control and Prevention. National Health and Nutrition Examination Survey.http://www.cdc.gov/nchs/nhanes.htm. Accessed November 21, 2013., (2013).

[27] Keller, K. L., Pietrobelli, A. & Faith, M. S. Genetics of food intake and body composition: lessons from twin studies. *Acta Diabetol* 40 Suppl 1, S95-100, (2003).

[28] Nan, C. et al. Heritability of body mass index in pre-adolescence, young adulthood and late adulthood. *Eur J Epidemiol* 27, 247-253, (2012).

[29] Lyon, H. N. & Hirschhorn, J. N. Genetics of common forms of obesity: a brief overview. *Am J Clin Nutr* 82, 215S-217S, (2005).

[30] Montague, C. T. et al. Congenital leptin deficiency is associated with severe early-onset obesity in humans. *Nature* 387, 903-908, (1997).

[31] O'Rahilly, S. & Farooqi, I. S. Human obesity: a heritable neurobehavioral disorder that is highly sensitive to environmental conditions. *Diabetes* 57, 2905-2910, (2008).
[32] Yang, W., Kelly, T. & He, J. Genetic epidemiology of obesity. *Epidemiol Rev* 29, 49-61, (2007).
[33] Rankinen, T. et al. The human obesity gene map: the 2005 update. *Obesity (Silver Spring)* 14, 529-644, (2006).
[34] Xia, Q. & Grant, S. F. The genetics of human obesity. *Ann N Y Acad Sci* 1281, 178-190, (2013).
[35] Heid, I. M. et al. Meta-analysis identifies 13 new loci associated with waist-hip ratio and reveals sexual dimorphism in the genetic basis of fat distribution. *Nat Genet* 42, 949-960, (2010).
[36] Benyshek, D. C. & Watson, J. T. Exploring the thrifty genotype's food-shortage assumptions: a cross-cultural comparison of ethnographic accounts of food security among foraging and agricultural societies. *Am J Phys Anthropol* 131, 120-126, (2006).
[37] Knowler, W. C., Pettitt, D. J., Saad, M. F. & Bennett, P. H. Diabetes mellitus in the Pima Indians: incidence, risk factors and pathogenesis. *Diabetes Metab Rev* 6, 1-27, (1990).
[38] Speakman, J. R. Thrifty genes for obesity, an attractive but flawed idea, and an alternative perspective: the 'drifty gene' hypothesis. *Int J Obes (Lond)* 32, 1611-1617, (2008).
[39] Flegal, K. M., Carroll, M. D., Ogden, C. L. & Curtin, L. R. Prevalence and trends in obesity among US adults, 1999-2008. *JAMA* 303, 235-241, (2010).
[40] Loos, R. J. & Yeo, G. S. The bigger picture of FTO: the first GWAS-identified obesity gene. *Nat Rev Endocrinol* 10, 51-61, (2014).
[41] Yang, J. et al. FTO genotype is associated with phenotypic variability of body mass index. *Nature* 490, 267-272, (2012).
[42] Qi, Q. et al. FTO genetic variants, dietary intake and body mass index: insights from 177,330 individuals. *Hum Mol Genet* 23, 6961-6972, (2014).
[43] Corella, D. et al. A high intake of saturated fatty acids strengthens the association between the fat mass and obesity-associated gene and BMI. *J Nutr* 141, 2219-2225, (2011).
[44] Dedoussis, G. V. et al. Does a short breastfeeding period protect from FTO-induced adiposity in children? *Int J Pediatr Obes* 6, e326-335, (2011).
[45] Roswall, N. et al. Association between Mediterranean and Nordic diet scores and changes in weight and waist circumference: influence of FTO and TCF7L2 loci. *Am J Clin Nutr* 100, 1188-1197, (2014).
[46] Figueiredo, J. C. et al. Genome-wide diet-gene interaction analyses for risk of colorectal cancer. *PLoS Genet* 10, e1004228, (2014).
[47] Andreasen, C. H. et al. Low physical activity accentuates the effect of the FTO rs9939609 polymorphism on body fat accumulation. *Diabetes* 57, 95-101, (2008).
[48] Cauchi, S. et al. Combined effects of MC4R and FTO common genetic variants on obesity in European general populations. *J Mol Med (Berl)* 87, 537-546, (2009).
[49] Jonsson, A. et al. Assessing the effect of interaction between an FTO variant (rs9939609) and physical activity on obesity in 15,925 Swedish and 2,511 Finnish adults. *Diabetologia* 52, 1334-1338, (2009).
[50] Lappalainen, T. J. et al. The common variant in the FTO gene did not modify the effect

of lifestyle changes on body weight: the Finnish Diabetes Prevention Study. *Obesity (Silver Spring)* 17, 832-836, (2009).

[51] Scott, R. A. et al. FTO genotype and adiposity in children: physical activity levels influence the effect of the risk genotype in adolescent males. *Eur J Hum Genet* 18, 1339-1343, (2010).

[52] Kilpelainen, T. O. et al. Physical activity attenuates the influence of FTO variants on obesity risk: a meta-analysis of 218,166 adults and 19,268 children. *PLoS Med* 8, e1001116, (2011).

[53] Grontved, A. & Hu, F. B. Television viewing and risk of type 2 diabetes, cardiovascular disease, and all-cause mortality: a meta-analysis. *JAMA* 305, 2448-2455, (2011).

[54] Ching, P. L. et al. Activity level and risk of overweight in male health professionals. *Am J Public Health* 86, 25-30, (1996).

[55] Qi, Q. et al. Television watching, leisure time physical activity, and the genetic predisposition in relation to body mass index in women and men. *Circulation* 126, 1821-1827, (2012).

[56] Watson, N. F. et al. Sleep duration and body mass index in twins: a gene-environment interaction. *Sleep* 35, 597-603, (2012).

[57] Sacks, F. M. et al. Comparison of weight-loss diets with different compositions of fat, protein, and carbohydrates. *N Engl J Med* 360, 859-873, (2009).

[58] Shai, I. et al. Weight loss with a low-carbohydrate, Mediterranean, or low-fat diet. *N Engl J Med* 359, 229-241, (2008).

[59] Vesely, J. M. & DeMattia, L. G. Obesity: long-term management after bariatric surgery. *FP Essent* 425, 29-33, (2014).

[60] Xu, M. et al. Genetic determinant for amino acid metabolites and changes in body weight and insulin resistance in response to weight-loss diets: the Preventing Overweight Using Novel Dietary Strategies (POUNDS LOST) trial. *Circulation* 127, 1283-1289, (2013).

[61] Zhang, X. et al. FTO genotype and 2-year change in body composition and fat distribution in response to weight-loss diets: the POUNDS LOST Trial. *Diabetes* 61, 3005-3011, (2012).

[62] Pan, Q. et al. Variation at the melanocortin 4 receptor gene and response to weight-loss interventions in the diabetes prevention program. *Obesity (Silver Spring)* 21, E520-526, (2013).

[63] Larsen, L. H. et al. Analyses of single nucleotide polymorphisms in selected nutrient-sensitive genes in weight-regain prevention: the DIOGENES study. *Am J Clin Nutr* 95, 1254-1260, (2012).

[64] McCaffery, J. M. et al. FTO predicts weight regain in the Look AHEAD clinical trial. *Int J Obes (Lond)* 37, 1545-1552, (2013).

[65] Moleres, A. et al. Obesity susceptibility loci on body mass index and weight loss in Spanish adolescents after a lifestyle intervention. *J Pediatr* 161, 466-470 e462, (2012).

[66] Hatoum, I. J. et al. Weight loss after gastric bypass is associated with a variant at 15q26.1. *Am J Hum Genet* 92, 827-834, (2013).

[67] Rinella, E. S. et al. Genome-wide association of single-nucleotide polymorphisms with weight loss outcomes after Roux-en-Y gastric bypass surgery. *J Clin Endocrinol Metab* 98, E1131-1136, (2013).

[68] Looney, S. M. & Raynor, H. A. Behavioral lifestyle intervention in the treatment of

obesity. *Health Serv Insights* 6, 15-31, (2013).

[69] Qi, L. Gene-diet interaction and weight loss. *Curr Opin Lipidol* 25, 27-34, (2014).

[70] Qi, L. Personalized nutrition and obesity. *Ann Med* 46, 247-252, (2014).

[71] Corella, D. & Ordovas, J. M. Can genotype be used to tailor treatment of obesity? State of the art and guidelines for future studies and applications. *Minerva Endocrinol* 38, 219-235, (2013).

In: Gene-Environment Interactions and Human Diseases ISBN: 978-1-63483-680-7
Editor: Lu Qi

Chapter 5

GENE-ENVIRONMENT INTERACTIONS ON LIPIDS

Dolores Corella
Genetic and Molecular Epidemiology Unit
Department of Preventive Medicine and Public Health
University of Valencia and CIBER OBN, Valencia, Spain

ABSTRACT

In recent years, numerous genome-wide association studies (GWAs) have provided us with knowledge of new genes associated with plasma lipid concentrations (i.e MLXIPL, SORT1, ANGPTL3, etc.) These new genes have been added to the broad spectrum of candidate genes that were already known (i.e., APOE, CETP, LPL, APOA1, etc.) to have an important influence on concentrations of different types of plasma lipids. There is a wide range of publications that have reported gene-environment interactions both for these classic genes and those discovered through GWAs. Although these interactions have included the analysis of various environmental factors such as tobacco smoking, physical activity, drugs, etc., the most commonly researched environmental factor has been the diet. Hence, there are hundreds of studies that have statistically significant interactions between polymorphisms in different genes and various components of the diet in determining plasma lipids and lipoproteins, not only in fasting lipids, but also in the postprandial state. In addition, lipidomics will offer the potential to identify new specific biomarkers. In this chapter, we shall review the main gene-environment studies on lipids and discuss the most important findings. In general we may say that, despite the fact that great heterogeneity still exists in the methodology and results obtained by these studies, the epidemiological design and the evidence level of these studies continue to improve by incorporating more randomized controlled intervention trials. The results of these studies are crucial to obtain a higher level of scientific evidence and in order to incorporate the findings of these gene-environmental interactions into personalized medicine.

1. Introduction

Plasma lipid concentrations are important biomarkers in many diseases, but they are mainly considered to be intermediate phenotypes of cardiovascular diseases [1, 2]. The key plasma lipids have been total cholesterol and triglycerides. These, being nonpolar lipids, need to be transported in the plasma associated with various lipoprotein particles. There are five major classes of plasma lipoproteins: chylomicrons, very-low-density lipoproteins (VLDL), intermediate-density lipoproteins (IDL), low-density lipoproteins (LDL), and high-density lipoproteins (HDL) [3]. Plasma lipoprotein particles contain different amounts of four major molecules: cholesterol, triglycerides, phospholipids and specific apoproteins (designated by an alphabetical nomenclature: A, B, C, D, E, F, H, J, L, etc.). Plasma lipids, draw medical attention when present in anomalous concentrations. Augmented or decreased levels largely occur because of alterations in the synthesis, degradation, and transport. High concentrations of LDL-cholesterol (LDL-C) are associated with a greater risk of cardiovascular events [4, 5], whereas high concentrations of HDL-C are, in general, considered to be a protective factor [6]. However, the situation has turned out to be much more complex than originally envisaged [8]. Regarding fasting triglycerides, which after several decades of being considered to play a minor role as a cardiovascular risk factor [9], are now re-emerging as an important risk factor [10]. This renewed interest has been driven by epidemiological evidence supporting raised triglycerides, remnant cholesterol, or triglyceride-rich lipoproteins as an independent cause of cardiovascular disease [10, 11]. On the other hand, some researchers have pointed out that these classic clinical biomarkers (triglycerides, total cholesterol, LDL-C and HDL-C) lack the necessary specificity and sensitivity and only increase significantly after serious dyslipidemia [12]. Therefore, sensitive biomarkers are needed for hyperlipidemia. Lipidomics and metabolomics offer the potential to identify new hyperlipidemia-specific biomarkers that would improve diagnosis and therapeutics at early stages [12, 13]. The alteration of the different plasma lipids and lipoproteins usually occurs in a number of combinations that have led to their classification into six different phenotypes: Type (Triglycerides/Chylomicrons); Type 2a (Cholesterol/LDL); Type 2b(Cholesterol and triglycerides/LDL and VLDL); Type 3 (Triglycerides and cholesterol/Remnants); Type 4 (Triglycerides/VLDL) and Type 5 (Triglycerides and cholesterol/VLDL and chylomicrons) [3].

Traditional environmental risk factors for alterations in plasma lipid concentrations include high saturated fat intake, high cholesterol intake, an unhealthy diet, low physical activity, tobacco smoking, etc. [14]. In addition, there is wide evidence of the genetic influence in the determination of these plasma lipid concentrations [15, 16]. The genetic influence could result from a single gene disorder (dominant or recessive) or from a polygenic alterations in multiple genes. The most outstanding single gene disorder is the influence of mutations in the LDL-C receptor gene (LDLR) in familial hypercholesterolemia, a well-characterized dominant disease, with high penetrance [17, 18]. The mutations in the LDLR result in high plasma levels of apolipoprotein B-containing lipoproteins (e.g., LDL and lipoprotein(a)). High plasma levels of lipoproteins increase the risk of cardiovascular events by as much as 20-fold if left untreated. It has been estimated that there are at least 20 million people with familial hypercholesterolemia worldwide [19]. Common variations in other multiples genes have also been significantly associated with different plasma lipid concentrations, although their influence is of a lesser magnitude [20, 21]. Moreover, various studies have shown that this genetic influence is not deterministic, but that it is modulated by environmental factors, giving rise to the concept of gene-environment interaction [22, 23]. Although various environmental factors (tobacco, physical activity, drugs, etc.) have been analyzed, the most commonly studied is diet. Hence, in this chapter we are going to focus mainly on gene-diet interactions as one of the main environmental factors interacting with genetic susceptibility in determining plasma lipids.

Though initially most results on gene-diet interactions come from observational studies [23], in recent years new data coming from dietary intervention studies, either including specific components of the diet or with patterns of whole diet, have been provided, so increasing the level of scientific evidence. What is more, recent studies have analyzed not only gene-diet interactions in plasma lipid concentrations but how these gene-diet interactions in determining plasma lipids also affect the incidence of cardiovascular diseases [24-26].

Over these years we have seen a spectacular development of technology applied to the analysis of the genome and that has increasingly enabled researchers to undertake more thorough and complex analyses of gene-diet interactions [27]. We have, therefore, moved on from studies which investigated the presence of genetic polymorphisms through primitive techniques based on restriction fragment length polymorphisms to studies in which high density arrays are used that allow us to analyze tens of thousands of polymorphisms and to investigate various metabolic pathways. Also over recent years, new omics (transcriptomics, proteomics, lipidomics, metabolomics, epigenomics) have been incorporated that are providing new knowledge by allowing us to integrate them into genomics and so have a more complete vision of the processes we are researching [28, 29]. In this chapter, we shall begin with a brief introduction to lipid metabolism, followed by another brief introduction to the technological development of the omics involved in this area of study and the methodology employed in studying gene-environment interactions, before going on to focus on the review of the main studies that have obtained the most noteworthy results on gene-environment interactions in determining plasma lipids. Among these, we shall provide a briefer analysis of those not centered on the diet, and will concentrate on those studies that have analyzed gene-diet interactions. Finally, we shall provide an overview of the results obtained and reflect upon the current situation and need for future research.

2. Lipid Metabolism

Lipid metabolism comprises complex biological pathways. Their homeostasis is achieved by the coordinated action of a large number of nuclear factors, binding proteins, apolipoproteins, enzymes, and receptors [30]. These pathways are also intertwined with energy metabolism and are subject to hormonal controls that are essential for adjustment to environmental and internal conditions. Lipid signaling pathways are especially complex. In the coming years, understanding the integrated lipidomic/metabolomic networks will constitute a key objective taking into account that the modulation lipid metabolism has huge therapeutic potential [13]. The liver plays a key role in lipid metabolism. Lipid metabolism is commonly subdivided into three components: the exogenous pathway, the endogenous pathway, and the reverse cholesterol transport [31]. Most of our knowledge about the relationship between plasma lipid concentrations and cardiovascular disease risk comes from studies in the fasting state. However, humans are almost continuously in the postprandial state. The increased awareness of the atherogenic properties of postprandial triglyceride-rich lipoproteins and the dramatic inter-individual variability in postprandial lipoprotein responses have propelled numerous studies on the gene-environment interactions modulating them [32].

The exogenous pathway starts with the intestinal absorption of triglycerides and cholesterol from dietary sources. Dietary fats represent about 30%–40% of the energy intake, mostly in the form of triglycerides. Under normal circumstances, the digestive system is very efficient in both digesting and absorbing triglycerides [30]. Cholesterol is another major dietary lipid absorbed in the enterocyte. Its homeostasis is maintained by a delicate equilibrium between dietary cholesterol absorption, de novo synthesis, and fecal excretion [3]. After absorption, triglycerides and cholesterol are re-esterified in the intestinal mucosal cells and then coupled with various apoproteins, phospholipids, and unesterified cholesterol into chylomicrons. These chylomicrons are secreted into lymph vessels and then enter the blood circulation and bind to the wall of capillaries in adipose and skeletal muscle tissue [30]. The apolipoproteins associated with nascent chylomicrons are apoB48, apoA-I, apoA-IV, apoC-II, and apoC-III. At the binding sites, the chylomicrons are hydrolyzed by lipoprotein lipase (LPL). Chylomicron remnants are removed from the plasma by hepatic receptors that bind to the apoE present in the particle. Subsequently, the bound remnants are catabolized by lysosomes. The liberated cholesterol is either converted into bile acids, excreted in bile, or incorporated into lipoproteins originated in the liver (VLDL) [31].

In the Endogenous Lipoprotein Metabolism, the liver is the central organ in this process beginning with the synthesis of VLDLs [34]. VLDL production is controlled primarily at the level of presecretory degradation. VLDL particles are precursors of intermediate-density lipoproteins and finally of LDL particles. Their predominant function is to supply cholesterol to cells with LDL receptors. LDL is known to be heterogeneous, comprising a number of distinct sub-populations defined on the basis of a number of characteristics, including particle density, size, charge, and chemical composition [35]. Thus, decreased size and increased density of LDL have been associated with increased cardiovascular disease risk [36].

Over 50 years ago, based on biochemical evidence, Glomset and Wright proposed that HDL are the major species transporting cholesterol from peripheral tissue to the liver for excretion, so-called reverse cholesterol transport [37]. Currently, this proposal is strongly supported by epidemiological studies showing a strong an inverse association between plasma HDLC concentrations and cardiovascular diseases [6]. HDL is synthesized by both the liver and the intestine. Unesterified cholesterol is esterified to cholesteryl ester by the lecithin-cholesterol acyltransferase (LCAT) and the small HDL3 particle becomes a larger HDL2 particle. HDL cholesteryl ester can be taken up selectively by the liver or by organs synthesizing steroid hormones or can also be selectively transferred to apoB-containing lipoproteins in exchange for triglycerides by the cholesteryl ester transfer protein (CETP) [38]. The end result is a system that allows the transfer of cholesterol through LDL to peripheral cells and its return to the liver through HDL, and that prevents excessive accumulation of cholesterol. However, the negative results of pharmacological trials increasing HDL-C levels are casting doubt over the beneficial role of increasing HDL [7]. Therefore, more specific measures of HDL metabolism are needed to reflect the beneficial effects of HDL particles [6, 7].

3. Analysis of the Genome, Environment and Phenotype

When we refer to gene-environment interactions in determining plasma lipid concentrations, we must bear in mind the difficulty of integrating the genotype, phenotype and environment [39].

3.1. The Phenotype

This difficulty is general for all diseases and, on some occasions, the definition of phenotype is very limited. In the case of plasma lipids, although the definition may have some methodological limitation as indicated above, at least for fasting total cholesterol, LDL-C, HDL-C and triglycerides, it seems a simpler and more standardized phenotype than others. Furthermore, as it involves widely available and relatively cheap measurement, traditional plasma lipid concentrations are phenotypes that are routinely measured in most studies that are carried out on cardiovascular risk factors. That has allowed researchers to carry out hundreds of studies specifically analyzing gene-environment interactions in determining plasma lipid concentrations, at least for the most conventional measurements [23, 40]. Other frequently analyzed fasting lipid phenotypes include particle size, density, composition, etc., [35, 36]. On the other hand, studies in the postprandial state analyze levels of postprandial lipids at pre-determined intervals throughout the postprandial period prior following, the ingestion of the selected meals [41]. Moreover, the enormous diversity of plasma lipids and their modifications make lipidomic analyses challenging. Lipidomics is the systems-level analysis of lipids and their interactions. It is a science still in its infancy but one that promises to revolutionize this field. The incorporation of lipidomics into new studies [42] will allow us to delve even deeper into the gene-environment interactions in determining those compounds (hundreds of molecular species have been reported in the human lipidome, covering the main lipid categories: fatty acyls, glycerolipids, glycerophospholipids, sphingolipids, sterol lipids, and prenol lipids) [13]. However, these determinations are still too expensive for intensive use in most studies and also greater standardization of measurements remains to be done for these new phenotypes [13].

3.2. The Genome

As far as measuring the genome is concerned, this can, in principle, be easily done as molecular biology has provided us with more accurate techniques for detecting the presence or absence of certain genetic variants. The most commonly studied genetic variants are the single nucleotide polymorphism (SNPs). Millions of SNPs are reported to be present in the genome, so it is interesting to study their groupings in blocks with high linkage disequilibrium (haplotypes). Sometimes, rather than analyzing various SNPs forming a haplotype, a single SNP chosen as a "tag" SNP, or a SNP representing the entire haplotype, is studied, so cutting the cost involved in its determination. Nevertheless, one limitation that we may find in studying SNPs is that sometimes those SNPs with statistically significant associations may not be the true causal variants, but mere genetic markers [40]. Thus, further studies on functionality have to be undertaken. Furthermore, in recent years high density genotyping arrays have been available that have allowed us to simultaneously determine thousands of genetic polymorphisms [43]. This ability has led to the so-called genome-wide association studies (GWAs) and to the discovery of new genes and SNPs associated with the different levels of plasma lipids. There are numerous GWAs in which dozens of those SNPs associated with different plasma lipids have been successfully identified [44-46]. What is more, the need to increase the statistical power of the associations has led to the formation of grand international consortiums for jointly analyzing the data of each cohort and so making new discoveries possible [47, 48]. Recently a number of meta-analyses including hundreds of thousands of individuals have been published identifying new gene variants associated with the lipid metabolism [48]. Also recently, genomic studies into lipids have incorporated the technologies of next generation sequencing for identifying new genetic variants associated with different lipid phenotypes, mainly identifying new low prevalence variants [49]. The next and necessary step will be to analyze the influence of the environment on the genetic effects reported. For some of the initial genes discovered through GWAs, we already have some results of gene-environment interactions, such as, for example the genes TCF7L2 transcription factor 7-like 2 (TCF7L2) and MLX Interacting Protein-Like (MLXIPL) [24, 25], but for other more recent ones, we are still at the research stage. This leads us to reflect upon the limitations involved in measuring environmental variables.

3.3. The Environment

Environmental factors are all those non-genetic factors to which an individual is exposed [22]. Some of the serious problems that have appeared in studying gene-environment interactions are the difficulty of reliably gathering data on environmental parameters and the complexity of the environment. Their complexity is huge including dozens of different factors such as tobacco smoking, physical activity, diet, drugs, air pollution, outdoor temperature, contaminants, stress, etc. Although tobacco, diet and physical activity, have been the most studied, we are aware that many others exist. Alcohol consumption has often been included in diet and this is how we are considering it in this review. Among these environmental factors the most directly involved in the genetic modulation of the effect on the different plasma lipid concentrations is diet, given our continuous exposure to it throughout our lives [50]. Thus, the most studied gene-environment interactions are those that take diet into account. These gene-diet interactions in determining plasma lipids have been crucial to the development of nutritional genomics [39, 50, 51]. For years, the common methods of gathering information on diet in observational studies have been criticized for their relative validity [50, 52] and higher evidence level is obtained from controlled randomized dietary intervention trials [52]. Moreover, foodstuffs are very complex mixtures of nutrients and it is possible that we are attributing an observed effect to a specific nutrient because we have some knowledge of it, even though the effect may be due to another food component that we do not know, or to one we have not paid attention to, for example, tomato intake. Various studies have shown it to have a protective effect against altered lipemia, oxidative stress and cardiovascular risk [53-55], but it is still unknown whether it is the vitamins, carotenoids, polyphenols, fiber, or other still unknown compounds that provide this protective effect [55]. These examples could be extended to many other dietary foods, as our knowledge of the action mechanisms is still very patchy. However, increasing importance is being placed on analyzing overall food patterns rather than focusing on specific foods or specific nutrients or components. This is certainly true in the case of the Mediterranean Diet, where we continually claim that "the whole is better than the sum of its parts" [56]. Both types of approaches (individual foodstuffs/nutrients or whole dietary patterns) are, therefore, necessary, in order to obtain general and specific knowledge. Furthermore, with the development of omics and their greater ease of application, it is important to integrate studies in which genetic variability and the levels of expression of candidate genes, are simultaneously analyzed [21, 57]. This combination of omics will open up new paths of knowledge on how environmental exposures may act and will provide a certain insight into the components of a specific diet that would merit more detailed investigation [58].

3.4. Other Omics in the Study of Gene-Environment Interactions

In addition to genomics, the recent advances in other omics are making it increasingly easier and cheaper to integrate those omics into gene-environment interaction studies applied to the lipid metabolism. Among these omics, transcriptomics provides us with knowledge of the transcriptome, either individually for each specific gene studied or analyzing the expression of various genes simultaneously on different scales. In this way we can investigate how exposure to different environmental factors affects the expression of all genes (genome-wide transcriptome) or the specific genes [59]. These studies of expression can be conducted either by analyzing intervention with whole diets (for example the Mediterranean diet as against a low fat diet) [60] or by administering specific foods (e.g., tomatoes, olive oil) or specific components of the diet (fiber, vitamins, etc.) [61]. Although initially these transcriptomic studies were carried out independently of other omics, in recent years, the general trend is to integrate them with other omics, mainly genomics [21] or also lipidomics/metabolomics and epigenomics [56, 58]. However, it must be borne in mind that one of the limitations of transcriptomics or epigenomics is that the transcriptome and the epigenome are not the same for all the cells of the organism, as is also true for genomics, but that the level of expression varies depending on the tissues analyzed, adding a little more difficulty to the investigation into these omics.

Despite this variability among tissues, we may say that the advances that perhaps have most contributed over recent years to the study of gene-environment interactions are those related with epigenetics/epigenomics [62]. This term is used to describe a variety of modifications to the genome that do not involve changes in the DNA sequence and can results in alteration of gene expression allowing for differential expression of common genetic information [62, 63]. It constitutes the missing link among genetics, the environment and disease [63]. One of the main advantages of the epigenomic study, unlike the variations in the genome, is that the epigenetic marks are reversible and may allow a rapid adaptation to the environment. There are 3 main categories of epigenetic marks: DNA methylation, histone modification and noncoding RNA. Several investigators have reported that several genes involved in the lipid metabolism are under epigenetic regulation, using a variety of in vivo and in vitro approaches [64, 65].

DNA methylation is a well characterized epigenetic modification of the genome. Most DNA methylation in humans occurs at cytosine–phosphate–guanine (CpG) dinucleotides and consists of the addition of a methyl group on position 5 of cytosine residues of the CpG island, providing marks in the genome by which genes are set to be transcriptionally activated or silenced [66]. Hypermethylation or hypomethylation of relevant islands have been associated with several phenotypes related with the lipid metabolism [64, 65]. As regards regulation by non-coding RNAs and although different classes of non-coding RNAs such as long non-coding RNAs, small nucleolar RNAs (snoRNAs) microRNA (miRNAs) have been characterized, miRNAs are currently the most important [67]. miRNAs are small (18-25 nucleotide) functional non-coding RNAs, that regulate gene expression of their target mRNA in a post-transcriptional manner and have emerged as crucial epigenetic regulators of many process related to lipid metabolism [68]. Numerous SNPs in miRNA target sites have also been described and shown allele-specific effects [68, 69]. Later in this chapter, we shall review in more detail the main studies that have shown gene-environment interactions by analyzing miRNA target site SNPs and their relationship with plasma lipid concentrations [26, 70]. Finally, the recent advances in Lipidomics/ Proteomics/Metabolomics are proving to be hugely important for increasing our knowledge on the lipid metabolism as well as for their integration with other omics.

4. Main Studies That Have Analyzed Gene-Environment Interactions in Determining Plasma Lipids

There are many studies that have analyzed various gene-environment interactions in determining plasma lipids. Most of them have focused on the diet, so it is these studies that we shall review in more detail here. First of all, we review the studies that have focused on tobacco smoking as an environmental factor, followed by studies that have analyzed physical activity and then the studies focusing on diet. Most of these studies are studies in which variations in the genome have been analyzed, although there are also a few that have undertaken their analyses incorporating other omics, mainly epigenomics.

4.1. Tobacco as an Environmental Factor

Tobacco smoking is an environmental factor that is relatively easy to measure. Normally its measurement is undertaken through questionnaires, although in some cases (the fewest), biomarkers of its consumption can be incorporated such as cotinine determinations, etc. [71]. Table 1 shows selected studies that have analyzed gene-smoking interactions in determining plasma lipids [72-79]. Plasma concentrations of HDL-C have been the lipid phenotype most influenced by the gene-smoking interactions analyzed. Although for the common genetic variants in the APOE, LPL APOC3 genes [72-75] as well as for the ABCG5/G8 genes [76-77], some studies have reported consistent gene-smoking interactions, the replication level is still low. The lack of significant interactions in some populations may simply be due to the fact that none exists for the variants and populations studied. However, another factor that has implications for power and effects is the range of smoking prevalence and amount smoked. Therefore, more individual studies and meta-analysis with even larger sample sizes are needed to establish how smoking may modify the effect of certain genetic variants on plasma lipid concentrations and the subsequent cardiovascular risk.

4.2. Physical Activity As an Environmental Factor

A report on an environment-wide association study (EWAS) showed that low physical activity is one of the main environmental factors contributing to all-cause mortality [80]. Although numerous studies have reported gene-physical activity interactions in determining several phenotypes [23, 81], replication for plasma lipids in still low. Physical activity is a much more complex factor to measure than tobacco smoking and that can give rise to problems of validity and accuracy. Normally in large population studies, physical activity is measured by means of validated questionnaires. However, in recent years a growing interest has been noted in incorporating direct measurements of physical activity using accelerometers [82, 83].

Table 2 shows selected studies that have analyzed gene-physical activity interactions in determining plasma lipids [84-91]. Although currently a great heterogeneity exists and results for specific genes have been inconsistent with few exceptions [84, 85], progress is likely to come when studies will be designed to truly address gene-physical activity interactions using randomized controlled trials.

4.3. Diet As an Environmental Factor

Over recent years, the number of publications that have described some gene-diet interaction between genetic polymorphism and dietary factors in determining plasma lipids has experienced exponential growth. Given the exhaustive nature of this list, we have opted to summarize the references of the main studies undertaken in Tables 3 and 4 and by way of example, dedicating more development in the text to the most important studies carried out by our research group. To present the studies, we shall take into account whether they are observational o interventional studies.

4.3.1. Gene-Diet Interactions in Observational Studies

One of the first studies to analyze gene-diet interactions on lipids was carried out by our research group on participants in the Framingham Study [92]. We found an APOE-alcohol interaction in determining LDL-C concentrations in men. Later, an APOE-alcohol interaction in determining HDL-C was described [93]. Moreover these gene-alcohol interactions were replicated in the Mediterranean population [94]. On the other hand, although, over the years, and encouraged by the results from initial studies, it has been affirmed that the E4 allele of the APOE polymorphism interacts with dietary saturated fat, so increasing LDL-C concentrations as well as CVD, more so in E4 individuals than in others (E2 and E3) [96], the evidence level of this interaction is still low. In terms of dietary fat, one of the first studies in determining plasma lipids was that reported by Prof Ordovas' group in Boston involving a polymorphism of the hepatic lipase (encoded by the LIPC gene) and total and saturated fat intake [97].

Table 1. Selected studies analyzing gene-smoking interactions in determining plasma lipids

Reference	Phenotype	Genetic factor/s	Description of the gene-smoking interaction
Peacock et al. [72]	Fasting plasma triglycerides, apoCIII, HDL-C, and apoAI concentrations	SNPs in LPL gene (HindIII and S447X) and in the APO AI-CIII gene cluster (G75A and C1100T)	In 315 healthy men and women from Iceland, they found that variations at both of the LPL and APOA1-APOC3 loci influenced levels of triglycerides, apoCIII, HDL-C, and apoAI, but these effects were strongly modulated by smoking and were different between men and women.
Corella et al. [73]	Fasting lipids (total cholesterol, HDL-C, LDL-C and triglycerides)	SNPs in the LPL and APOC3 gene loci (LPL-HindIII, LPL-S447X, and APOC3-SstI) and the APOE common polymorphism	In a healthy Spanish population (n = 1,029), a significant gene-gene-smoking interactions was found (interaction HindIII-APOE-smoking, P = 0.019), indicating that smoking abolished the increase in HDL-C levels observed in E4/H(-) subjects.
Lee et al. [74]	Fasting lipids (total cholesterol, HDL-C, LDL-C and triglycerides)	LPL (s447X) and APOE common polymorphisms	A gene-smoking interaction was found in 4,058 subjects from a representative sample of the Singapore population (Chinese, Malays, and Asian Indians). The association between the X447 allele HDL-C concentration was modulated by APOE genotype in males and cigarette smoking in females.
Fiegenbaum et al. [75]	Fasting lipids (total cholesterol, HDL-C, LDL-C and triglycerides)	APOC3 and APOE common polymorphisms	In a southern Brazilian population of European descent (n=673) a gene-smoking interactions was found involving the APOC3 polymorphism in determining HDL-C concentrations in women with the E3/E3 genotype.
Junyent et al. [76]	Plasma HDL-C	(SNPs) at ABCG5/G8 (i18429G>A, i7892T>C, Gln604GluC>G, 5U145A>C, Tyr54CysA>G, Asp19HisG>C, i14222A>G, and Thr400LysC>A)	In the Boston Puerto Rican Health Study (243 men and 602 women) a significant gene-smoking interaction was found. Carriers of the minor alleles at ABCG5/G8 (Gln604GluC>G, Asp19HisG>C, i14222A>G) SNPs displayed lower concentrations of HDL-C only if they were smokers (P = 0.001-0.025).
García-Rios et al. [77]	Fasting lipids (total cholesterol, HDL-C, LDL-C and triglycerides)	(SNPs) at ABCG5 (i7892A>G, i18429C>T, Gln604GluC>G, i11836G>A) and five at ABCG8 (5U145T>G, Tyr54CysA>G, Asp19HisG>C, i14222T>C, and Thr400LysG>T)	In 500 subjects with genetic diagnosis of familial hypercholesterolemia from Spain, a gene-smoking interaction was found. Carriers of the minor alleles at ABCG5 (i7892A>G, i18429C>T, i11836G>A) SNPs displayed significantly lower HDL-C, higher TC and higher TG respectively, only in smokers

Reference	Phenotype	Genetic factor/s	Description of the gene-smoking interaction
Dumitrescu et al. [78]	Plasma lipids (total cholesterol, HDL-C, LDL-C and triglycerides)	49 SNP in genes associated with lipids (PCSK9, ANGPTL3 CELSR2/PSRC/SORT1, GALNT2, APOB, GCKR, ABCG8, HMGCR, TIMD4, MLXIPL, LPL, TRIB1, TTC39B, ABCA1, FADS1, APOA1/C3/A4/A5, MMAB-MVK, HNF1A, LIPC, CETP, LCAT, LIPG, ANGPTL4, LDLR, CILP2/PBX4/NCA, BCL3, APOE/C1/C4, TOMM40,MAFB, HNF4A and PLTP)	Gene-smoking interactions in determining plasma lipids were tested in over 41,000 racially/ethnically diverse samples with lipid levels from the Population Architecture Using Genomics and Epidemiology (PAGE) study. No significant SNP-smoking interactions were detected after correction for multiple comparisons. The most significant interaction was TTC39B rs471364 -smoking in determining HDL-C levels among Mexican Americans/Hispanics
Grammer et al. [79]	LDL diameters, oxidized LDL, and lipoprotein-associated phospholipase A2	APOE (Alleles E2, E3 and E4)	In 3263 participants of the LUdwigshafen RIsk and Cardiovascular Health study, tobacco smoking interacted significantly with the common APOE polymorphisms in determining lipid phenotypes. Smokers with the E4 allele had reduced average LDL diameters, elevated oxidized LDL, and lipoprotein-associated phospholipase A2.

Table 2. Selected studies analyzing gene-physical activity interactions in determining plasma lipids

Reference	Phenotype	Genetic factor/s	Description of the gene-physical activity interaction
Corella et al. [84]	Fasting lipids (total cholesterol, HDL-C, LDL-C and triglycerides)	APOE polymorphism (E2,E3 and E4 alleles)	In a healthy Mediterranean population (396 men and 513 women) a gene physical activity in determining HDL-C concentrations in males was found. In sedentary men the E2 allele was associated with higher HDL-C. Conversely, in active men, the opposite effect was observed, with E4 carriers showing the highest HDL-C concentrations.
Bernstein et al. [85]	Fasting lipids (total cholesterol, HDL-C, LDL-C and triglycerides)	APOE polymorphism (E2,E3 and E4 alleles)	In 1708 randomly selected men and women from Switzerland a gene-physical activity in determining plasma lipids was observed in males. Among men, increased high-intensity activity had greater protective effects in E4 carriers, increasing HDL-C and decreasing triglycerides.
Bernstein et al. [86]	Fasting lipids (total cholesterol, HDL-C, LDL-C and triglycerides)	CETP polymorphisms	In 1720 randomly selected European men and women Switzerland, no statistically significant (CETP-physical activity interactions in determining lipid profile were detected in either gender.
Wilund et al. [87]	Fasting lipids (total cholesterol, HDL-C, LDL-C and triglycerides) and HDL-C subfractions measured by nuclear magnetic resonance (NMR)	Common polymorphisms in CETP, APOE and LPL genes	Middle- to older-aged men and women with at least 1 lipoprotein-lipid risk factor underwent 6 months of endurance exercise training while on a low-fat diet. CETP genotype may contribute to the interindividual differences in plasma HDL-C subfraction changes occurring with endurance exercise training. No effect was detected for APOE or LPL genotypes.
Ferreira et al. [88]	Postprandial lipemia	APOE polymorphism (E2,E3 and E4 alleles)	Intervention study with a 500 Kcal test in 30 men depending on the APOE genotype. Exercise was effective in attenuating postprandial lipemia in E2 subjects. E4 subjects showed no attenuation.
Brinkley et al. [89]	Fasting triglycerides, VLDL, and HDL-C	LIPC -514C>T, rs1800588	Intervention study with physical activity in 76 subjects. LIPC -514C>T variant significantly affected training-induced anti-atherogenic changes in VLDL-TG, VLDL particles, and HDL-C
Smith et al. [90]	Fasting HDL and LDL-C. Particle size and particle concentration	LIPG polymorphisms (rs2000813 and rs6507931)	In 539 men and 584 women participating in the GOLDN study some LIPG-physical activity interactions were found for HDL and some LDL measures
Halverstadt et al. [91]	HDL-C, particle size and particle concentrations	LIPG polymorphism (rs2000813)	Intervention study with exercise training in 83 subjects. LIPG genotype was associated with the response of HDL-C and its subfractions to exercise training.

Table 3. Selected observational studies analyzing gene-diet interactions in determining plasma lipids

Reference	Study population	Phenotype	Genetic variant (s)	Gene-diet interaction
Corella et al. [92]	Cross-sectional analysis in participants in the Framingham Offspring Study (1014 men and 1133 women)	LDL-C concentrations	*APOE (alleles E2, E3, E4)*	Alcohol intake interacted with the APOE SNP in determining LDL-C in men. In E2 subjects, LDL-C was significantly lower in drinkers than in nondrinkers but was significantly higher in drinkers than in nondrinkers in E4 subjects.
Djoussé et al. [93]	Cross-sectional design in the National Heart, Lung, and Blood Institute Family Heart Study (670 men and women)	HDL-C concentrations	*APOE (alleles E2, E3, E4)*	Alcohol intake interacted with the APOE SNP in determining HDL-C concentrations. The increase in HDL associated with alcohol was stronger in subjects without the epsilon4 allele than in E4 carriers.
Corella et al. [94]	Nested case-control study in the EPIC-Spain cohort (534 incident coronary heart disease cases and 1123 controls)	HDL-C and LDL-C	*APOE (alleles E2, E3, E4)*	Alcohol intake interacted with the APOE SNP in determining LDL-C concentrations. In E2 LDL-C was significantly lower in drinkers than in nondrinkers. This observation was also detected when analyzing coronary heart disease incidence. An interaction effect with saturated fat intake was also reported.
Campos et al. [95]	Cross-sectional study in 420 randomly selected free-living Costa Ricans	VLDL and HDL-C	*APOE (alleles E2, E3, E4)*	The APOE genotype interacted with saturated fat in determining VLDL and HDL-C concentrations (higher VLDL and lower HDL-C in E2 carriers with a high fat)
Ordovas et al. [97]	Cross-sectional analysis in participants in the Framingham Offspring Study (1020 men and 1110 women)	HDL-C concentrations and HDL particle size	*LIPC (-514C>T)*	The -514C>T LIPC polymorphism interacted with dietary fat in determining HDL-related measures. T allele was associated with significantly greater HDL-C concentrations and large HDL size only in subjects consuming <30% of energy from fat.

Table 3. (Continued)

Reference	**Study population**	**Phenotype**	Genetic variant (s)	**Gene-diet interaction**
Tai et al. [98]	Cross-sectional study in an Asian population (1324 Chinese, 471 Malays and 375 Asian Indians)	Fasting triglycerides and HDL-C	*LIPC (-514C>T)*	The -514C>T LIPC polymorphism interacted with dietary fat in determining TG concentration and the HDL-C/TG ratio (P = 0.001) in the overall sample.
Tai et al. [99]	Cross-sectional study in participants in the Framingham Offspring Study (1003 men and 1103 women)	Plasma triglycerides and apoC-III concentrations	PPARA (L162V polymorphism)	Carriers of the 162V allele had significantly lower plasma triglycerices and apoC-III concentrations when consuming a high PUFA diet, in which *n-3* and *n-6* fatty acids and seem to have a similar role.
Chan et al. [100]	Cross-sectional study among Participants in the 1998 Singapore National Health Survey: 1964 men and 2284 women.	Plasma lipids	PPARA (V227A polymorphism)	In carriers of the A227 allele, increasing PUFA intake was associated with lower HDL-C concentrations. In those who were homozygous for the common allele, this association was much weaker.
Volcik et al. [101]	Cross-sectional study among participants in the biethnic Atherosclerosis Risk in Communities (ARIC) Study (10134 whites and 3480 African Americans)	Plasma lipids	PPARA (L162V and 3'UTR C→T polymorphisms)	A significant interaction between the *PPARA* 3'UTR C→T SNP and LC *n–3* fatty acids on total cholesterol and LDL-C concentrations in African American participants. No significant interaction with the *PPARA* L126V SNP.
Lai et al. [102]	Cross-sectional study among participants in the Framingham Offspring cohort (1001 men and 1147 women)	Plasma lipids, remnant-like particle concentrations, and lipoprotein particle size	APOA5 SNPs: –1131T>C, –3A>G, IVS3+476G>A, and 1259T>C and 56C>G	Significant gene-diet interactions between the -1131T>C polymorphism and total PUFA intake were found in determining fasting triglycerides, remnants and particle size. However, the PUFA-APOA5 interactions were specific for dietary *n-6* fatty acids. No significant interaction with *n-3* fatty acids were found

Reference	Study population	Phenotype	Genetic variant (s)	Gene-diet interaction
Lu et al. [103]	Cross-sectional investigation in the Doetinchem Cohort Study (3575 subjects)	Plasma lipids	*FADS* cluster (rs174546, rs482548, and rs174570 polymorphisms)	Significant associations between rs174546 genotypes and total and non-HDL-cholesterol concentrations in the group with a high intake of n-3 PUFA but not in the low intake group
Baik et al. [104]	Prospective cohort of 5314 Korean men and women	Plasma lipids	LPL polymorphisms (Ser447Stop)	Significant interactive effects of the LPL polymorphism and alcohol consumption and unsaturated fat on follow-up measurement of HDL-C concentrations.
Li et al. [105]	Cross-sectional analysis in 780 diabetic men who participated in the Health Professionals Follow-Up Study	HDL-C concentrations	CETP (TaqIB polymorphism)	Significant CETP-fat intake (animal fat, saturated and monounsaturated) interactions were found in determining HDL-C concentrations.
Corella et al. [106]	Cross-sectional analysis in 4210 participants in the PREDIMED Study	Plasma lipids	CETP polymorphisms (Taq1B and -4,502C>T)	No statistically significant interactions of these SNPs (independently or as diplotype) with alcohol, dietary fat, and adherence to the Mediterranean in determining HDL-C were found.
Richardson et al. [107]	Cross-sectional study in two replication cohorts: The Framingham study (1259 men and 1352 women) and the GOLDN Study (481 men and 513 women	Anthropometrics, lipids and glucose	Seven SNPs in the *PLIN4* gene: rs884164, rs1609717, rs7250947, rs8887, rs8102428, rs892158, and rs11673616.	*N-3* fatty acids modulate the associations between the rs8887 SNP and anthropometrics. rs884164 showed interaction with both n3 and n6 PUFA modulating anthropometric and lipid phenotypes.
Richardson et al. [70]	Cross-sectional study and meta-analysis including 28.000 individuals from ten cohorts participating in the CHARGE consortium.	Plasma lipids	LPL (rs13702 SNP; a microRNA binding site SNP)	The rs13702 minor allele in the LPL gene that disrupts a microRNA recognition element seed site for the miR-410, interacted with PUFA in determining triglycerides. Replicated in a meta-analysis including 9 populations

Table 3. (Continued)

Reference	Study population	Phenotype	Genetic variant (s)	Gene-diet interaction
Corella et al. [26]	Cross-sectional analysis in 7187 participants in the PREDIMED at baseline	Plasma lipids	LPL (rs13702 SNP; a microRNA binding site SNP)	The LPL rs13702T>C polymorphism was strongly associated with lower triglycerides in C allele carriers and interacted synergistically with dietary monounsaturated and unsaturated fat intake decreasing triglycerides.
Ortega-Azorín et al. [25]	Cross-sectional analysis in 7166 participants in the PREDIMED at baseline	Plasma lipids	MLXIPL polymorphism (rs3812316, C771G; Gln241His)	The MLXIPL-rs3812316 was associated with lower baseline triglycerides. This association was modulated by baseline adherence to the Mediterranean diet. When adherence to Mediterranean diet was high, the protection was stronger.

Table 4. Selected dietary intervention studies analyzing gene-diet interactions in determining plasma lipids

Reference	Study population	Phenotype	Genetic variant (s)	Gene-diet interaction
Jansen et al. [111]	Intervention study (three consecutive diets, each of 4 weeks' duration) in 41 healthy male subjects	Changes in cholesterol, LDL-C and ApoB concentrations	*APOA4 (Thr437Ser)*	LDL-C response to dietary fat is influenced by the 347Ser mutation of APOA4. Carriers of the 347Ser allele presented a greater decrease in LDL-C when they were switched from the SFA to the NCEP type 1 diet than homozygous the 347Thr allele.
Delgado-Lista et al. [112]	Intervention study in 88 normolipidemic young men who were given a fatty meal	Postprandial lipids	*ABCA1 (rs2575875, rs4149272 and rs2230806)*	Major allele homozygotes for ABCA1 polymorphisms have a lower postprandial response as compared to minor allele carriers
Pérez-Martínez [113]	Intervention study in 88 subjects (were given a fat-rich meal)	Postprandial lipids	*MC4R* (rs12970134)	Individuals carrying the G/G genotype displayed a higher postprandial response of triglycerides (total cholesterol and large TRL-triglycerides than did carriers of the minor A-allele.
Pérez-Martínez [114]	Intervention study in two independent samples. 88 healthy Spanish men and 271 healthy US subjects (men and women) underwent an oral-fat-load test	Postprandial lipids	*(PLIN1, 6209T-->C; PLIN4, 11482G-->A; PLIN5, 13041A-->G; and PLIN6, 14995A-->T)*	In both studies the presence of the minor C and A alleles at PLIN1 and PLIN4, respectively, were associated with a lower postprandial response that may result in lower atherogenic risk.
Moreno-Luna et al. [115]	Intervention study in 88 APOE E3/E3 subjects (were given a fat-rich meal)	Postprandial lipids	*APOA5 (1131T>C, c.-3A>G, c.56C>G, IVS3+476G>A, and c.1259T>C)*	The presence of the APOA5*2 and APOA5*3 haplotypes was associated with a higher postprandial response that could be involved in the higher cardiovascular risk associated with the 56G and -1131C alleles

Table 4. (Continued)

Reference	Study population	Phenotype	Genetic variant (s)	Gene-diet interaction
Zhang et al. [116]	Intervention study in 734 overweight or obese adults who were randomly assigned to one of 4 diets in the Pounds Lost Trial	Changes in total cholesterol, LDL-C and HDL-C	*APOA5* (rs964184)	The APOA5 rs964184 polymorphism interacted with and dietary fat intake in the determination of changes in total cholesterol, LDL-C and HDL-C at 2-years. In the low-fat intake group, carriers of the risk allele (G allele) exhibited greater reductions in total cholesterol and LDL-C cholesterol than did noncarriers.
Qi et al. [117]	Intervention study in 732 overweight/obese adults from a 2-year randomized weight-loss trial (POUNDS LOST), and replicated in 171 overweight/obese adults from an independent trial (DIRECT)	Changes in plasma lipids	CETP (rs3764261)	Gene-diet interactions were found and replicated. These results suggested that the CETP rs3764261 CC genotype might derive greater effects on raising HDL-cholesterol and lowering triglycerides in a low-carbohydrate/high-fat weight-loss diet instead of a low-fat diet
Brahe et al. [118]	Intervention study in 841 subjects participating in the DIOGENES trial. 8 weeks on the low-energy diet and 6-month ad libitum weight maintenance diet were analyzed	Changes in plasma lipids	240 SNPs in 42 candidate genes	After adjusting for multiple testing, a SNP-dietary protein interaction effect on triglycerides was identified for the LPIN1 polymorphism (rs4315495), with a decrease in triglycerides per A-allele/protein unit.
Irvin et al. [119]	Intervention study in 40 participants from the GOLDN Study (changes were analyzed between fasting and 3.5 hours after a high-fat meal)	Change in plasma concentration of 35 fatty acids and 11 sterols	SNPs (from the Affymetrix 6.0 array) and methylation markers (from the Illumina Methyl 450 array)	Two SNPs (rs12247017 and rs12240292) in the sorbin and SH3 domain containing 1 (SORBS1) gene were associated with b-Sitosterol after correction for multiple testing.

Reference	**Study population**	**Phenotype**	**Genetic variant (s)**	**Gene-diet interaction**
Frazier-Wood et al. [120]	Intervention study in 817 men and women from the GOLDN study after ingestion of an oral fat load.	Change in Chylomicrons	26 SNPs in the LRP-1 gene	Eleven LRP-1 SNPs were significantly associated with the change in chylomicron concentration
An et al. [121]	Intervention study in 817 men and women from the GOLDN study after ingestion of an oral fat load	Changes in postprandial lipids	Genome-wide association study	Several loci in determining postprandial response were identified. The most relevant was the CDH13 locus non-HDL response to dietary high-fat intake challenge
Gomez-Delgado et al. [122]	Intervention study in 507 metabolic syndrome patients selected from the CORDIOPREV trial	Postprandial lipids and changes in triglycerides	TNFA (rs1800629, and rs1799964)	The rs1800629 SNP interacted with intervention with Mediterranean diet to influence triglyceride metabolism
Corella et al. [26]	Randomized controlled intervention trial with Mediterranean Diet versus control diet in 7187 participants in the PREDIMED at baseline	Plasma lipids	LPL (rs13702 SNP; a microRNA binding site SNP)	The LPL rs13702T>C polymorphism interacted with the dietary intervention (Mediterranean diet versus control group) for changes in triglycerides at 3 years in the randomized trial.

This work was crucial due to its pioneering nature and for advancing nutrigenetics. It showed the heterogeneity of the genetic effects on HDL-C concentrations and particle size depending on the amount of total fat intake in the Framingham study. Later studies on this genetic modulation in a South Asian population also showed a statistically significant interaction between the LIPC polymorphism (-514 C > T) and total fat intake on plasma triglycerides [98]. Table 3 provides summary of these and of other selected studies focusing on observational gene-diet interactions on lipids [25, 26, 70, 92-95, 97-107]. In general, heterogeneity in the results of gene-diet interaction still exist and more replication in needed. However, in recent years some studies have included independent cohorts for replication of finding [70] and have meta-analyzed data. In addition, genetic risk scores including dozens of polymorphisms [108] are being gradually incorporated in the gene-diet interaction studies on plasma lipids [109], increasing the amount of genes analyzed, although the number of published studies is still scarce in comparison with additive scores in studies of gene-diet interactions in determining body-weight [110].

4.3.2. Gene-Diet Interactions in Intervention Studies

Observational studies have many limitations and bias that result in a low level of scientific evidence for the associations found. Thus, more intervention studies are needed to better characterize the gene-diet interactions [52]. Taking into account the great difficulties in conducting intervention studies, the number of adequately powered intervention studies analyzing gene-diet interactions is still low. Moreover, there are hardly any studies that have later examined these interactions and the few that have done so, have not managed to replicate the results. Table 4 [26, 111-122] summarizes selected interventional gene-diet interaction studies in determining plasma lipids. In these studies, both interventions with specific foods or with whole dietary patterns (such as the Mediterranean diet) have been considered. In this regard, it is motivating to note the interesting results obtained by our group in the PREDIMED randomized intervention trial analyzing the interaction effects of the intervention with Mediterranean diet (rich in unsaturated fat) with a genetic polymorphism in the LPL gene located in a the microRNA-410 binding site [26]. In this study linking genetics and epigenetics, we replicated previous findings obtained in a previous observational meta-analysis [70] and extended results. We detected a statistically significant gene-diet interaction between the LPL-rs13702 polymorphism and the dietary intervention for changes in triglycerides at 3 years in the randomized trial. These results were consistent with the gene-diet interaction observed for triglycerides at baseline (Table 3). Thus, among participants allocated to the Mediterranean diet intervention groups, carriers of the variant allele had greater reductions in triglycerides than did noncarriers. However, in the control group (low-fat diet group), no significant decreases in triglycerides in carriers of the variant allele were observed [26]. Furthermore, we observed an effect of this gene-diet interaction in determining stroke incidence. Thus, only among participants assigned to a Mediterranean diet intervention, supplemented with extra-virgin olive oil or nuts, did carriers of the variant allele have a statistically significant lower stroke risk than did homozygous subjects for the common allele. However, those carriers of the protective variant allele who were assigned to the low-fat group did not experience the same level of protection from this allele, and the statistical association became nonsignificant. These results are encouraging in the field of nutrigenetics and although currently, firm conclusions in general cannot be established in the field of gene-diet interactions on lipids, it is crucial to undertake new intervention studies aimed at replicating previously reported interactions in order to increase their evidence level and so make the clinical or preventive applications in cardiovascular genomics possible.

SUMMARY AND PERSPECTIVES

In this chapter, we have reviewed the main gene-environment interactions in determining lipid metabolism focusing basically on gene-diet interactions, as they have been the ones most commonly studied. If we look back from the current perspective where we enjoy the huge advances in the technological development of omics, we realize that, despite the fact that the early studies were undertaken with very primitive technology and provided very limited analyses, they did open the way for the knowledge that we have today. We have learnt from both their correct findings and their mistakes over these years of experience, and all this baggage together with the huge development and lowering of costs that is going to take place in omics technology along with the possibility of incorporating next generation sequencing, leads us to believe in a very promising future. The results of the first studies on gene-environment interactions allowed us to see that the influence of the main genes discovered in the lipid metabolism was not deterministic. The association of the risk alleles with detrimental phenotypes could be modified by different environmental factors, especially the diet. Although the scientific evidence of the initial studies was not very high, being mainly observational studies with little replication, in recent years interest in replicating findings has increased and methodological aspects have been improved. Furthermore, the need to undertake more randomized dietary intervention trials has been accepted in order to minimize bias and increase the level of scientific evidence. Outstanding among these studies is the PREDIMED study, which is a randomized controlled intervention trial with Mediterranean diet versus a control diet in the primary prevention of cardiovascular diseases. Here, not only have gene-diet interactions in determining plasma lipid concentrations been reported on, but also the effect of these gene-diet interactions in determining the incidence of cardiovascular diseases. This has allowed us to have a more complete vision of the interactions that may affect not only the intermediate phenotypes of cardiovascular diseases, such as plasma lipid concentrations, but also their repercussions on the final hard event. The PREDIMED study, therefore, is a good example to follow in order to undertake future studies, essential for increasing the level of evidence on gene-environment interactions and for allowing us to use the results of these studies in personalized medicine. We should, however, not forget that we need to evolve from the simple scenarios examined (i.e., one single dietary component, SNP, and risk factor) to more realistic situations involving multiple interactions, not only with diet, but with other environmental factors interacting jointly in a holistic manner. To investigate this complexity we need to both apply and integrate the latest advances in omics technologies and to improve and automate the measuring and validation of diet and the other environmental factors through better computational developments. To do so, biomedical information technology and bioinformatics will be crucial to progress over the coming years in this discipline.

Key Facts

1. The first studies on gene-environment interactions were very important because they allowed us to know that the association between the most relevant genes in lipid metabolism and plasma lipids was not deterministic, but that it could be modulated by environmental factors
2. Hundreds of gene-environment interactions in determining plasma lipid concentrations have been described including physical activity, tobacco smoking, drugs, etc. However, the most studied interactions are those that take diet into account, as these were crucial to the development of nutritional genomics.
3. The development of omics technology over the last ten years has been spectacular. This huge improvement has gradually been incorporated into reports on lipid gene-environment interactions. This work has now moved on from using the more costly and time-consuming primitive methodologies that only allow a few polymorphisms to be analyzed to high-throughput methodologies that can analyze tens of thousands of polymorphisms.
4. Despite this great technological improvement, we still rely on older methodologies in order to measure environmental variables, so posing a limitation. Hence, in the coming years we shall have to work more on automating and increasing the validity and accuracy of measuring diet, physical activity and other more complex environmental variables.
5. Although for many genetic variants there is a great heterogeneity and a lack of results replication in the gene-environment interactions observed in the initial studies, for others, replication of results has indeed been obtained in various populations.
6. Most studied undertaken in the field of lipid gene-environment interactions have been observational studies so there is a need to increase the number of randomized controlled intervention trials in order to raise the level of evidence.
7. More longitudinal studies are required to provide us with knowledge on the evolution of gene-environment interactions, not only for lipids, but also for incidence of cardiovascular disease.
8. Incorporating high-density genotyping and improving computational efficiency in genome-wide-interaction analyses will provide us with new gene-environment interactions
9. Future studies in this field have to incorporate not only genomics, but also the new omics such as transcriptomics, epigenomics, lipidomics, metabolomics so that we may arrive at an integrated analysis of gene-environmental interactions
10. As more powerful computational tools become available, a more holistic vision will have to be adopted in studies, simultaneously analyzing the complex interactions between the different environmental factors.

References

[1] Rached, FH. et al. An overview of the new frontiers in the treatment of atherogenic dyslipidemias. *Clin Pharmacol Ther*. 96, 57-63 (2014).

[2] Fischer, K. et al. Biomarker profiling by nuclear magnetic resonance spectroscopy for the prediction of all-cause mortality: an observational study of 17,345 persons. *PLoS Med.* 11, e1001606 (2014).
[3] Griffin, BA. et al. Metabolism of VLDL and LDL subclasses. *Curr Opin Lipidol.* 5, 200-206 (1994).
[4] Bruckert, E. et al. Evidence supporting primary prevention of cardiovascular diseases with statins: Gaps between updated clinical results and actual practice. *Arch Cardiovasc Dis.* 107, 188-200 (2014).
[5] Ridker PM. LDL cholesterol: controversies and future therapeutic directions. *Lancet.* 384, 607-617 (2014).
[6] Rader, DJ. et al. HDL and cardiovascular disease. *Lancet.* 384, 618-625 (2014).
[7] Kingwell, BA. et al. HDL-targeted therapies: progress, failures and future. *Nat Rev Drug Discov.* 13, 445-464 (2014).
[8] Wilkins, JT. et al. Coronary heart disease risks associated with high levels of HDL cholesterol. *J Am Heart Assoc.* 3, e000519 (2014).
[9] Bansal, S. et al. Fasting compared with nonfasting triglycerides and risk of cardiovascular events in women. *JAMA.* 298, 3-309 (2007).
[10] Nordestgaard, BG. et al. Triglycerides and cardiovascular disease. *Lancet.* 384, 9943-626 (2014).
[11] Iso, H. et al. Fasting and non-fasting triglycerides and risk of ischemic cardiovascular disease in Japanese men and women: The Circulatory Risk in Communities Study (CIRCS). *Atherosclerosis.* 237, 361-8 (2014).
[12] Chen, H. et al. Metabolomics in dyslipidemia. *Adv Clin Chem.* 66, 101-19 (2014).
[13] Quehenberger, O1. et al. The human plasma lipidome. *N Engl J Med.* 365, 19-1812 (2011).
[14] Farrell, TC. et al. The primary prevention of cardiovascular disease: nurse practitioners using behaviour modification strategies. *Can J Cardiovasc Nurs.* 24, 8-15 (2014).
[15] Holdt, LM. et al. From genotype to phenotype in human atherosclerosis--recent findings. *Curr Opin Lipidol.* 24, 410-418 (2013).
[16] Keenan, TE. et al. Genetics of lipid traits and relationship to coronary artery disease. *Curr Cardiol Rep.* 15, 396 (2013).
[17] Varghese, MJ. Familial hypercholesterolemia: A review. *Ann Pediatr Cardiol.* 7, 107-17 (2014).
[18] Knowles, JW.et al. Reducing the burden of disease and death from familial hypercholesterolemia: A call to action. *Am Heart J.* 168, 807-811 (2014).
[19] Watts, GF. et al. Integrated guidance on the care of familial hypercholesterolaemia from the International FH Foundation. *Int J Cardiol.* 171, 309-325 (2014).
[20] Bryant, EK. et al. A multiethnic replication study of plasma lipoprotein levels-associated SNPs identified in recent GWAS. PLoS One. 8, e63469 (2013).
[21] Yao, C. et al. Integromic Analysis of Genetic Variation and Gene Expression Identifies Networks for Cardiovascular Disease Phenotypes. *Circulation.* (in press) (2014).
[22] Corella, D. et al. Integration of environment and disease into 'omics' analysis. *Curr Opin Mol Ther.* 7, 569-576 (2005).
[23] Parnell, LD. et al. CardioGxE, a catalog of gene-environment interactions for cardiometabolic traits. *BioData Min.* 7, 21 (2014).

[24] Corella, D. et al. Mediterranean diet reduces the adverse effect of the TCF7L2-rs7903146 polymorphism on cardiovascular risk factors and stroke incidence: a randomized controlled trial in a high-cardiovascular-risk population. *Diabetes Care.* 36, 3803-3811 (2013).

[25] Ortega-Azorín, C. et al. Amino acid change in the carbohydrate response element binding protein is associated with lower triglycerides and myocardial infarction incidence depending on level of adherence to the Mediterranean diet in the PREDIMED trial. *Circ Cardiovasc Genet.* 7, 49-58 (2014).

[26] Corella, D. et al. MicroRNA-410 regulated lipoprotein lipase variant rs13702 is associated with stroke incidence and modulated by diet in the randomized controlled PREDIMED trial. *Am J Clin Nutr.* 100, 719-731 (2014).

[27] Phillips, CM. Nutrigenetics and metabolic disease: current status and implications for personalised nutrition. *Nutrients.* 5, 32-57 (2013).

[28] Gibson, G. Wellness and health omics linked to the environment: the WHOLE approach to personalized medicine. *Adv Exp Med Biol.* 799, 1-14 (2014).

[29] Kuo, TC. et al. 3Omics: a web-based systems biology tool for analysis, integration and visualization of human transcriptomic, proteomic and metabolomic data. *BMC Syst Biol.* 7, 64 (2013).

[30] Dashty, M. A quick look at biochemistry: lipid metabolism. *J Diabet Metab* 5, 1 (2014).

[31] Kwiterovich, PO Jr. The metabolic pathways of high-density lipoprotein, low-density lipoprotein, and triglycerides: a current review. *Am J Cardiol.* 86, 5L-10L (2000).

[32] Chan, DC. et al. Postprandial hypertriglyceridemia and cardiovascular disease: current and future therapies. *Curr Atheroscler Rep.* 15, 309 (2013).

[33] Huff, MW. Dietary cholesterol, cholesterol absorption, postprandial lipemia and atherosclerosis. *Can. J. Clin. Pharmacol.* 10, A26–32 (2003).

[34] Havel, RJ. et al. Quantification of triglyceride transport in blood plasma: a critical analysis. *Fed. Proc.* 34, 2250–2257 (1975).

[35] Diffenderfer, MR. et al. The composition and metabolism of large and small LDL. *Curr Opin Lipidol.* 25, 221-6 (2014).

[36] Hoogeveen, RC. et al. Small dense low-density lipoprotein-cholesterol concentrations predict risk for coronary heart disease: the Atherosclerosis Risk In Communities (ARIC) study. *Arterioscler Thromb Vasc Biol.* 34, 1069-1077 (2014).

[37] J.A. Glomset, J.L. Wright. Some properties of a cholesterol esterifying enzyme in human plasma. *Biochim Biophys Acta*, 89, 266–276 (1964).

[38] Levinson, SS. et al. Implications of reverse cholesterol transport: Recent studies. *Clin Chim Acta.* 439C, 154-161 (2015).

[39] Ordovas, JM. et al. Nutritional genomics. *Annu Rev Genomics Hum Genet.* 5, 71-118 (2004).

[40] Ordovas, JM. et al. Nutrient-gene interactions in lipoprotein metabolism - an overview. *Forum Nutr.* 60, 102-9 (2007).

[41] Purcell, R. et al. High-fat meals rich in EPA plus DHA compared with DHA only have differential effects on postprandial lipemia and plasma 8-isoprostane F2α concentrations relative to a control high-oleic acid meal: a randomized controlled trial. *Am J Clin Nutr.* 100, 1019-28 (2014).

[42] Stegemann, C. et al. Lipidomics profiling and risk of cardiovascular disease in the prospective population-based Bruneck study. *Circulation.* 129, 18-1821 (2014).

[43] Nievergelt, CM. et al. Marine Resiliency Study Investigators, Schork NJ. Chip-based direct genotyping of coding variants in genome wide association studies: utility, issues and prospects. *Gene*. 540, 104-109 (2014).

[44] Kathiresan, S. et al. A genome-wide association study for blood lipid phenotypes in the Framingham Heart Study. *BMC Med Genet*. 8, S17 (2007).

[45] Kathiresan, S. et al. Six new loci associated with blood low-density lipoprotein cholesterol, high-density lipoprotein cholesterol or triglycerides in humans. *Nat Genet.* 40, 189-97 (2008).

[46] Kathiresan, S. et al. Common variants at 30 loci contribute to polygenic dyslipidemia. *Nat Genet*. 41, 56-65 (2009).

[47] Teslovich, TM. et al. Biological, clinical and population relevance of 95 loci for blood lipids. *Nature.* 466, 7307-7707 (2010).

[48] Global Lipids Genetics Consortium. et al. Discovery and refinement of loci associated with lipid levels. *Nat Genet.* 45, 1274-83 (2013).

[49] Lange, LA. et al. Whole-exome sequencing identifies rare and low-frequency coding variants associated with LDL cholesterol. *Am J Hum Genet.* 94, 2-233 (2014).

[50] Corella, D. et al. Single Nucleotide Polymorphisms that influence lipid metabolism: Interaction with Dietary Factors. *Annu Rev Nutr*. 25, 341-90 (2005).

[51] Corella, D. et al. Nutrigenomics in cardiovascular medicine. *Circ Cardiovasc Genet.* 2, 637-651 (2009).

[52] Tucker, KL. et al. Quantifying diet for nutrigenomic studies. *Annu Rev Nutr.* 33, 349-371 (2013).

[53] Vilahur, G. et al. Intake of cooked tomato sauce preserves coronary endothelial function and improves apolipoprotein A-I and apolipoprotein J protein profile in high-density lipoproteins. *Transl Res*. S1931-5244(14)00420-4 (2014).

[54] Arranz, S. et al. Influence of olive oil on carotenoid absorption from tomato juice and effects on postprandial lipemia. *Food Chem*. 168, 203-210 (2015).

[55] Perveen, R. et al. Tomato (Solanum lycopersicum) carotenoids & lycopenes chemistry; Metabolism, absorption, nutrition and allied health claims- A comprehensive review. *Crit Rev Food Sci Nutr*. (in press) (2013).

[56] Corella, D. et al. How does the Mediterranean diet promote cardiovascular health? Current progress toward molecular mechanisms: gene-diet interactions at the genomic, transcriptomic, and epigenomic levels provide novel insights into new mechanisms. *Bioessays.* 36, 526-37 (2014).

[57] Chen, C. et al. Genetic dissection of blood lipid traits by integrating genome-wide association study and gene expression profiling in a porcine model. *BMC Genomics.* 14, 848 (2013).

[58] Voros, S. et al. Precision phenotyping, panomics, and system-level bioinformatics to delineate complex biologies of atherosclerosis: Rationale and design of the "Genetic Loci and the Burden of Atherosclerotic Lesions" study. *J Cardiovasc Comput Tomogr.* 8, 442-451 (2014).

[59] Siemelink, MA. et al. Biomarkers of coronary artery disease: the promise of the transcriptome. *Curr Cardiol Rep*. 16, 513 (2014).

[60] Castañer, O. et al. In vivo transcriptomic profile after a Mediterranean diet in high-cardiovascular risk patients: a randomized controlled trial. *Am J Clin Nutr.* 98, 845-53 (2013).

[61] Camargo, A. et al. Marin C, Perez-Martinez P, Uceda M, Lopez-Miranda J, Perez-Jimenez F. Gene expression changes in mononuclear cells in patients with metabolic syndrome after acute intake of phenol-rich virgin olive oil. BMC Genomics. 11, 253 (2010).

[62] Jiménez-Chillarón, JC. et al. The role of nutrition on epigenetic modifications and their implications on health. *Biochimie*. 94, 2242-2263 (2012).

[63] Baccarelli, A. et al. Environmental exposures, epigenetics and cardiovascular disease. *Curr Opin Clin Nutr Metab Care*. 15, 323-9 (2012).

[64] Meaney S. Epigenetic regulation of cholesterol homeostasis. *Front Genet*. 5, 311 (2014).

[65] Irvin, MR. et al. Epigenome-wide association study of fasting blood lipids in the Genetics of Lipid-lowering Drugs and Diet Network study. *Circulation*. 130, 7-565 (2014).

[66] Zaina S. Unraveling the DNA methylome of atherosclerosis. *Curr Opin Lipidol*. 25, 148-53 (2014).

[67] Fernández-Hernando, C. et al. MicroRNAs in metabolic disease. *Arterioscler Thromb Vasc Biol*. 33, 178-85 (2013).

[68] Novák, J. et al. MicroRNAs involved in the lipid metabolism and their possible implications for atherosclerosis development and treatment. *Mediators Inflamm*. 2014, 275867 (2014).

[69] Hausser, J. et al. Identification and consequences of miRNA-target interactions-beyond repression of gene expression. *Nat Rev Genet*. 15, 599-612 (2014).

[70] Richardson, K. et al. Gain-of-function lipoprotein lipase variant rs13702 modulates lipid traits through disruption of a microRNA-410 seed site. *Am J Hum Genet*. 92, 1-5 (2013).

[71] Delgado, G. et al. Cotinine as a marker for risk prediction in the LUdwigshafen Risk and Cardiovascular Health Study. *Respir Physiol Neurobiol*. S1569-9048(14)00280-8,(2014).

[72] Peacock, RE. et al. Variation at the lipoprotein lipase and apolipoprotein AI-CIII gene loci are associated with fasting lipid and lipoprotein traits in a population sample from Iceland: interaction between genotype, gender, and smoking status. Peacock RE, Temple A, Gudnason V, Rosseneu M, Humphries SE. *Genet Epidemiol*. 14, 265-82 (1997).

[73] Corella, D. et al. Associations of LPL and APOC3 gene polymorphisms on plasma lipids in a Mediterranean population: interaction with tobacco smoking. *J Lipid Res*. 43, 416-27 (2002).

[74] Lee, J. et al. The lipoprotein lipase S447X polymorphism and plasma lipids: interactions with APOE polymorphisms, smoking, and alcohol consumption. *J Lipid Res*. 45, 1132-9 (2004).

[75] Fiegenbaum, M. et al. Association between plasma lipid parameters and APOC3 genotypes in Brazilian subjects: effect of gender, smoking and APOE genotypes. *Clin Chim Acta*. 380, 175-181 (2007).

[76] Junyent, M. et al. The effects of ABCG5/G8 polymorphisms on plasma HDL cholesterol concentrations depend on smoking habit in the Boston Puerto Rican Health Study. *J Lipid Res*. 50, 565-73 (2009).

[77] Garcia-Rios, A. et al. Genetic variations at ABCG5/G8 genes modulate plasma lipids concentrations in patients with familial hypercholesterolemia. *Atherosclerosis.* 210, 486-92 (2010).

[78] Dumitrescu, L. et al. No evidence of interaction between known lipid-associated genetic variants and smoking in the multi-ethnic PAGE population. *Hum Genet.* 132, 1427-31 (2013).

[79] Grammer, TB. et al. Smoking, apolipoprotein E genotypes, and mortality (the Ludwigshafen RIsk and Cardiovascular Health study). *Eur Heart J.* 34, 1298-1305 (2013).

[80] Patel, CJ. et al. Systematic evaluation of environmental and behavioural factors associated with all-cause mortality in the United States National Health and Nutrition Examination Survey. *Int J Epidemiol* 42, 1795-1810 (2013).

[81] Lee, YC. et al. A database of gene-environment interactions pertaining to blood lipid traits, cardiovascular disease and type 2 diabetes. *J Data Mining Genomics Proteomics* 2, 106 (2011).

[82] Pedišić, Z. et al. Accelerometer-based measures in physical activity surveillance: current practices and issues. *Br J Sports Med.* (in press), (2014).

[83] Gorman, E. et al. Accelerometry analysis of physical activity and sedentary behavior in older adults: a systematic review and data analysis. *Eur Rev Aging Phys Act.* 11, 35-49 (2014).

[84] Corella, D. et al. Environmental factors modulate the effect of the APOE genetic polymorphism on plasma lipid concentrations: ecogenetic studies in a Mediterranean Spanish population. *Metabolism.* 50, 936-44 (2001).

[85] Bernstein, MS. et al. Physical activity may modulate effects of ApoE genotype on lipid profile. *Arterioscler Thromb Vasc Biol.* 22, 133-40 (2002).

[86] Bernstein, MS. et al. No physical activity x CETP 1b.-629 interaction effects on lipid profile. *Med Sci Sports Exerc.* 35, 1124-9 (2003).

[87] Wilund, KR. et al. Changes in high-density lipoprotein-cholesterol subfractions with exercise training may be dependent on cholesteryl ester transfer protein (CETP) genotype. *Metabolism.* 51, 774-8 (2002).

[88] Ferreira, AP. et al. The effect of aerobic exercise intensity on attenuation of postprandial lipemia is dependent on apolipoprotein E genotype. Atherosclerosis. 229, 139-44 (2013).

[89] Brinkley, TE. et al. Hepatic lipase gene -514C>T variant is associated with exercise training-induced changes in VLDL and HDL by lipoprotein lipase. *J Appl Physiol.* 111, 1871-6 (2011).

[90] Smith, CE. et al. Physical inactivity interacts with an endothelial lipase polymorphism to modulate high density lipoprotein cholesterol in the GOLDN study. *Atherosclerosis.* 206, 500-4 (2009).

[91] Halverstadt, A. et al. High-density lipoprotein-cholesterol, its subfractions, and responses to exercise training are dependent on endothelial lipase genotype. *Metabolism.* 52, 1505-11 (2003).

[92] Corella, D. et al. Alcohol drinking determines the effect of the APOE locus on LDL-cholesterol concentrations in men: the Framingham Offspring Study. *Am J Clin Nutr.* 73, 736-45 (2001).

[93] Djoussé, L. et al. Apolipoprotein E polymorphism modifies the alcohol-HDL association observed in the National Heart, Lung, and Blood Institute Family Heart Study. *Am J Clin Nutr.* 80, 1639-44 (2004).
[94] Corella, D. et al. Saturated fat intake and alcohol consumption modulate the association between the APOE polymorphism and risk of future coronary heart disease: a nested case-control study in the Spanish EPIC cohort. *J Nutr Biochem.* 22, 487-94 (2011).
[95] Campos, H. et al. Gene-diet interactions and plasma lipoproteins: role of apolipoprotein E and habitual saturated fat intake. *Genet Epidemiol.* 20, 117-128 (2001).
[96] Minihane, AM. et al. ApoE genotype, cardiovascular risk and responsiveness to dietary fat manipulation. *Proc Nutr Soc.* 66, 183-97 (2007).
[97] Ordovas, JM. et al. Dietary fat intake determines the effect of a common polymorphism in the hepatic lipase gene promoter on high-density lipoprotein metabolism: evidence of a strong dose effect in this gene-nutrient interaction in the Framingham Study. *Circulation.* 106, 18-2315 (2002).
[98] Tai, ES. et al. Dietary fat interacts with the -514C>T polymorphism in the hepatic lipase gene promoter on plasma lipid profiles in a multiethnic Asian population: the 1998 Singapore National Health Survey. *J Nutr.* 133, 3399-408 (2003).
[99] Tai, ES. et al. Polyunsaturated fatty acids interact with the PPARA-L162V polymorphism to affect plasma triglyceride and apolipoprotein C-III concentrations in the Framingham Heart Study. *J Nutr.* 135, 397-403 (2005).
[100] Chan, E. et al. The V227A polymorphism at the PPARA locus is associated with serum lipid concentrations and modulates the association between dietary polyunsaturated fatty acid intake and serum high density lipoprotein concentrations in Chinese women. *Atherosclerosis.* 187, 309-15 (2006).
[101] Volcik, KA. et al. Peroxisome proliferator-activated receptor [alpha] genetic variation interacts with n-6 and long-chain n-3 fatty acid intake to affect total cholesterol and LDL-cholesterol concentrations in the Atherosclerosis Risk in Communities Study. *Am J Clin Nutr.* 87, 1926-31 (2008).
[102] Lai, CQ. et al. Dietary intake of n-6 fatty acids modulates effect of apolipoprotein A5 gene on plasma fasting triglycerides, remnant lipoprotein concentrations, and lipoprotein particle size: the Framingham Heart Study. *Circulation.* 113, 2062-70 (2006).
[103] Lu, Y. et al. Dietary n-3 and n-6 polyunsaturated fatty acid intake interacts with FADS1 genetic variation to affect total and HDL-cholesterol concentrations in the Doetinchem Cohort Study. *Am J Clin Nutr.* 92, 258-65 (2010).
[104] Baik, I. et al. A lipoprotein lipase gene polymorphism interacts with consumption of alcohol and unsaturated fat to modulate serum HDL-cholesterol concentrations. *J Nutr.* 143, 1618-25 (2013).
[105] Li, TY. et al. Interaction between dietary fat intake and the cholesterol ester transfer protein TaqIB polymorphism in relation to HDL-cholesterol concentrations among US diabetic men. *Am J Clin Nutr.* 86, 1524-9 (2007).
[106] Corella, D. et al. Gene-environment interactions of CETP gene variation in a high cardiovascular risk Mediterranean population. *J Lipid Res.* 51, 2798-807 (2010).
[107] Richardson, K. et al. The PLIN4 Variant rs8887 Modulates Obesity Related Phenotypes in Humans through Creation of a Novel miR-522 Seed Site. *PLoS One.* 6, e17944 (2011).

[108] Warren, H. et al. Genetic prediction of quantitative lipid traits: comparing shrinkage models to gene scores. *Genet Epidemiol.* 38, 72-83 (2014).
[109] Walia, GK. et al. Association of common genetic variants with lipid traits in the Indian population. *PLoS One.* 9, e10168 (2014).
[110] Casas-Agustench, P. et al. Saturated Fat Intake Modulates the Association between an Obesity Genetic Risk Score and Body Mass Index in Two US Populations. *J Acad Nutr Diet.* 114, 1954-66 (2014).
[111] Jansen, S. et al. Effect of 347-serine mutation in apoprotein A-IV on plasma LDL cholesterol response to dietary fat. *Arterioscler Thromb Vasc Biol.* 17, 1532-1538 (1997).
[112] Delgado-Lista, J. et al. ABCA1 gene variants regulate postprandial lipid metabolism in healthy men. *Arterioscler Thromb Vasc Biol.* 30, 1051-7 (2010).
[113] Perez-Martinez, P. et al. A variant near the melanocortin-4 receptor gene regulates postprandial lipid metabolism in a healthy Caucasian population. *Br J Nutr.* 106, 468-471 (2011).
[114] Perez-Martinez, P. et al. Postprandial triacylglycerol metabolism is modified by the presence of genetic variation at the perilipin (PLIN) locus in 2 white populations. *Am J Clin Nutr.* 87, 744-52 (2008).
[115] Moreno-Luna, R. et al. Two independent apolipoprotein A5 haplotypes modulate postprandial lipoprotein metabolism in a healthy Caucasian population. *J Clin Endocrinol Metab.* 92, 2280-5 (2007).
[116] Zhang, X. et al. APOA5 genotype modulates 2-y changes in lipid profile in response to weight-loss diet intervention: the Pounds Lost Trial. *Am J Clin Nutr.* 96, 917-22 (2012).
[117] Qi, Q. et al. CETP genotype and changes in lipid levels in response to weight-loss diet intervention: gene-diet interaction analysis in the POUNDS LOST and DIRECT randomized trials. *J Lipid Res.* (in press), (2014).
[118] Brahe, LK. et al. Influence of SNPs in nutrient-sensitive candidate genes and gene-diet interactions on blood lipids: the DiOGenes study. *Br J Nutr.* 110, 5-790 (2013).
[119] Irvin, MR. et al. Genomics of post-prandial lipidomic phenotypes in the Genetics of Lipid lowering Drugs and Diet Network (GOLDN) study. *PLoS One.* 9, e99509 (2014).
[120] Frazier-Wood, AC. et al. The association between LRP-1 variants and chylomicron uptake after a high fat meal. *Nutr Metab Cardiovasc Dis.* 23, 1154-8 (2013).
[121] An, P. et al. Genome-wide association studies identified novel loci for non-high-density lipoprotein cholesterol and its postprandial lipemic response. *Hum Genet.* 133, 919-930 (2014).
[122] Gomez-Delgado, F. et al. Polymorphism at the TNF-alpha gene interacts with Mediterranean diet to influence triglyceride metabolism and inflammation status in metabolic syndrome patients: From the CORDIOPREV clinical trial. *Mol Nutr Food Res.* 58, 1519-27 (2014).

In: Gene-Environment Interactions and Human Diseases
Editor: Lu Qi
ISBN: 978-1-63483-680-7

Chapter 6

GENE-ENVIRONMENT INTERACTIONS ON TYPE 2 DIABETES

***Min Xu*[1] *and Lu Qi*[2,3,*]**

[1]Shanghai Clinical Center for Endocrine and Metabolic Diseases, Shanghai Institute of Endocrine and Metabolic Diseases, Ruijin Hospital Affiliated to Shanghai Jiao Tong University School of Medicine, Shanghai, China

[2]Department of Epidemiology, School of Public Health and Tropical Medicine, Tulane University, New Orleans, LA, US

[3]Department of Nutrition, Harvard School of Public Health, Boston, MA, US

ABSTRACT

Rate of type 2 diabetes (T2D) has climbed sharply over the past decades and become a major public health threat worldwide. Compelling evidence shows that a shift from principally nutritionally dense tradition diet and more active lifestyle to more energy-dense (Western-pattern) diet and more sedentary lifestyle plays pivotal role in promoting epidemic of T2D. With revolutionary advances in genotyping and sequencing technology, large-scale genome-wide association studies have made great strides in unravelling the genetic basis of T2D, and facilitating new efforts to investigate gene-environment interaction. A large group of studies have been performed to assess interactions between genetic factors and lifestyle factors, mostly focusing on candidate genes and being conducted in small-sized observational settings. Recently, data from large scale, prospective cohorts and randomized trial are emerging. Methodology development is crucial in detection gene-environment interaction on T2D. In addition, it remains a major challenge to translate the findings into public health and medical practice.

* Dr. Lu Qi, Department of Epidemiology, School of Public Health and Tropical Medicine, Tulane University, New Orleans, Louisiana 70112. Telephone: 504-988-7259; Fax: 504-988-3548; Email address: lqi1@tulane.edu; luqi@hsph.harvard.edu.

1. INTRODUCTION

Diabetes is one of the leading causes of morbidity and mortality, and has become one of the leading health threats in most of the developed countries and many developing countries. Globally, the number of diabetes patients has been predicated to be 552 million by 2030 [1]. Type 2 diabetes (T2D) accounts for more than 90% of diabetes, in which insulin resistance and beta cell dysfunction are the major pathogenesis basis. It is widely acknowledged that genetic components, environmental factors, and their interactions, determine an individual's risk for T2D.

Transition from a traditional lifestyle characterized by healthy, nutrient-rich foods intake and high-level physical activity, to an "obesogenic" pattern characterized by increased consumption of energy-dense foods and sedentary lifestyle has been noted to be paralleled with rapid increase of T2D worldwide. Such coincidence suggests that dramatic changes in environment factors may play a key role in driving the epidemic of T2D. Appealing evidence from numerous epidemiological studies has associated environment factors such as higher consumption of sugar-sweetened beverages and fried food, lower level of physical activity, and shorter sleep duration with T2D. On the contrary, intensive lifestyle modifications such as increasing physical activity and intakes of healthy diets can postpone the onset or reduce the risk of developing diabetes [2].

In the past ten years, great progress has been achieved in identifying novel genetic variants robustly associated with T2D, mainly through the successful application of genome-wide association studies (GWAS) [3]. The number of the confirmed T2D susceptibility loci has been around 70 [4, 5]. It has been noted for long time that susceptibility to environmental risk factors and response to clinical interventions varies considerably between individuals, indicating potential interactions between the genetic factors and environmental factors [6, 7].

The dramatic progress in genetic research provides high possibility to comprehensively explore the potential roles of gene-environmental interaction in pathogenesis, prevention and treatment of T2D.

In the present chapter, we will briefly introduce the major environmental risk factors, including lifestyle and dietary intakes, and findings from the genetic research of T2D. We will particularly address the studies of gene-environment interactions on T2D or related metabolic risk factors in observational settings or randomized controlled trials.

2. ENVIRONMENTAL FACTORS IN THE DEVELOPMENT OF TYPE 2 DIABETES

T2D is one of the major metabolic disorders closely related to environment, such as diet and lifestyle. A large body of data from epidemiological studies has consistently shown that dietary factors, physical inactivity, and long sedentary time, among many others, are related to risk of T2D.

Several dietary factors have been linked to diabetes risk. For example, high intakes of sugar-sweetened beverages, fried-foods, red meats or processed meats, refined grains, and trans- and saturated fat have been associated with increased risk of T2D [8-11], whereas high consumption of whole grains, fruits, vegetables, legumes, nuts, coffee, and moderate alcohol reduce the risk [8, 10, 12, 13]. Specific dietary patterns defined by combinations of different foods and nutrients have also been related to diabetes risk [10, 14].

Obesity is thus far the most important, modifiable risk factor for T2D. Physical activity is a well-established protective factor on T2D [15]; whereas television watching and other sedentary behaviors have been related to increased risk of the disease [16]. Smoking and other components of the social and built environment have also been identified as risk factors [17]. In addition, environmental exposures during early development, such as nutritional environment during both pre- and postnatal life is increasingly recognized as a potentially critical contributor to the risk of T2D in later life [18].

3. Genetic Risk Factors for Type 2 Diabetes

Before GWAS was first introduced to this filed, candidate-gene based association analysis and linkage analysis are the predominant methods in genetic mapping on complex diseases including T2D. The *PPARG* gene, which encodes the nuclear receptor PPAR-γ, was first identified using the candidate gene approach [19-21]. Candidate gene analyses have also identified variants in several other genes associated with T2D risk, including *IRS1*, *KCNJ11*, *WFS1*, *HNF1A*, and *HNF1B* [22, 23 24]. The *TCF7L2* gene was initially identified to be associated with risk of T2D using linkage analysis in Caucasians [25]; and such association has been consistently confirmed across various populations [26-28].

Since 2007, application of GWAS in population studies has made great strides in unraveling the genetic basis of T2D. The first GWAS of T2D identified two new diabetes loci: *HHEX* and *SLC30A8* [29]. More diabetes-associated loci were identified through larger-scale meta-analysis of GWAS. The first round of meta-analysis by the DIAGRAM consortium identified 6 novel loci [30], and the second round meta-analysis [31] identified 12 additional loci. Another GWAS meta-analysis including 21 cohort data of European decent conducted by the Glucose and Insulin-Related Traits Consortium (MAGIC) identified 5 T2D related loci [32]. To date, in total more than 70 loci have been identified in Caucasians through GWAS.

GWAS of T2D conducted in other ethnic groups have led to identification of additional loci. In fact, the majority of the diabetes-associated loci are shared by different ethnicity groups such as between Caucasians and Asians. However, population specificity of genetic predisposition may also exist [33], due to difference in linkage disequilibrium (LD) structure in different populations [34]. Several GWASs of T2D have been performed in non-Caucasian populations and contributed to identification of new loci [35-41]. For example, meta-analysis of eight GWASs in East Asian populations identified 8 new T2D loci [35]. *KCNQ1* was thought to be an Asian-specific susceptibility gene for T2D when it was first detected in Japanese [40, 41] and followed by multiple replication studies in other Asian populations. Recent large-scale meta-analysis combining GWASs in Caucasians reported an independent signal of *KCNQ1* rs231362 [31], which is different from the previously reported single neucleotide polymorphisms (SNPs) in Asian populations (rs2237892 [40], rs2237895 [37], rs2237897 [41], rs163182 [42]). The minor allele frequency (MAF) of rs231362 in Caucasians is 0.48, which is much higher than 0.08 for rs2237892 and 0.05 for rs2237897.

Of note, most of the risk alleles for T2D loci share a consistent pattern of decreasing frequencies along human migration from sub-Saharan Africa to East Asia [43]. Such difference is hypothesized to be due to the promotion of energy storage and usage appropriate to inconsistent energy intake. Recently, Mahajan et al. conducted a meta-analysis including 26,488 cases and 83,964 controls of European, East Asian, South Asian and Mexican and Mexican American ancestry. They observed a significant excess in the directional consistency of T2D risk alleles across ancestry groups and identified seven new T2D susceptibility loci. The study emphasizes the importance of cross-ethnic analyses in discovering new susceptible genes [44], and improving fine-mapping resolution at the T2D susceptibility loci [44].

4. Gene-Environment Interaction on Type 2 Diabetes

The genetic effects of individual variants or loci are generally modest. The odds ratios (ORs) associated with each risk allele usually range from 1.1 to 1.3 [45]. In addition, when all the identified genetic variants are considered together, they only account for a small proportion of the disease risk, indicating huge 'missing heritability'. It is believed that at least part of the missing heritability could be explained by interactions between the genetic variants and environmental factors. However, most of the previous genetic discovery studies did not consider the potential modification effect of environmental factors.

In 1962, Neel proposed the thrifty gene hypothesis to explain the rapid increase of diabetes and obesity in modern society [46]. According to this hypothesis, the diabetes or obesity-predisposing genotypes would have been advantageous in the early evolutionary history because these genotypes promote fat deposition in adipocytes which protect human beings to survive at the period of famine (positive selection). However, such genotypes become disadvantageous in modern society when over-nutrition has become popular [46]. If the hypothesis is true, it may provide a judicious model to account for the gene-environment interaction. However, evidence directly supporting the hypothesis is largely lacking. Recently, Ayub et al. tried to test the hypothesis at 65 loci associated with T2D in samples of African, European, and East Asian ancestry. The results failed to demonstrate that positive selection has a powerful influence driving the prevalence of T2D risk alleles [47].

Even though, studies attempting to test the gene-environment interaction have provided preliminary evidence to support that such interplays between the genetic and environmental factors may exist.

The gene-environment interactions for diabetes-related traits have been reported since the mid-1990s [48]; the majority of these studies were performed in relatively small cohorts, case-control studies, and clinical trials. By 2014, only one study has modelled gene-lifestyle interactions for the majority of the GWAS-identified loci with diabetes incidence [49], a handful of other interaction studies have explored subsets of these variants in relation to diabetes (several selected studies are shown in Table 1).

4.1. Observational Studies

Most of the previous studies on gene-environment interactions were performed in observational settings such as case-control and cohort studies. Early studies were often focused on the selected genetic variants in candidate genes, while recent studies concentrated more on the GWAS-identified variants.

4.1.1. Gene-Diet Interaction

Lifestyle and diet habits are important modifiable risk factors for diabetes. Several studies have assessed interactions between SNPs in the GWAS-identified genes, such as *TCF7L2, SLC30A8, GIPR, HNF1B,* and *FTO*, and dietary factors in relation to risk of T2D (Table 1).

The *TCF7L2* gene is by far the strongest susceptibility locus associated with T2D. Testing interactions between *TCF7L2* variants and dietary factors, such as whole grains, fat, protein, glycemic index, and Mediterranean diet (MedDiet), and lifestyle factors have attracted extensive attention [50-53]. The *TCF7L2* rs7903146 genotype was reported to modify the inverse association between whole-grain intake and diabetes risk (P for interaction = 0.016) in the European Prospective Investigation into Cancer and Nutrition (EPIC)-Potsdam cohort. In the CC carriers, whole-grain intake was significantly associated with a reduced diabetes risk (hazard ratio [HR] for 50 g portion/day = 0.86; 95% confidence interval [CI]: 0.75, 0.99); whereas the T-allele attenuated the associations to null (HR=1.08; 95% CI: 0.96, 1.23) [51]. The *TCF7L2* rs7903146 variant was also found to modify associations between fiber intakes and T2D in a case-control study derived from the Malmö Diet and Cancer Study (MDCS) [53], which including 1,649 diabetes cases and 5,216 non-diabetic controls.

The genetic association with T2D risk increased with higher intake of dietary fiber, with ORs ranging from 1.24 to 1.56 from the lowest to highest quintile of fiber intake (P for interaction = 0.049).

The MedDiet includes proportionally high consumption of olive oil, legumes, unrefined cereals, fruits, and vegetables, moderate to high consumption of fish, moderate consumption of dairy products (mostly as cheese and yogurt), moderate wine consumption, and low consumption of meat and meat products. In a prospective study of 7,018 individuals with a median follow-up of 4.8 years, Corella et al. found that adherence to the MedDiet reduced the increase in blood levels of fasting glucose and lipids associated with *TCF7L2* rs7903146 [52].

Zinc transporter-8 gene (*SLC30A8*) is a robust T2D-associated genetic locus that has been widely validated across different populations. The gene encodes ZnT-8, which transports zinc from the cytoplasm into insulin secretory vesicles in pancreatic beta cells, an important step in insulin synthesis and secretion. Zinc is an essential trace element found in most foods that facilitates catalytic, structural, and transcriptional actions [54]. In epidemiology studies, plasma zinc concentrations have been associated with impaired glucose regulation (IGR) and T2D risk. Recently, investigators of the Cohorts for Heart and Aging Research in Genetic Epidemiology (CHARGE) consortium [55, 56] conducted two large-scale gene-diet interaction studies. In one study [56], 14 cohorts were included to assess the interaction of 20 genetic variants related to glycemic traits and zinc metabolism with dietary zinc intake (food sources) and 5 cohorts were included to assess the interaction with total zinc intake (food sources and supplements) on fasting glucose levels among individuals of European ancestry without diabetes.

Table 1. Selected studies regarding gene-lifestyle interaction on type 2 diabetes

Loci	SNPs	Study design	Sample size	Environmental	Main findings
Observational					
Candidate gene					
PPARG	Pro12Ala and 1431C > T	Cohort	4,676 participants, general population	Fat intake [110]	Dietary fat intake modified the association of variations at the *PPARG* locus with type 2 diabetes risk. High fat intake was associated with an increased type 2 diabetes risk among ProPro and CC homozygotes, but not in Ala and T carriers.
IRS1	rs2943641	Cohort	15,227 women and 9,614 men aged 45-74 y; 12 y of follow-up, 1,567 incident T2D cases	Carbohydrate and fat intake [59]	The interactions of *IRS1* rs2943641 interacted with carbohydrate and fat intakes on incident type 2 diabetes was in a sex-specific fashion. Among men, the T allele was associated with decreased risk in the lowest tertile of fat intake (P-trend = 0.01, P-interaction = 0.02). Among women, the T allele was associated with decreased risk only in the lower tertiles of carbohydrate intake.
GWAS identified gene					
TCF7L2	rs7903146	Case-control	2,318 type 2 diabetes, 25,230 control subjects	Whole-grain intake [51]	The *TCF7L2* rs7903146 T-allele modified the inverse association between whole-grain intake and type 2 diabetes risk (P for interaction =0.016). Whole-grain intake was inversely associated with type 2 diabetes risk among CC homozygote carriers; while not in the T-allele carriers.
FTO and MC4R	*FTO*-rs9939609, *MC4R*-rs17782313	Case-control	3,430 type 2 diabetic patients, 3,622 control subjects	Mediterranean Diet [111]	The associations of the *FTO*-rs9939609 and the *MC4R*-rs17782313 with type 2 diabetes depended on a high adherence to the Mediterranean diet. When adherence to the Mediterranean diet was low, carriers of the variant alleles had higher type 2 diabetes risk than wild-type subjects.
SLC30A8	rs13266634	Case-control	785 type 2 diabetic patients, 793 control subjects	Plasma zinc [57]	Each 10-µg/dl increment of plasma zinc was associated with 22% lower odds of type 2 diabetes in TT genotype carriers, 17% lower odds in CT genotype carriers, and 7% lower odds in CC genotype carriers.

Loci	SNPs	Study design	Sample size	Environmental	Main findings
SLC30A8	rs11558471	Meta Cohort	Up to 45,821 participants	Total and dietary zinc intake [56]	A nominally significant interaction between total zinc intake and the *SLC30A8* rs11558471 variant on fasting glucose levels (β-coefficient ± SE per A allele for 1 mg/day of greater total zinc intake: −0.0017 ± 0.0006 mmol/L, summary interaction P value = 0.005); which suggests a stronger inverse association between total zinc intake and fasting glucose in individuals carrying the glucose-raising A allele compared with individuals who do not carry it.
GIPR	rs10423928	Cohort	24,840 participants	Carbohydrate and fat intakes [112]	The *GIPR* TT-genotype carriers within the highest carbohydrate quintile were at 23% decreased T2D risk as compared to that in the lowest quintile. AA-genotype carriers in the highest fat quintile were at 69% decreased type 2 diabetes risk as compared to the lowest fat quintile.
HNF1B	rs4430796	Cross-sectional	16,003 population-based cohort of initially nondiabetic middle-aged adults.	Physical activity [113]	The *HNF1B* rs4430796 variant modified associations of physical activity with type 2 diabetes. The A allele was associated with decreased risk of diabetes in sedentary individuals (hazard rate ratio [HRR] = 0.85; 95% CI 0.74–0.96; P = 0.011), with a contrasting effect in physically active individuals (HRR = 1.10; 95% CI 1.03–1.18; P = 0.007).
Genetic risk score	10 SNPs	A prospective, nested, case-control study of	1,196 diabetic and 1,337 nondiabetic men.	Western dietary pattern [14]	A more Western dietary pattern significantly increased risk of T2D only among those with a high genetic risk score. Secondary analysis suggested the interaction was attributable to the red and processed meat component of the Western diet.
Genetic risk score	49 SNPs	Cohort; Prospective analysis	12,403 incident T2D cases;	Lifestyle [49]	The relative effect of a T2D genetic risk score is larger in younger individuals (p for interaction = 1.20 x 10-4); participants who were leaner, both in terms of body mass index (p for interaction = 1.5x10-3) and waist circumference (p for interaction = 7.49x10-9). The 10-y cumulative incidence of T2D rose from 0.25% to 0.89% across extreme quartiles of the genetic score in normal weight individuals, compared to 4.22% to 7.99% in obese individuals.

Table 1. (Continued)

Loci	SNPs	Study design	Sample size	Environmental	Main findings
Clinical trials					
IRS1	rs294364	2 years of weight loss diet intervention	738 overweight adults	Carbohydrate diet [114]	At 6 months, participants with the IRS1 SNP rs294364 risk-conferring CC genotype had greater decreases in weight loss than those without this genotype in the highest-carbohydrate diet group; whereas the genetic effect was not significant in participants assigned to the lowest-carbohydrate diet group (P for interaction=0.03).
TCF7L2	rs12255372	Intervention study	3,548 participants	Lifestyle [50]	The diabetes risk was higher in the placebo group (HR=1.81; 95%CI 1.19-2.75) than in the lifestyle-intervention groups (reduced intakes of total fat and saturated fat and increased intake of fiber; moderate exercise for at least 30 minutes per day; HR=1.24; 0.73-2.12).
PPM1K	rs1440581	2 years of weight loss diet intervention	734 overweight adults	Fat diet [100]	In the energy restricted high-fat diet group, the C allele was related to less weight loss (adjusted means: TT, -6.7kg; CT, -5.8kg; CC, -4.1 kg, respectively), whereas no significant genetic effect was observed in the low-fat diet group at 6 months of intervention. The C allele carrier had smaller decreases in serum insulin and HOMA-IR values than those without this allele in high-fat diet group, while an opposite effect was observed in participants assigned to the low-fat diet group.

A nominally significant yet biologically plausible interaction was observed between *SLC30A8* (rs11558471) and total zinc intake. Higher total zinc intake was found to attenuate the glucose-raising effect of the rs11558471 *SLC30A8* (zinc transporter) variant. Shan et al. recently assessed the interaction of *SLC30A8* rs13266634 with plasma zinc levels associated with T2D in 1,796 Chinese: 218 newly diagnosed IGR patients, 785 newly diagnosed T2D patients, and 793 individuals with normal glucose tolerance (NGT) (P for interaction = 0.01) [57]. In all the study samples, each 10-μg/dl higher plasma zinc level was associated with a decreased risk of T2D with multivariate-adjusted OR of 0.87 (95% CI 0.85-0.90). The associations was more prominent in the TT homozygotes, each 10-μg/dl increment of plasma zinc was associated with 22% (OR, 0.78; 95% CI, 0.72-0.85) lower odds of T2D; while 17% (0.83; 0.80-0.87) and 7% (0.93; 0.90-0.97) lower odds were observed in the CT and CC genotype carriers, respectively.

In addition, the CHARGE consortium recently examined whether the established glucose- and insulin-associated loci modified the relationships of whole grains, zinc, and magnesium, and overall dietary patterns with fasting blood glucose and insulin concentrations. Besides the tentative evidence of interactions for variants in *SLC30A8* and total zinc intake on fasting glucose concentrations, it was also found that higher whole-grain intake was associated with less reduction in fasting insulin in those with the insulin-raising allele of rs780094 (*GCKR*) [58]. However, no interaction between the selected genetic variants and dietary magnesium or dietary patterns was observed. In another large prospective cohort study from southern Sweden, significant interactions between the *IRS1* and *GIPR* variants, respectively, with dietary fat and carbohydrate intakes on risk of incident T2D were recently reported [55, 59].

The majority of available gene-diet interaction studies have focused on single nutrient or food. However, the potential methodological or conceptual limitations need to be addressed. First, in a "real world," no such people eat a single nutrient or food. Thus, the "single nutrient/food" approach may not adequately account for complicated nutrient interactions in free-living populations.

The cumulative effects of complex of nutrients/foods intake, such as dietary patterns or quality indexes, may be more appropriate to be used in testing gene-diet interactions. In addition, to use combined genetic effect in the gene-environmental interaction tests is a reasonable and effective way, especially when the individual genetic variation effect is minor. A genetic risk score summing allelic counts for the established susceptibility loci from GWAS has been widely used to represent the overall genetic predisposition. Even though it is less informative at the biological level, the genetic score is a powerful method for demonstrating an interaction. Qi et al. [14] assessed whether the established genetic variants, mainly from GWASs, modified dietary patterns in predicting diabetes risk. They calculated a genetic risk score using a simple count method, assuming that each SNP is independently associated with T2D risk under an additive genetic model. It was found that the Western dietary pattern was significantly associated with an increased risk of T2D only among those with a high genetic risk score. Secondary analysis suggested the interaction was largely attributable to the red and processed meat component of the Western diet. They concluded that genetic predisposition may synergistically interact with a Western dietary pattern in determining diabetes risk in men.

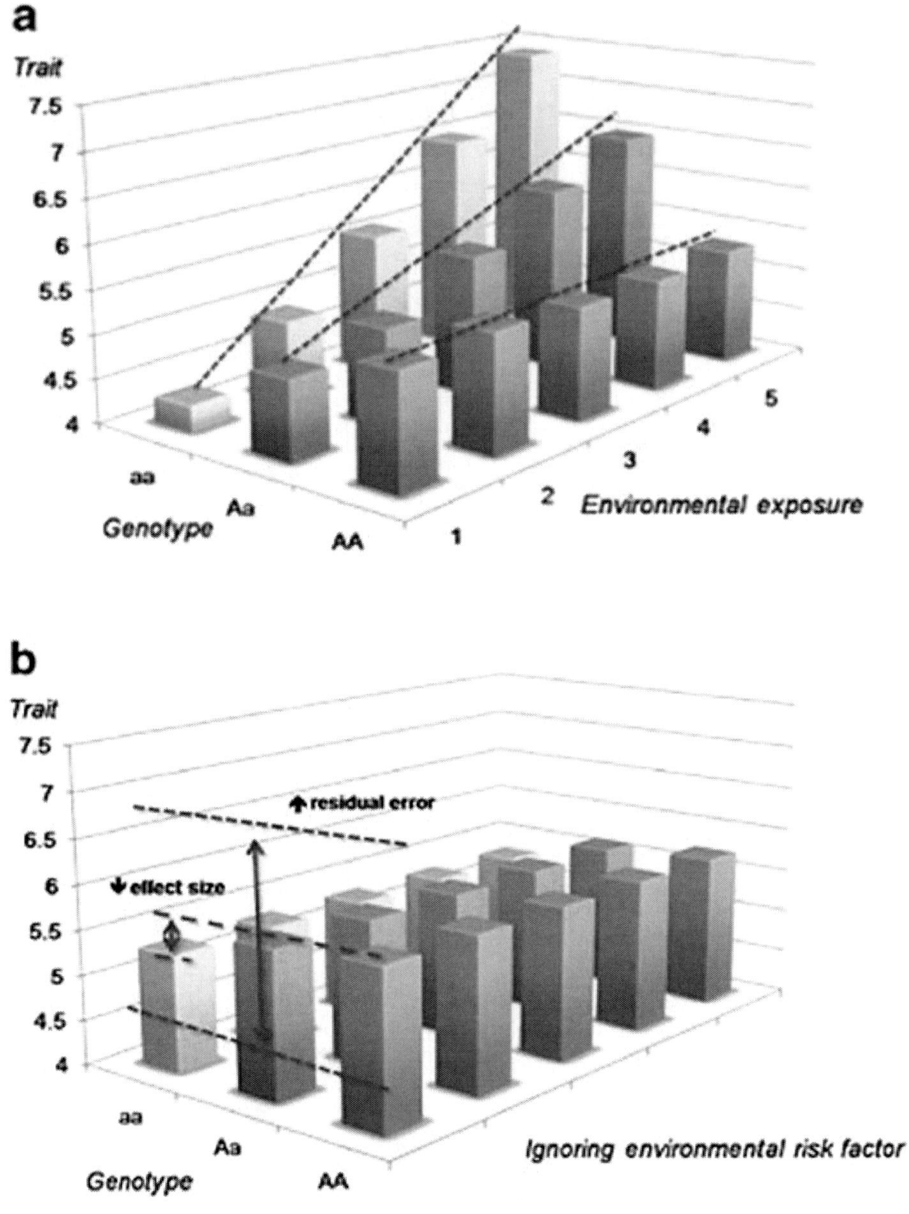

*Adopted from Franks PW. Curr Diab Rep 2011; 11:552–561.

Figure 1. Gene-environmental interaction in diseases. a, Genetic effects on [disease] traits that differ in magnitude [and sometimes direction] across environmental contexts. b, Shows the same data as in Panel a, but under the assumption that the genetic effects are equal across the spectrum of the environmental exposure (i.e., no interactions exist).

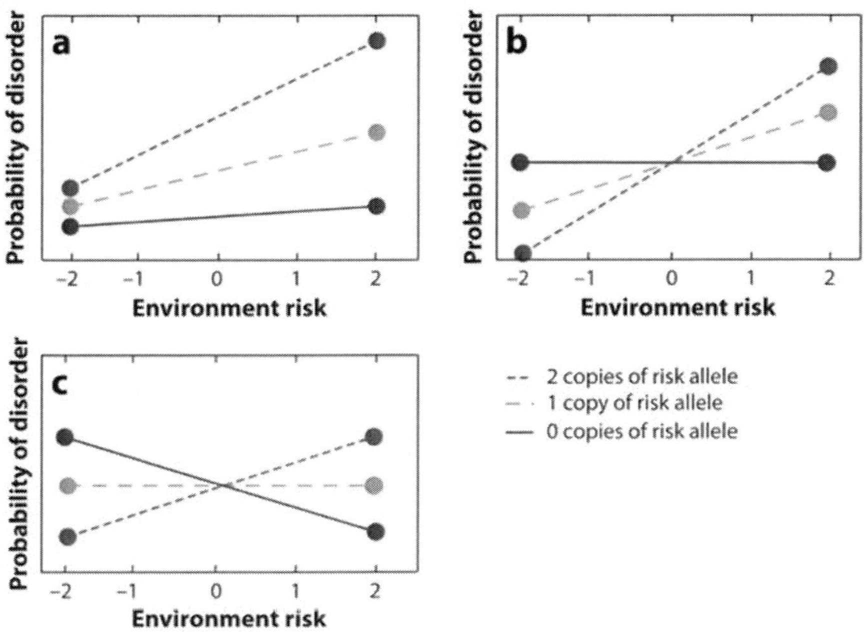

*Adopted from Dick DM. Annu Rev Clin Psychol. 2011; 7: 383–409.

Figure 2. Different types of gene-environment interactions. a. Fan-shaped interactions, in which the influence of genotype is greater in one environmental context than in another; b. and c. Cross-over interactions, in which the same individuals who are most adversely affected by negative environments may also be those who are most likely to benefit from positive environments.

4.1.2. Gene-Physical Activity Interaction

It is well established that physical activity reduces T2D risk. However, the extent of protection conferred by physical activity may differ according to the genetic factors, suggesting potential interaction between the genetic variation and physical activity on the disease risk [60].

Several previous studies examined the interaction of physical activity with candidate genes such as *PPARG, SLC2A2 (GLUT2)* and *ABCC8 (SUR1)* [61]. Among these candidate-gene based studies, the most extensively examined variant is *PPARG* Pro12Ala. *PPARG* is one of the earliest identified and confirmed genes related to T2D risk [5], and encodes an essential transcription factor involved in lipid and glucose metabolism, fatty acid transport, and adipocyte differentiation. Moreover, *PPARG* is demonstrated to be the target for thiazolidinediones, insulin-sensitizing drugs that are used to treat T2D [62]. It was found that the Ala12 allele was related to a favorable diabetes-related traits response to physical activity either in the exercise training interventions or habitual physical activity in observational studies [63-67].

Many recent studies have focused on GWAS-identified variants for interaction analyses. The *FTO* variants are related to both obesity and T2D [68]. In addition, many other obesity-related genetic variants also affect risk of T2D or its risk factors [69]. It was found that physical activity might attenuate the influence of *FTO* variants on obesity risk [70] in a meta-analysis of 218,166 adults and 19,268 children. The association of the *FTO* risk allele with the odds of obesity is attenuated by 27% in physically active adults [70]. No such interaction was found in children and adolescents.

The approach of genetic risk score-based analyses have been successful in demonstrating an attenuation in the effect of body mass index (BMI) susceptibility score on BMI with increasing levels of physical activity in 20,000 individuals from the UK and in ~12,000 individuals from the US [71, 72]. Recently, in a prospective cohort of initially nondiabetic middle-aged adults, Brito et al. [64] tested the interaction of 17 GWAS-identified SNPs with physical activity on incident T2D. They reported a significant interaction for a SNP in *HNF1B*, and the interaction on IGR and incident diabetes remained significant after correction for multiple testing (*P* for interaction = 0.015 and 0.007, respectively). More recently, in the ARIC study including 8,101 men and women aged 45-64 years at baseline, a statistically significant interaction was found between physical activity and genetic risk score on incident diabetes. The genetic risk score was created based on 65 SNPs. Specifically, the association between physical activity and T2D incidence was weaker in those with higher T2D genetic risk (HR=1.04; P =0.56) than those with lower genetic risk (HR =0.82; P =0.05). Furthermore, the genetic risk scores for insulin resistance (IR), insulin secretion, fasting insulin (FI) and fasting glucose were also created. The results suggest that the observed interaction was principally driven by genetic risk to IR and was most pronounced in women [73]. However, in a recent meta-analysis of 54,884 individuals from 22 cross-sectional studies did not find an interaction between 5 GWAS-identified variants (*TCF7L2 and GCKR, ADCY5, GIPR, and VPS13C*) and physical activity on 2-hour glucose levels obtained during an oral glucose tolerance test [74].

Langenberg et al. reported a large gene-lifestyle interaction study on T2D [49]. In the InterAct study including 12,403 incident T2D cases and a representative sub-cohort of 16,154 individuals from a cohort of 340,234 European participants with 3.99 million person-years of follow-up, the relative effect of a T2D genetic risk score was greater in younger than older individuals (P for interaction = 1.20×10^{-4}). Relative genetic risk (per standard deviation [4.4 risk alleles]) was also greater in participants who were leaner, both in BMI (P for interaction = 1.5×10^{-3}) and waist circumference (P for interaction = 7.5×10^{-9}). The 10-y cumulative incidence of T2D rose from 0.25% to 0.89% across the extreme quartiles of the genetic score in normal weight individuals, compared to 4.22% to 7.99% in obese individuals. However, no significant interactions were observed between the genetic score and sex, diabetes family history, physical activity, or dietary habits assessed by a Mediterranean diet score.

4.1.3. Gene-Prenatal Nutrition Interaction

It has been acknowledged that early life development may play a pivotal role in affecting later life health. According to the thrifty phenotype hypothesis, malnutrition during fetal development might lead to poor fetal and infant growth, and predispose to adverse health outcomes [75]. Fetal adaptation to an intrauterine malnutrition environment may lead to permanent changes that influence insulin secretion and resistance, and risk of adulthood T2D. Previous studies have shown consistent association between low birth weight, an indicator of intrauterine malnutrition status, and T2D risk [76, 77].

There is evidence suggesting that the association of prenatal environmental factors with risk of developing T2D may be modified by genetic variants [78]. For example, the *PPARG* Pro12Ala polymorphism significantly modulated the association between prenatal exposure to famine and impaired glucose tolerance and T2D. Several studies analyzed the GWAS-identified variants and provided further evidence to the gene-prenatal nutrition interactions

[79-81]. Li et al., examined the interactions between genetic susceptibility to obesity or T2D, evaluated on the basis of 32 BMI-associated variants or 35 T2D-associated variants, and birth weight in relation to adulthood risk of T2D. They included two nested case-control studies of 2,591 T2D cases and 3,052 healthy controls from the Nurses' Health Study and Health Professional Follow-up Study [82]. A significant interaction was observed between birth weight and obesity genetic risk score, in which the obesity genetic scores showed a stronger association with T2D risk in individuals with low birth weight, with the multivariable-adjusted OR was 2.55 (95% CI 1.34-4.84) by comparing extreme quartiles of the obesity genetic score; whereas the OR was 1.27 (1.04-1.55) among individuals with birth weight > 2.5 kg (P for interaction = 0.017). The T2D related GRS, however, did not interact with birth weight. The data suggest that low birth weight and genetic susceptibility to obesity may synergistically affect adulthood risk of T2D.

In a Chinese cohort [83], Hong et al. also reported the obesity genetic risk score showed significant interaction with birth weight in relation to obesity (P for interaction < 0.001), in which the obesity genetic risk score calculated on six SNPs associated with obesity risk in Chinese. The genetic effect appeared to be more pronounced in individuals with normal range of birth weight (25-75%) than those with either low (< 25%) or high (> 75%) birth weight. The data suggest that birth weight may significantly influence individual's genetic susceptibility to obesity.

4.2. Gene-Diet/Lifestyle Interaction in Clinical Trials

Results from the observational studies usually have several limitations. For example, observational studies are likely to be confounded by measured and unmeasured factors and prone to reverse causation. Randomized intervention trials may act as an alternative approach for testing gene-diet/lifestyle interaction in prospective manner [84]. Randomization minimizes the potential confounding effects that may seriously bias gene-lifestyle/diet interactions. Specifically, in a randomized diet intervention trial, dietary factors are usually precisely defined and interventions are prescribed. In addition, a unique strength to gene-diet/lifestyle interaction in randomized intervention trials is that, the results may provide more direct evidence to instruct genetic-targeted diet/lifestyle modifications in future public health and clinical practice [84]. However, most of the existing clinical trials are in relative small size; the power for detection of moderate gene-diet/lifestyle interactions is therefore a major challenge.

The Diabetes Prevention Program (DPP) is thus far the largest randomized controlled trial of lifestyle modification and metformin for diabetes prevention. Gene-lifestyle interactions on incident T2D have been extensively tested in this trial, and significant interactions between lifestyle intervention and variants at *TCF7L2* [85], *WFS1* [86], other established T2D-related loci [87, 88], established glucose- and insulin-associated loci [89], common and rare variants at the *MC4R* [90] and *SLC80A8* loci [91], as well as other candidate loci [92-94] have been reported. For example, in the DPP, a stronger association between the TT genotype of *TCF7L2* rs12255372 and the diabetes risk was found in the placebo group (HR=1.81; 95%CI 1.19-2.75) than in the lifestyle intervention groups, which was characterized as increased intake of fiber; moderate exercise for at least 30 minutes per day and reduced intakes of total fat and saturated fat and; HR=1.24; 0.73-2.12). Although test of interaction was not statistically significant (P > 0.10), these data suggest that healthy diet/lifestyle intervention may attenuate the genetic effects. In another clinical trial, Tübingen Lifestyle Intervention Program (TULIP), which tested exercise and diet intervention with decreased intake of fat and increased intake of fibers (> 15 g fiber per 1,000 kcal), it was found that the CC genotype of *TCF7L2* rs7903146 was significantly related to greater weight loss in participants with high fiber intake, but not in those with low fiber intake [95].The Preventing Overweight Using Novel Dietary Strategies (Pounds Lost) is a clinical trial aimed to assess the possible advantage for reduced-calorie weight-loss diets that emphasizes different macronutrient contents, including in total of 811 overweight (BMI ≥ 25kg/m^2) and obese (BMI ≥ 30kg/m^2) adults who were randomly assigned to 1 of 4 diets varying in fat, protein and carbohydrates for 2 years [96]. In this trial, the participants got an average of 6 kg weight loss at the first 6 months of intervension, and began to regain weight after 12 months until by 2 years, weight loss remained similar in various diets groups. In the Pounds Lost trial, the T2D-associated *TCF7L2* SNP rs12255372 significantly interacted with fat intake on changes in BMI, total fat mass, and trunk fat mass (all P < 0.05) at 6 months, with non-significant larger decreases for TT (risk genotype) carriers on a low-fat diet [97]. The *FTO* SNP rs1558902 was also found to modify the dietary protein on 2-year changes in fat-free mass, total percentage of fat mass, and total-, visceral-, and superficial adipose tissue mass in the Pounds Lost trial [98].

Several other T2D-associated genetic variants have also been found to modify diet interventions on weight change and improvement of insulin resistance. Also in the Pounds Lost trial, the significant interactions between the *IRS1* SNP rs294364, and the amino acid metabolites related genotype *PPM1K* rs1440581 and dietary intake in relation to changes in weight loss and insulin resistance were reported, respectively [99, 100]. At 6 months, participants with the *IRS1* SNP rs294364 risk-conferring CC genotype had greater decrease in weight loss than those without this genotype in the highest-carbohydrate diet group; whereas the genetic effect was not significant in participants assigned to the lowest-carbohydrate diet group (P for interaction=0.03). The gene-diet interaction was attenuated at 2 years due to weight regain. Branched amino acids (BCAAs) or aromatic amino acids (AAAs) were recently found to be related to obesity, insulin sensitivity and T2D risk [101-103] and weight loss induced by or gastric bypass surgery or behavioral/dietary intervention trial [104, 105]. SNP rs1440581 (*PPM1K*), which was firstly identified by the amino acid metabolites GWAS, was found to modify dietary fat on weight loss and changes in insulin resistance [100]. In the energy restricted high-fat diet group, the C allele was related to less weight loss (adjusted means: TT, -6.7kg; CT, -5.8kg; CC, -4.1 kg, respectively, $P = 0.001$ in an additive pattern), whereas no significant genetic effect was observed in the low-fat diet group ($P = 0.53$), at 6 months of intervention. The C allele carrier had smaller decreases in serum insulin and homeostasis model of assessment for insulin resistance index (HOMA-IR) values than those without this allele in high-fat diet group (both $P = 0.02$), while an opposite effect was observed in participants assigned to the low-fat diet group ($P = 0.02$ and 0.04, respectively).

These data highlight that gene-diet/lifestyle interaction analyses in the randomized clinical trials may generate data reflecting direct response to dietary or/and lifestyle interventions. However, how to validate the findings of gene-diet/lifestyle interactions remains a major challenge, considering relatively small size in most of the existing trials and studies are generally different in designs. Replication, functional exploration, and translation of the findings into personalized diet/lifestyle interventions remain the chief challenges.

SUMMARY AND PERSPECTIVES

A large group of environmental factors, especially dietary intake and lifestyle, have been related to diabetes risk. In the past ten years, momentous progress in genetic studies has been achieved. These findings provide unprecedented opportunities to investigate gene-environment interaction on diabetes risk. Previous studies on gene-environment interaction have provided preliminary but promising evidence. It is notable that most of the previous studies are generally limited by lack of replication. In addition, because of the observational nature of these studies, confounding and reverse causation are inevitable. Improvement in measuring lifestyle factors in population studies is another major challenge. Study shows that moderate decreases in the accuracy of measurement of environmental factors may lead to a 20-fold reduction in statistical power to detect an interaction [106]. Moreover, conventional statistical approaches are underpowered for detection of gene-environment interactions. Hence, a range of new methods have been proposed to address these challenges, especially on a genome-wide scale [107-109]. Breakthrough in the field will heavily rely on advances in methodology development and collective efforts of well-designed, prospective cohorts with comprehensive lifestyle information. Large scale collaborations to detect gene-diet/lifestyle interactions in the randomized controlled trials are urgently needed.

It remains debatable how to translate statistically significant interactions into biological interactions. Individual risk-based, or 'personalized', diet and lifestyle interventions have long been proposed as an attractive target in future prevention or treatment of T2D. The findings of gene-environment interactions would inform future development of personalized interventions based on genetic profile that are better tailored to individuals' needs. Advance in high-throughput genotyping technology is facilitating offer of direct-to-consumer genetic testing services; and raises great hope that genetic testing will pave the way to personalized prevention and medicine. Curbing the epidemic of T2D calls not only for changes in lifestyle but also in policy and social environment. In the following years, public health practice will not be able to ignore the emerging information from research of gene-environment interactions.

Key Facts

1. Diabetes has become epidemic, largely due to transition of environment and lifestyle during the past decades;
2. Both environment factors and genetic factors may independently affect diabetes risk, and interact with each other;
3. Genetic research has made great strides in identification of a group of variants associated with diabetes risk; however, these genetic variants only account for a small proportion of the disease risk;

4. The majority of the studies of gene-environment interaction on diabetes was performed in observational settings, and the findings are limited by lack of replication;
5. Methodology development is crucial in detection gene-environment interaction on T2D;
6. It remains a major challenge to translate the findings into public health and medical practice.

ACKNOWLEDGMENT

Dr Qi is supported by NIH grants from the National Heart, Lung, and Blood Institute (HL071981, HL034594, HL126024), the National Institute of Diabetes and Digestive and Kidney Diseases (DK091718, DK100383, DK078616), the Boston Obesity Nutrition Research Center (DK46200), and United States – Israel Binational Science Foundation Grant 2011036. Dr Qi was a recipient of the American Heart Association Scientist Development Award (0730094N).

No conflict of interest

REFERENCES

[1] Global Diabetes Plan 2011-2021. http://www.idf.org/global-diabetes-plan-2011-2021.
[2] Diabetes Prevention Program Research, G. REDUCTION IN THE INCIDENCE OF TYPE 2 DIABETES WITH LIFESTYLE INTERVENTION OR METFORMIN. *The New England journal of medicine* 346, 393-403, doi:10.1056/NEJMoa012512 (2002).
[3] Visscher, Peter M., Brown, Matthew A., McCarthy, Mark I. & Yang, J. Five Years of GWAS Discovery. *American Journal of Human Genetics* 90, 7-24, doi:10.1016/j.ajhg.2011.11.029 (2012).
[4] Replication, D. I. G. et al. Genome-wide trans-ancestry meta-analysis provides insight into the genetic architecture of type 2 diabetes susceptibility. *Nature genetics* 46, 234-244, doi:10.1038/ng.2897 (2014).
[5] Morris, A. P. et al. Large-scale association analysis provides insights into the genetic architecture and pathophysiology of type 2 diabetes. *Nature genetics* 44, 981-990, doi:10.1038/ng.2383 (2012).
[6] Bouchard, C. Genomic predictors of trainability. *Exp Physiol* 97, 347-352, doi:10.1113/expphysiol.2011.058735 (2012).
[7] Bouchard, C. et al. The response to long-term overfeeding in identical twins. *The New England journal of medicine* 322, 1477-1482, doi:10.1056/NEJM199005243222101 (1990).
[8] Hu, F. B. Globalization of Diabetes: The role of diet, lifestyle, and genes. *Diabetes Care* 34, 1249-1257, doi:10.2337/dc11-0442 (2011).

[9] Malik, V. & Hu, F. Sweeteners and Risk of Obesity and Type 2 Diabetes: The Role of Sugar-Sweetened Beverages. *Curr Diab Rep* 12, 195-203, doi:10.1007/s11892-012-0259-6 (2012).

[10] Ley, S. H., Hamdy, O., Mohan, V. & Hu, F. B. Prevention and management of type 2 diabetes: dietary components and nutritional strategies. *The Lancet* 383, 1999-2007, doi:10.1016/S0140-6736(14)60613-9 (2014).

[11] Pan, A. et al. Changes in Red Meat Consumption and Subsequent Risk of Type 2 Diabetes: Three Cohorts of US Men and Women. *JAMA internal medicine* 173, 10.1001/jamainternmed.2013.6633, doi:10.1001/jamainternmed.2013.6633 (2013).

[12] Ding, M., Bhupathiraju, S. N., Chen, M., van Dam, R. M. & Hu, F. B. Caffeinated and Decaffeinated Coffee Consumption and Risk of Type 2 Diabetes: A Systematic Review and a Dose-Response Meta-analysis. *Diabetes Care* 37, 569-586, doi:10.2337/dc13-1203 (2014).

[13] Cahill, L. E. et al. Fried-food consumption and risk of type 2 diabetes and coronary artery disease: a prospective study in 2 cohorts of US women and men. *The American Journal of Clinical Nutrition* 100, 667-675, doi:10.3945/ajcn.114.084129 (2014).

[14] Qi, L., Cornelis, M. C., Zhang, C., van Dam, R. M. & Hu, F. B. Genetic predisposition, Western dietary pattern, and the risk of type 2 diabetes in men. *The American Journal of Clinical Nutrition* 89, 1453-1458, doi:10.3945/ajcn.2008.27249 (2009).

[15] Sigal, R. J., Kenny, G. P., Wasserman, D. H., Castaneda-Sceppa, C. & White, R. D. Physical Activity/Exercise and Type 2 Diabetes: A consensus statement from the American Diabetes Association. *Diabetes Care* 29, 1433-1438, doi:10.2337/dc06-9910 (2006).

[16] Forouhi, N. G. & Wareham, N. J. The EPIC-InterAct Study: A Study of the Interplay between Genetic and Lifestyle Behavioral Factors on the Risk of Type 2 Diabetes in European Populations. *Current Nutrition Reports* 3, 355-363, doi:10.1007/s13668-014-0098-y (2014).

[17] Jaddoe, V. W. V. et al. Fetal Exposure to Parental Smoking and the Risk of Type 2 Diabetes in Adult Women. *Diabetes Care* 37, 2966-2973, doi:10.2337/dc13-1679 (2014).

[18] Li, Y., Qi, Q., Workalemahu, T., Hu, F. B. & Qi, L. Birth Weight, Genetic Susceptibility, and Adulthood Risk of Type 2 Diabetes. *Diabetes Care* 35, 2479-2484, doi:10.2337/dc12-0168 (2012).

[19] Altshuler, D. et al. The common PPARgamma Pro12Ala polymorphism is associated with decreased risk of type 2 diabetes. *Nature genetics* 26, 76-80, doi:10.1038/79216 (2000).

[20] Gouda, H. N. et al. The association between the peroxisome proliferator-activated receptor-gamma2 (PPARG2) Pro12Ala gene variant and type 2 diabetes mellitus: a HuGE review and meta-analysis. *American journal of epidemiology* 171, 645-655, doi:10.1093/aje/kwp450 (2010).

[21] Deeb, S. S. et al. A Pro12Ala substitution in PPARgamma2 associated with decreased receptor activity, lower body mass index and improved insulin sensitivity. *Nature genetics* 20, 284-287, doi:10.1038/3099 (1998).

[22] Ahlqvist, E., Ahluwalia, T. S. & Groop, L. Genetics of type 2 diabetes. *Clinical chemistry* 57, 241-254, doi:10.1373/clinchem.2010.157016 (2011).

[23] Rung, J. et al. Genetic variant near IRS1 is associated with type 2 diabetes, insulin resistance and hyperinsulinemia. *Nature genetics* 41, 1110-1115, doi:10.1038/ng.443 (2009).

[24] Almind, K. et al. Aminoacid polymorphisms of insulin receptor substrate-1 in non-insulin-dependent diabetes mellitus. *Lancet* 342, 828-832 (1993).

[25] Grant, S. F. et al. Variant of transcription factor 7-like 2 (TCF7L2) gene confers risk of type 2 diabetes. *Nature genetics* 38, 320-323, doi:10.1038/ng1732 (2006).

[26] Lyssenko, V. et al. Mechanisms by which common variants in the TCF7L2 gene increase risk of type 2 diabetes. *The Journal of clinical investigation* 117, 2155-2163, doi:10.1172/JCI30706 (2007).

[27] Saxena, R. et al. Common single nucleotide polymorphisms in TCF7L2 are reproducibly associated with type 2 diabetes and reduce the insulin response to glucose in nondiabetic individuals. *Diabetes* 55, 2890-2895, doi:10.2337/db06-0381 (2006).

[28] Helgason, A. et al. Refining the impact of TCF7L2 gene variants on type 2 diabetes and adaptive evolution. *Nature genetics* 39, 218-225, doi:10.1038/ng1960 (2007).

[29] Sladek, R. et al. A genome-wide association study identifies novel risk loci for type 2 diabetes. *Nature* 445, 881-885, doi:http://www.nature.com/nature/journal/v445/n7130/suppinfo/nature05616_S1.html (2007).

[30] Zeggini, E. et al. Meta-analysis of genome-wide association data and large-scale replication identifies additional susceptibility loci for type 2 diabetes. *Nature genetics* 40, 638-645, doi:10.1038/ng.120 (2008).

[31] Voight, B. F. et al. Twelve type 2 diabetes susceptibility loci identified through large-scale association analysis. *Nature genetics* 42, 579-589, doi:10.1038/ng.609 (2010).

[32] Dupuis, J. et al. New genetic loci implicated in fasting glucose homeostasis and their impact on type 2 diabetes risk. *Nature genetics* 42, 105-116, doi:10.1038/ng.520 (2010).

[33] Kato, N. Insights into the genetic basis of type 2 diabetes. *Journal of diabetes investigation* 4, 233-244, doi:10.1111/jdi.12067 (2013).

[34] Kato, N. Ethnic differences in genetic predisposition to hypertension. *Hypertens Res* 35, 574-581 (2012).

[35] Qi, L. et al. Genetic variants at 2q24 are associated with susceptibility to type 2 diabetes. *Human Molecular Genetics* 19, 2706-2715, doi:10.1093/hmg/ddq156 (2010).

[36] Shu, X. O. et al. Identification of New Genetic Risk Variants for Type 2 Diabetes. *PLoS Genetics* 6, e1001127, doi:10.1371/journal.pgen.1001127 (2010).

[37] Tsai, F.-J. et al. A Genome-Wide Association Study Identifies Susceptibility Variants for Type 2 Diabetes in Han Chinese. *PLoS Genetics* 6, e1000847, doi:10.1371/journal.pgen.1000847 (2010).

[38] Li, H. et al. A Genome-Wide Association Study Identifies GRK5 and RASGRP1 as Type 2 Diabetes Loci in Chinese Hans. *Diabetes* 62, 291-298, doi:10.2337/db12-0454 (2013).

[39] Yamauchi, T. et al. A genome-wide association study in the Japanese population identifies susceptibility loci for type 2 diabetes at UBE2E2 and C2CD4A-C2CD4B. *Nature genetics* 42, 864-868, doi:http://www.nature.com/ng/journal/v42/ n10/abs/ng.660.html#supplementary-information (2010).

[40] Yasuda K, M. K., Horikawa Y, Hara K, Osawa H, Furuta H, Hirota Y, Mori H, Jonsson A, Sato Y, Yamagata K, Hinokio Y, Wang HY, Tanahashi T, Nakamura N, Oka Y,

Iwasaki N, Iwamoto Y, Yamada Y, Seino Y, Maegawa H, Kashiwagi A, Takeda J, Maeda E, Shin HD, Cho YM, Park KS, Lee HK, Ng MC, Ma RC, So WY, Chan JC, Lyssenko V, Tuomi T, Nilsson P, Groop L, Kamatani N, Sekine A, Nakamura Y, Yamamoto K, Yoshida T, Tokunaga K, Itakura M, Makino H, Nanjo K, Kadowaki T, Kasuga M. Variants in KCNQ1 are associated with susceptibility to type 2 diabetes mellitus. *Nat Genet*. 40, 1092-1097 (2008).

[41] Unoki H, T. A., Kawaguchi T, Hara K, Horikoshi M, Andersen G, Ng DP, Holmkvist J, Borch-Johnsen K, Jørgensen T, Sandbaek A, Lauritzen T, Hansen T, Nurbaya S, Tsunoda T, Kubo M, Babazono T, Hirose H, Hayashi M, Iwamoto Y, Kashiwagi A, Kaku K, Kawamori R, Tai ES, Pedersen O, Kamatani N, Kadowaki T, Kikkawa R, Nakamura Y, Maeda S. SNPs in KCNQ1 are associated with susceptibility to type 2 diabetes in East Asian and European populations. *Nature genetics* 40, 1098-1102 (2008).

[42] Cui B, Z. X., Xu M, Guo T, Zhu D, Chen G, Li X, Xu L, Bi Y, Chen Y, Xu Y, Li X, Wang W, Wang H, Huang W, Ning G. A genome-wide association study confirms previously reported loci for type 2 diabetes in Han Chinese. *PLoS One* 6, e22353 (2011).

[43] Chen, R. et al. Type 2 Diabetes Risk Alleles Demonstrate Extreme Directional Differentiation among Human Populations, Compared to Other Diseases. *PLoS Genetics* 8, e1002621, doi:10.1371/journal.pgen.1002621 (2012).

[44] Replication, D. I. G. & Meta-analysis Consortium, A. G. E. N. T. D. C. S. A. T. D. C. M. A. T. D. C. T. D. G. E. b. N.-g. s. i. m.-E. Genome-wide trans-ancestry meta-analysis provides insight into the genetic architecture of type 2 diabetes susceptibility. *Nature genetics* 46, 234-244, doi:10.1038/ng.2897 (2014).

[45] Morris, A. P. et al. Large-scale association analysis provides insights into the genetic architecture and pathophysiology of type 2 diabetes. *Nature genetics* 44, 981-990, doi:10.1038/ng.2383 (2012).

[46] Neel, J. V. Diabetes mellitus: a "thrifty" genotype rendered detrimental by "progress"? *American journal of human genetics* 14, 353-362 (1962).

[47] Ayub, Q. et al. Revisiting the Thrifty Gene Hypothesis via 65 Loci Associated with Susceptibility to Type 2 Diabetes. *American Journal of Human Genetics* 94, 176-185, doi:10.1016/j.ajhg.2013.12.010 (2014).

[48] Franks, P. W., Pearson, E. & Florez, J. C. Gene-Environment and Gene-Treatment Interactions in Type 2 Diabetes: Progress, pitfalls, and prospects. *Diabetes Care* 36, 1413-1421, doi:10.2337/dc12-2211 (2013).

[49] Langenberg, C. et al. Gene-Lifestyle Interaction and Type 2 Diabetes: The EPIC InterAct Case-Cohort Study. *PLoS Medicine* 11, e1001647, doi:10.1371/journal.pmed.1001647 (2014).

[50] Florez, J. C. et al. TCF7L2 Polymorphisms and Progression to Diabetes in the Diabetes Prevention Program. *The New England journal of medicine* 355, 241-250, doi:10.1056/NEJMoa062418 (2006).

[51] Eva Fisher, H. B., Andreas Fritsche, Frank Doering, Hans-Georg Joost and Matthias B. Schulze. Whole-grain consumption and transcription factor-7-like 2 (TCF7L2) rs7903146: gene–diet interaction in modulating type 2 diabetes risk. *British Journal of Nutrition* 101, 478-481 (2009).

[52] Corella, D. et al. Mediterranean Diet Reduces the Adverse Effect of the TCF7L2-rs7903146 Polymorphism on Cardiovascular Risk Factors and Stroke Incidence: A randomized controlled trial in a high-cardiovascular-risk population. *Diabetes Care* 36, 3803-3811, doi:10.2337/dc13-0955 (2013).

[53] Hindy, G. et al. Role of TCF7L2 risk variant and dietary fibre intake on incident type 2 diabetes. *Diabetologia* 55, 2646-2654, doi:10.1007/s00125-012-2634-x (2012).

[54] Prasad, A. S. Zinc: mechanisms of host defense. *J Nutr* 137, 1345-1349 (2007).

[55] Sonestedt, E. et al. Genetic Variation in the Glucose-Dependent Insulinotropic Polypeptide Receptor Modifies the Association between Carbohydrate and Fat Intake and Risk of Type 2 Diabetes in the Malmö Diet and Cancer Cohort. *The Journal of Clinical Endocrinology & Metabolism* 97, E810-E818, doi:doi:10.1210/jc.2011-2444 (2012).

[56] Kanoni, S. et al. Total Zinc Intake May Modify the Glucose-Raising Effect of a Zinc Transporter (SLC30A8) Variant: A 14-Cohort Meta-analysis. *Diabetes* 60, 2407-2416, doi:10.2337/db11-0176 (2011).

[57] Shan Z, B. W., Zhang Y, Rong Y, Wang X, Jin Y, Song Y, Yao P, Sun C, Hu FB, Liu L. Interactions between zinc transporter-8 gene (SLC30A8) and plasma zinc concentrations for impaired glucose regulation and type 2 diabetes. *Diabetes* 63, 1796-1803 (2014).

[58] Nettleton, J. A. et al. Interactions of Dietary Whole-Grain Intake With Fasting Glucose– and Insulin-Related Genetic Loci in Individuals of European Descent: A meta-analysis of 14 cohort studies. *Diabetes Care* 33, 2684-2691, doi:10.2337/dc10-1150 (2010).

[59] Ericson, U. et al. Sex-specific interactions between the IRS1 polymorphism and intakes of carbohydrates and fat on incident type 2 diabetes. *The American journal of clinical nutrition* 97, 208-216, doi:10.3945/ajcn.112.046474 (2013).

[60] Franks, P. Gene × Environment Interactions in Type 2 Diabetes. *Curr Diab Rep* 11, 552-561, doi:10.1007/s11892-011-0224-9 (2011).

[61] Kilpeläinen, T. O. et al. Physical activity modifies the effect of SNPs in the SLC2A2 (GLUT2) and ABCC8 (SUR1) genes on the risk of developing type 2 diabetes. Vol. 31 (2007).

[62] Spiegelman, B. M. PPAR-gamma: adipogenic regulator and thiazolidinedione receptor. *Diabetes* 47, 507-514, doi:10.2337/diabetes.47.4.507 (1998).

[63] Adamo, K. B. et al. Influence of Pro12Ala peroxisome proliferator-activated receptor γ2 polymorphism on glucose response to exercise training in type 2 diabetes. *Diabetologia* 48, 1503-1509, doi:10.1007/s00125-005-1827-y (2005).

[64] Brito, E. C. et al. Previously Associated Type 2 Diabetes Variants May Interact With Physical Activity to Modify the Risk of Impaired Glucose Regulation and Type 2 Diabetes: A Study of 16,003 Swedish Adults. *Diabetes* 58, 1411-1418, doi:10.2337/db08-1623 (2009).

[65] Franks, P. W. et al. Does peroxisome proliferator-activated receptor γ genotype (Pro12ala) modify the association of physical activity and dietary fat with fasting insulin level? *Metabolism - Clinical and Experimental* 53, 11-16, doi:10.1016/j.metabol.2003.08.005.

[66] Ruchat, S. M. et al. Improvements in glucose homeostasis in response to regular exercise are influenced by PPARG Pro12Ala variant: results from the HERITAGE Family Study. *Diabetologia* 53, 679-689, doi:10.1007/s00125-009-1630-2 (2010).

[67] Nelson, T. L. et al. Association of the peroxisome proliferator–activated receptor γ gene with type 2 diabetes mellitus varies by physical activity among non-Hispanic whites from Colorado. *Metabolism - Clinical and Experimental* 56, 388-393, doi:10.1016/j.metabol.2006.10.022.

[68] Frayling, T. M. et al. A common variant in the FTO gene is associated with body mass index and predisposes to childhood and adult obesity. *Science* 316, 889-894 (2007).

[69] Speliotes, E. K. et al. Association analyses of 249,796 individuals reveal 18 new loci associated with body mass index. *Nat Genet* 42, 937-948, doi: 10.1038/ng.686 (2010).

[70] Kilpeläinen, T. O. et al. Physical Activity Attenuates the Influence of FTO Variants on Obesity Risk: A Meta-Analysis of 218,166 Adults and 19,268 Children. *PLoS Medicine* 8, e1001116, doi:10.1371/journal.pmed.1001116 (2011).

[71] Qi, Q. et al. Television Watching, Leisure-Time Physical Activity and the Genetic Predisposition in Relation to Body Mass Index in Women and Men. Circulation 126, 1821-1827, doi:10.1161/CIRCULATIONAHA.112.098061 (2012).

[72] Li, S. et al. Physical Activity Attenuates the Genetic Predisposition to Obesity in 20,000 Men and Women from EPIC-Norfolk Prospective Population Study. *PLoS Medicine* 7, e1000332, doi:10.1371/journal.pmed.1000332 (2010).

[73] Klimentidis, Y., Chen, Z., Arora, A. & Hsu, C.-H. Association of physical activity with lower type 2 diabetes incidence is weaker among individuals at high genetic risk. *Diabetologia* 57, 2530-2534, doi:10.1007/s00125-014-3380-z (2014).

[74] Scott, R. A. et al. No Interactions Between Previously Associated 2-Hour Glucose Gene Variants and Physical Activity or BMI on 2-Hour Glucose Levels. *Diabetes* 61, 1291-1296, doi:10.2337/db11-0973 (2012).

[75] Hales, C. N. & Barker, D. J. The thrifty phenotype hypothesis. *British medical bulletin* 60, 5-20 (2001).

[76] Barker, D. J. et al. Type 2 (non-insulin-dependent) diabetes mellitus, hypertension and hyperlipidaemia (syndrome X): relation to reduced fetal growth. *Diabetologia* 36, 62-67 (1993).

[77] Whincup, P. H. et al. Birth weight and risk of type 2 diabetes: a systematic review. *Jama* 300, 2886-2897 (2008).

[78] de Rooij, S. R. et al. The effects of the Pro12Ala polymorphism of the peroxisome proliferator-activated receptor-gamma2 gene on glucose/insulin metabolism interact with prenatal exposure to famine. *Diabetes care* 29, 1052-1057 (2006).

[79] Pulizzi, N. et al. Interaction between prenatal growth and high-risk genotypes in the development of type 2 diabetes. *Diabetologia* 52, 825-829 (2009).

[80] van Hoek, M., Langendonk, J. G., de Rooij, S. R., Sijbrands, E. J. G. & Roseboom, T. J. Genetic Variant in the IGF2BP2 Gene May Interact With Fetal Malnutrition to Affect Glucose Metabolism. *Diabetes* 58, 1440-1444, doi:10.2337/db08-1173 (2009).

[81] Grarup N, R. C., Andersson EA, Andersen G, Nielsen AL, Albrechtsen A, Clausen JO, Rasmussen SS, Jørgensen T, Sandbaek A, Lauritzen T, Schmitz O, Hansen T, Pedersen O. Studies of association of variants near the HHEX, CDKN2A/B, and IGF2BP2 genes with type 2 diabetes and impaired insulin release in 10,705 Danish subjects: validation and extension of genome-wide association studies. *Diabetes* 56, 3105-3111 (2007).

[82] Li, Y., Qi, Q., Workalemahu, T., Hu, F. B. & Qi, L. Birth weight, genetic susceptibility, and adulthood risk of type 2 diabetes. *Diabetes care* 35, 2479-2484, doi:10.2337/dc12-0168 (2012).

[83] Hong, J. et al. Genetic susceptibility, birth weight and obesity risk in young Chinese. *International journal of obesity* 37, 673-677, doi:10.1038/ijo.2012.87 (2013).

[84] Qi, L. Gene-Diet Interactions in Complex Disease: Current Findings and Relevance for Public Health. *Current nutrition reports* 1, 222-227, doi:10.1007/s13668-012-0029-8 (2012).

[85] Florez, J. C. et al. TCF7L2 polymorphisms and progression to diabetes in the Diabetes Prevention Program. *The New England journal of medicine* 355, 241-250, doi:10.1056/NEJMoa062418 (2006).

[86] Florez, J. C. et al. Testing of diabetes-associated WFS1 polymorphisms in the Diabetes Prevention Program. *Diabetologia* 51, 451-457, doi:10.1007/s00125-007-0891-x (2008).

[87] Moore, A. F. et al. Extension of type 2 diabetes genome-wide association scan results in the diabetes prevention program. *Diabetes* 57, 2503-2510, doi:10.2337/db08-0284 (2008).

[88] Hivert, M. F. et al. Updated genetic score based on 34 confirmed type 2 diabetes Loci is associated with diabetes incidence and regression to normoglycemia in the diabetes prevention program. *Diabetes* 60, 1340-1348, doi:10.2337/db10-1119 (2011).

[89] Florez, J. C. et al. Effects of genetic variants previously associated with fasting glucose and insulin in the Diabetes Prevention Program. *PLoS One* 7, e44424, doi:10.1371/journal.pone.0044424 (2012).

[90] Pan, Q. et al. Variation at the melanocortin 4 receptor gene and response to weight-loss interventions in the diabetes prevention program. *Obesity* (Silver Spring) 21, E520-526, doi:10.1002/oby.20459 (2013).

[91] Billings, L. K. et al. The INFLUENCE of rare Genetic variation in SLC30A8 on diabetes incidence and beta-cell function. *J Clin Endocrinol Metab*, jc20132378, doi:10.1210/jc.2013-2378 (2014).

[92] Mather, K. J. et al. Common variants in genes encoding adiponectin (ADIPOQ) and its receptors (ADIPOR1/2), adiponectin concentrations, and diabetes incidence in the Diabetes Prevention Program. *Diabet Med* 29, 1579-1588, doi:10.1111/j.1464-5491.2012.03662.x (2012).

[93] Jablonski, K. A. et al. Common variants in 40 genes assessed for diabetes incidence and response to metformin and lifestyle intervention in the diabetes prevention program. *Diabetes* 59, 2672-2681, doi:10.2337/db10-0543 (2010).

[94] Moore, A. F. et al. The association of ENPP1 K121Q with diabetes incidence is abolished by lifestyle modification in the diabetes prevention program. *J Clin Endocrinol Metab* 94, 449-455, doi:10.1210/jc.2008-1583 (2009).

[95] Heni, M., Herzberg-Schafer, S., Machicao, F., Haring, H. U. & Fritsche, A. Dietary fiber intake modulates the association between variants in TCF7L2 and weight loss during a lifestyle intervention. *Diabetes care* 35, e24, doi: 10.2337/dc11-2012 (2012).

[96] Sacks, F. M. et al. Comparison of weight-loss diets with different compositions of fat, protein, and carbohydrates. *The New England journal of medicine* 360, 859-873, doi: 10.1056/NEJMoa0804748 (2009).

[97] Mattei, J., Qi, Q., Hu, F. B., Sacks, F. M. & Qi, L. TCF7L2 genetic variants modulate the effect of dietary fat intake on changes in body composition during a weight-loss intervention. *The American journal of clinical nutrition* 96, 1129-1136, doi:10.3945/ajcn.112.038125 (2012).

[98] Zhang, X. et al. FTO Genotype and 2-Year Change in Body Composition and Fat Distribution in Response to Weight-Loss Diets: The POUNDS LOST Trial. *Diabetes* 61, 3005-3011, doi: 10.2337/db11-1799 (2012).

[99] Qi, Q. et al. Insulin receptor substrate 1 gene variation modifies insulin resistance response to weight-loss diets in a 2-year randomized trial: the Preventing Overweight Using Novel Dietary Strategies (POUNDS LOST) trial. *Circulation* 124, 563-571, doi: 10.1161/CIRCULATIONAHA.111.025767 (2011).

[100] Xu, M. et al. Genetic Determinant for Amino Acid Metabolites and Changes in Body Weight and Insulin Resistance in Response to Weight-Loss DietsClinical Perspective The Preventing Overweight Using Novel Dietary Strategies (POUNDS LOST) Trial. *Circulation* 127, 1283-1289, doi: 10.1161/CIRCULATIONAHA.112.000586 (2013).

[101] Huffman KM, Shah SH, Stevens RD, Bain JR, Muehlbauer M, Slentz CA, Tanner CJ, Kuchibhatla M, Houmard JA, Newgard CB, Kraus WE. . Relationships between circulating metabolic intermediates and insulin action in overweight to obese, inactive men and women. *Diabetes Care.* 32, 1678-1683, doi: 10.2337/dc08-2075 (2009).

[102] Newgard CB, An J, Bain JR, Muehlbauer MJ, Stevens RD, Lien LF, Haqq AM, Shah SH, Arlotto M, Slentz CA, Rochon J, Gallup D, Ilkayeva O, Wenner BR, Yancy WS, Jr., Eisenson H, Musante G, Surwit RS, Millington DS, Butler MD, Svetkey LP. A branched-chain amino acid-related metabolic signature that differentiates obese and lean humans and contributes to insulin resistance. *Cell Metab.* 9, 311-326, doi: 10.1016/j.cmet.2009.02.002 (2009).

[103] Wang TJ, Larson MG, Vasan RS, Cheng S, Rhee EP, McCabe E, Lewis GD, Fox CS, Jacques PF, Fernandez C, O'Donnell CJ, Carr SA, Mootha VK, Florez JC, Souza A, Melander O, Clish CB, Gerszten RE. . Metabolite profiles and the risk of developing diabetes. *Nat Med.* 17, 448-453, doi: 10.1038/nm.2307 (2011).

[104] Laferrere B, Reilly D, Arias S, Swerdlow N, Gorroochurn P, Bawa B, Bose M, Teixeira J, Stevens RD, Wenner BR, Bain JR, Muehlbauer MJ, Haqq A, Lien L, Shah SH, Svetkey LP, Newgard CB. Differential metabolic impact of gastric bypass surgery versus dietary intervention in obese diabetic subjects despite identical weight loss. *Sci Transl Med.* 3, 80re82, doi: 10.1126/scitranslmed.3002043 (2011).

[105] Shah SH, Crosslin DR, Haynes CS, Nelson S, Turer CB, Stevens RD, Muehlbauer MJ, Wenner BR, Bain JR, Laferrère B, Gorroochurn P, Teixeira J, Brantley PJ, Stevens VJ, Hollis JF, Appel LJ, Lien LF, Batch B, Newgard CB, Svetkey LP. Branched-chain amino acid levels are associated with improvement in insulin resistance with weight loss. *Diabetologia.* 55, 321-330, doi: 10.1007/s00125-011-2356-5 (2012).

[106] Moffitt, T. E., Caspi, A. & Rutter, M. Strategy for investigating interactions between measured genes and measured environments. *Archives of general psychiatry* 62, 473-481, doi:10.1001/archpsyc.62.5.473 (2005).

[107] Paré, G., Cook, N. R., Ridker, P. M. & Chasman, D. I. On the Use of Variance per Genotype as a Tool to Identify Quantitative Trait Interaction Effects: A Report from the Women's Genome Health Study. *PLoS Genetics* 6, e1000981, doi:10.1371/journal.pgen.1000981 (2010).

[108] Zhang, Y. & Liu, J. S. Bayesian inference of epistatic interactions in case-control studies. *Nature genetics* 39, 1167-1173, doi:http://www.nature.com/ng/journal/v39/n9/suppinfo/ng2110_S1.html (2007).

[109] Kooperberg, C. & LeBlanc, M. Increasing the power of identifying gene × gene interactions in genome-wide association studies. *Genetic epidemiology* 32, 255-263, doi:10.1002/gepi.20300 (2008).

[110] Lamri, A. et al. Dietary fat intake and polymorphisms at the PPARG locus modulate BMI and type 2 diabetes risk in the D.E.S.I.R. prospective study. *Int J Obes* (Lond) 36, 218-224, doi:10.1038/ijo.2011.91 (2012).

[111] Ortega-Azorin, C. et al. Associations of the FTO rs9939609 and the MC4R rs17782313 polymorphisms with type 2 diabetes are modulated by diet, being higher when adherence to the Mediterranean diet pattern is low. *Cardiovascular diabetology* 11, 137, doi:10.1186/1475-2840-11-137 (2012).

[112] Sonestedt, E. et al. Genetic variation in the glucose-dependent insulinotropic polypeptide receptor modifies the association between carbohydrate and fat intake and risk of type 2 diabetes in the Malmo Diet and Cancer cohort. *The Journal of clinical endocrinology and metabolism* 97, E810-818, doi:10.1210/jc.2011-2444 (2012).

[113] Brito, E. C. et al. Previously associated type 2 diabetes variants may interact with physical activity to modify the risk of impaired glucose regulation and type 2 diabetes: a study of 16,003 Swedish adults. *Diabetes* 58, 1411-1418, doi:10.2337/db08-1623 (2009).

[114] Qi, Q. et al. Insulin receptor substrate 1 (IRS1) gene variation modifies insulin resistance response to weight-loss diets in a two-year randomized trial. *Circulation* 124, 563-571, doi:10.1161/CIRCULATIONAHA.111.025767 (2011).

In: Gene-Environment Interactions and Human Diseases
Editor: Lu Qi
ISBN: 978-1-63483-680-7

Chapter 7

CANCER HEALTH DISPARITIES: CONTRIBUTIONS FROM TUMOR BIOLOGY AND GENE-ENVIRONMENT INTERACTIONS

Tiffany A. Wallace and Stefan Ambs
Laboratory of Human Carcinogenesis,
Center of Cancer Research, National Cancer Institute (NCI),
National Institutes of Health (NIH), Bethesda, MD, US

ABSTRACT

Cancer incidence and clinical outcomes show great variations across nations and between population groups. These variations are largely explained by differences in age distribution, diet and lifestyle, access to health care, cultural barriers, environmental exposures, and ancestry-related factors. The global pattern of cancer is dynamic due to fluctuations in the age distribution of populations, improvements in cancer prevention and early detection in affluent countries, and rapid diet and lifestyle changes worldwide. Some population groups (i.e., developed versus non-developed countries or different racial/ethnic groups) experience a disproportionally high cancer burden. For example, cancers caused by infections are significantly more common in developing countries, and they over-proportionally affect immigrant populations in the United States and other countries.

Additionally, in the United States and the Caribbean, cancer incidence and mortality rates are higher in populations of African ancestry when compared with other population groups. The causes of these disparities are multi-faceted and may include tumor biological and genetic factors, and their interaction with the environment. In this chapter, we will discuss the magnitude and causes of global cancer health disparities and will, with a focus on prostate and breast cancers, evaluate how environmental, genetic and tumor biological factors contribute to existing cancer incidence and clinical outcome disparities amongst population groups.

INTRODUCTION

Large global disparities in cancer incidence, prevalence, and mortality are evident for almost all cancer sites [1-4]. It is predicted that most of the future cancer incidence, morbidity and mortality will occur in the developing world. Currently, developing nations are burdened with cancers that are due to infectious diseases like cervical and liver cancer, but in the future these nations will be increasingly burdened with non-communicable cancers because of a rising cancer risk associated with the adaptation of a Western style diet, use of tobacco and alcohol, and a decrease in physical activity.

Breast and prostate cancer are more common in Europe or North American than in less affluent countries. Low incidence rates for these diseases are reported from parts of Africa and Asia, with the lowest rates occurring in China [1-5]. To date, it remains unresolved whether the low reported occurrence rates for breast and prostate cancer in Africa are factually correct or an artifact of under-reporting due to lack of screening and inadequate cancer surveillance. Notably, cancer mortality from breast and prostate cancer is quite high in some parts of Africa, with rates similar to those observed in Europe and North America [1, 2]. Nevertheless, incidence rates for these two cancers are increasing in parts of the world, including Africa and Asia, and are most likely attributed to changes in diet, lifestyle, and reproductive factors.

Many studies have explored the causes of the large global variations in cancer incidence and mortality between different population groups. It is clear the contributing risk factors are multifaceted. While risk factors including socioeconomic status, lifestyle exposures, access to healthcare, racial and ethnic discrimination, language and cultural barriers, and delayed disease diagnosis are well described, they do not fully explain the observed differences in cancer incidence, morbidity, and mortality that is found amongst various population groups. Therefore, it is suggested that possibly unrecognized ancestry-related factors or interethnic differences in environmental exposures might influence tumor biology and contribute to cancer health disparities. Likely, it is a complex interaction between biological/genetic, modifiable, and/or environmental factors that play a role in these documented cancer health disparities. Elucidating the interplay of these factors may help explain to some extent the observations of cancer health disparities seen on an international scale.

ENVIRONMENTAL AND BEHAVIORAL FACTORS IN CANCER HEALTH DISPARITIES

Modifiable environmental exposures have been suggested to be a major risk factor for common cancers across different population groups. This is supported by marked differences of cancer rates amongst various countries and migration studies that show immigrants tend to acquire the cancer rates of their new home country within a few generations [4, 6-9]. These studies suggest that ancestral genetic differences between human populations are not a major cancer cause, which is consistent with other reports [10]. If ancestry differences were largely responsible for global variations in cancer rates, one would expect that variations in cancer rates would persist in immigrant populations, as genetics-based differences in cancer susceptibility do not change quickly. Alternatively, others argue that heritable factors could be responsible for some of the population differences in cancer rates. A Scandinavian study reported that a considerable proportion of the risk to develop common cancers, like prostate and breast cancer, is attributed to heritable factors [11]. Subsequently, genome-wide association studies (GWAS) of cancer have identified numerous susceptibility loci for prostate, breast and other cancers that significantly influence cancer risk in the general population and are thought to substantially contribute to the global cancer burden [12-18]. Several of these studies report that cancer risk loci may affect some population groups differently than others [12, 17-22]. For example, deleterious 8q24 variants seem to confer a higher risk for prostate cancer in men of African ancestry than men of European ancestry [12, 19], as also suggested by admixture mapping [23]. As more and more of these cancer susceptibility loci are identified, further exploration is needed to identify the underlying biologically relevant variants to improve cancer risk modeling in populations with diverse ancestry.

While genetic predisposition to cancer is well documented, most cancer risk-conferring genetic variants do not alone cause cancer, but instead interact with harmful environmental exposures and thereby increase both the effects of these exposures and the odds of developing cancer. For example, a recent study observed that pesticide exposure modifies the association of 8q24 cancer susceptibility variants with prostate cancer [24]. Another study found that the interaction of household lead exposure with a single nucleotide polymorphism (SNP) in a zinc finger protein, JAZF1, was associated with prostate cancer in African American men [25]. These examples linking genetic factors and environmental exposures to the risk of developing common cancers may help us to better understand some cancer health disparities.

GENETIC AND TUMOR BIOLOGY CONTRIBUTIONS TO CANCER HEALTH DISPARITIES

As mentioned previously, genetic factors can significantly contribute to cancer susceptibility, and these susceptibility loci can vary significantly across population groups. Perhaps the seminal paper demonstrating population-specific risk variants for prostate cancer was a study conducted by Freedman et al. in 2006. This study conducted a whole-genome admixture scan of 1,597 African-American prostate cancer cases and 873 disease-free controls and identified 8q24 as a prostate cancer risk locus in African-American men [23]. The design and results of this study provided evidence that the 8q24 locus confers an increased disease risk among men of West African ancestry when compared with men of European ancestry. These data strongly suggest that genetic differences by ancestry can explain some of the observed differences in disease frequency between population groups defined by race/ethnicity. Other admixture studies have supported this hypothesis. For example, in two studies investigating United States (U.S.) Hispanic/Latino women, it was found that higher European ancestry is associated with an increased risk for breast cancer in this population group [26, 27]. Prokunina-Olsson and colleagues described a germ-line polymorphism that encodes a new interferon, termed interferon λ 4 (IFNL4). The variant allele is a deletion polymorphism (termed ΔG; rs368234815) that commonly occurs in subjects of African ancestry, but not in subjects from other population groups. Only subjects with this deletion can express functional interferon lambda 4 protein, which inhibits clearance of HCV, increasing the risk of liver cancer among carriers and thereby affecting population groups of African ancestry more than other population groups [28]. Thus, findings from these and other investigations suggest that population differences in genetic ancestry can lead to population differences in cancer susceptibility.

Observed differences in gene expression among population groups due to common genetic variations have also been identified [29-31]. Two studies investigated gene expression variations between individuals with European ancestry and individuals with African ancestry (Nigeria) using lymphoblastoid cell lines [30, 31]. These authors assessed the enrichment of biological processes and pathways by genes that are systematically differentially expressed by race/ethnicity. Notably, processes related to antimicrobial humoral response, chemokine and cytokine-mediated inflammation, histamine H1 receptor-mediated signaling pathway, toll-receptor signaling pathway, and the vascular endothelial growth factor signaling pathway were enriched. These results suggest differences in population genetics between people of European and West African ancestry cause gene expression differences affecting host immune response, inflammation and chemotaxis, and angiogenesis. Consistently, these findings correspond with other gene expression profiling studies [32, 33], and while preliminary, raise the possibility that differences in common genetic variations among population groups could lead to population group-selective alterations in cancer-related pathways that control host response, inflammation, and tumor angiogenesis [33, 34].

Functional genetic variants in key inflammatory and immune-related genes can show large frequency differences between population groups [33, 35-38]. Interestingly, many of these genes have important functions in tumor biology [39, 40]. One may speculate that population-specific functional genetic polymorphisms are perhaps most common in genes that regulate the immune system and host defense and developed as a necessary adaptation of the host to become resistant to infections unique to a given environment. Other population-specific genetic variations may arise in the course of human adaptation to diet and eco-regions [41]. There are several examples of functional genetic polymorphisms that are frequent in populations of African descent because they confer resistance to parasitic infections but may increase the susceptibility to other diseases including cancer. Trypanolytic apolipoprotein L1 variations are significantly more common in African populations and in African-Americans because they are thought to confer resistance to trypanosome infections [42]. However, while seemingly protective, these variations have also been linked to the development of severe kidney disease in African-Americans [42].

Another example of functional genetic polymorphisms is SNPs in the *NOS2* and Duffy antigen (*DARC*) genes that evolved in Africa to resist malaria infections [43-45]. Interestingly, both genes may have key functions in tumor biology [46-48]. The *DARC* polymorphism abolishes Duffy antigen expression which is the erythrocyte receptor for the *Plasmodium vivax* malaria parasite [45]. However, this antigen also serves as a chemokine receptor and reservoir (e.g., for IL-8 and MCP-1), and specifically interacts with the tumor suppressor KAI1 [48, 49]. In summary, there is evidence that population genetics can lead to population differences in disease susceptibility. Common genetic variations may influence cancer-related pathways and cancer incidence and mortality in a population-specific manner. However, the relative contribution of these genetic factors to common cancers is still poorly understood.

POPULATION DIFFERENCES IN RESPONSE TO CANCER THERAPY: DO THEY EXIST?

Minority populations remain underrepresented in clinical trials, despite the critical importance of race/ethnic diversity in clinical trial participation. The lack of diversity in clinical trials makes it difficult to assess whether the efficacy of cancer therapies is the same or similar for African-American, Asian-American, Hispanic/Latino, Native American, or European-American patients [50]. Analyses of the scientific literature and a focused assessment of clinical trial data strongly suggest that only modest survival differences are evident for African-American and European-American patients when treated comparably for similar-stage cancers [51-53]. The findings from these studies indicate that therapy outcomes are similar for these two patient groups and conclude that differences in disease stage at presentation should be the primary target of interventions designed to reduce survival heath disparities [51]. Other studies in support of these conclusions emphasize that some persistent differences in survival between African-Americans and European-Americans could be caused by the deleterious effects of co-morbidities that over-proportionally affect African-Americans [54-60].

However, studies suggesting that therapy outcomes are similar amongst population groups do not go unchallenged. Multiple studies in recent years, including findings from randomized clinical trials, report yet unexplained differences in survival between African-American and European-American cancer patients [61-73]. Studies investigating differences in tumor biology and pharmacogenomics have offered some explanation to the source of these treatment disparities. Pharmacogenomics studies investigate the influence of genetic variation on drug response in patients by correlating gene expression or polymorphisms with a drug's efficacy or toxicity. Evidence supports that differences in activity of drug metabolizing enzymes, such as cytochrome P450 and dihydropyrimidine dehydrogenase (DPYD), may potentially modulate pharmacokinetics and affects the amount of cancer drug reaching target organs and effective therapeutic levels. Specific SNPs in these enzymes occur in varying frequencies among distinct populations including Asian and African-American cancer patients and may account for the occurrence of unwanted side effects and/or reduced effectiveness observed with the use of anti-cancer drugs [74-77]. Another study has shown that allelic differences in AT-rich interaction domain (ARID) genes that occur with greater frequency among pediatric acute lymphoblastic leukemia patients from diverse ethnic populations has been linked with poor response to anti-leukemic drugs and increased risk of clinical relapse [78]. Furthermore, there is the observation that African-Americans encounter more toxic side effects from anthracycline-based therapies than European-Americans or Hispanic/Latinos [79, 80], perhaps explaining why these patients are more likely to terminate their chemotherapy prematurely [81]. The combination of the decreased efficacy and increase in toxic side effects may explain some of the population differences found in response to cancer therapy. Furthermore, genotypes that affect a patient's response to anthracycline-based therapies also have been described [82, 83]. As such, researchers have started to examine genetic contributions in the response to therapy and how genetic factors may lead to race/ethnic differences in either the benefits or unwanted side effects of chemotherapy [84]. These studies provide evidence that tumor biology and genetics may contribute to the cancer treatment disparities observed amongst population groups.

Lastly, population differences in therapy response and survival may also relate to stress, chronic depression and lack of social support [85]. Stress has been linked to an increased risk of developing cancer because it influences the immune response and can lead to immune dysfunction [85-88]. Chronic stress can enhance the release of interleukin-6, an inflammatory cytokine with a key function in cancer progression [89, 90]. Thus, it is reasonable to assume that stress and other bio-behavioral factors may affect disease outcomes and could contribute to cancer health disparities. Still, there is a deficiency of data examining the relationship between biological and bio-behavioral factors and health disparities in survival.

PROSTATE CANCER DISPARITIES AMONGST AFRICAN-AMERICAN MEN

The public health burden of prostate cancer is substantial. The largest U.S. cancer health disparity exists in prostate cancer with African-American men having the highest incidence and mortality rates [5]. Many factors have been postulated to contribute to this disparity, such as access to care, attitudes toward care, socioeconomic and educational disparities, differences in type and aggressiveness of treatment, tumor growth rates and disease aggressiveness, location and histopathological variables of the tumor, genetic predisposition, and dietary fat intake [53, 91-98]. There is evidence from clinical studies that prostate cancer outcomes are similar for African-American and European-American men when the disease is organ-confined [58, 99], but African-Americans appear to have worse outcomes when the disease is advanced [63-65, 72]. Furthermore, prostate tumors tend to develop earlier as a clinical disease in African-American men than in other men [97].

Modifiable and environmental risk factors are known to contribute to the majority of prostate tumors globally [100]. One risk factor that has been widely studied between population groups is dietary patterns [101-103]. African-Americans have a lower overall fruit and vegetable consumption, with the caveat that they may consume a higher frequency of specific types of fruits and vegetables (e.g., cruciferous vegetables) that are considered to be protective against cancer [102, 103]. In addition to the assessment of fruit and vegetable consumption, meat and animal fat intake have been identified as a candidate prostate cancer risk factor for African-Americans [104, 105]. Lastly, it is well-established that African-Americans have higher rates of vitamin D deficiency than European-Americans. Given that vitamin D has been associated with having anti-cancer properties, it has been hypothesized that this deficiency puts African-Americans at an increased risk of cancer [106]. While the relationship between vitamin D and cancer has been controversial in past years, recent large prevention trials support the hypothesis that vitamin D decreases the risk of clinically relevant prostate cancers [107, 108].

Nutrition is associated with cancer not just through diet, but additionally through the effects of obesity. A high calorie intake combined with a sedentary lifestyle can lead to obesity. Obesity is a risk factor for cancer and has been associated with an increased prostate cancer-specific mortality and treatment failure [109-113]. African-Americans have the highest proportion of a body mass index of greater than 30, which is considered a threshold for being obese [114]. About 37% percent of African-American men fall into this category versus 32% of the entire U.S. male population. Thus, it is possible that the high obesity rate among African-American men has some contribution to the higher disease incidence and mortality among them. Studies have investigated the effects of obesity on cancer and found that dietary factors and obesity may influence prostate tumor biology by modifying gene expression patterns [115-117]. Furthermore, gene signatures in prostate tumors may develop in response to nutrition and lifestyle intervention [115].

Prostate gland biology is androgen-dependent. A racial disparity in tumor androgen receptor expression in men with a localized disease has been observed [118]. Furthermore, several epidemiological studies report African-American men have modestly higher blood testosterone levels than European-American men [119-121]. However, the biological significance of this observation has since become uncertain because blood male sex hormone levels were not associated with the risk of prostate cancer in prospective studies and in a meta-analysis [122]. It was also hypothesized that prostate glands from African-American men may have an increased availability of male sex hormones, increased androgen receptor signaling, and an increased potential of cell growth and tumor development. However, studies have shown that prostatic tissue, testosterone and dihydrotestosterone levels are not different between African-Americans and European-Americans [123, 124]. Prostate size is also similar in both groups of men, suggesting that an enhanced androgenic effect is not present in the gland of African-American men [125]. Interestingly though, serum estrogen levels do differ significantly by race/ethnicity with the highest levels being detected in African-American men and in a group of men from Tobago [126, 127]. This is of interest because estrogen may induce inflammation-induced oxidative/nitrosative stress in the prostate gland [128]. In addition, a higher proportion of African-American than European-American men may have chronic inflammation in their non-cancerous prostate, as observed in one study [129]. Given that prostatic tissue inflammation and prostatitis are risk factors for prostate cancer [130, 131], prevalent inflammation in the prostate gland of African-American males may put these men at an increased prostate cancer risk.

In addition to modifiable and environmental risk factors, numerous genetic studies suggest a significant influence of common genetic variations in prostate disease development [12, 14, 19]. Many studies have examined the possibility of low penetrance genes contributing to the excessive burden of prostate cancer in African-American men. While common genetic variations that are associated with prostate cancer mortality have yet to be identified [132], genetic variations associated with risk have been described. As previously detailed above, the best described risk locus for prostate cancer to date is located at 8q24. Several additional GWAS identified other prostate cancer susceptibility loci (i.e., 11q13, 17q12, 19q33, and Xp11) and have been validated in studies of men with African ancestry [20, 133]. Because of the importance of inflammation in disease development, researchers also evaluated the relationship between prostate cancer and common genetic variations in genes comprising inflammation and host defense pathways. These studies yielded preliminary evidence suggesting that low penetrance genetic variability in these pathways may influence cancer risk in African-Americans and other population groups [134-141].

Prostate tumors in African-Americans tend to be more aggressive at diagnosis than they are in European-Americans [142]. This observation generated an interest in understanding the biology of these tumors. One study by Powell et al. addressed this question by analyzing autopsy material from prostates of 1056 African-American and European-American men who did not have a prior diagnosis of prostate cancer. Interestingly, Powell and coworkers observed that the cancer prevalence was similar between the two race/ethnicity groups at this subclinical stage. However, amongst the men with clinically detectable prostate tumors, African-American men seemed to have faster growing and more aggressive tumors, when compared with European-American men [97]. Additional research will have to investigate further whether mechanisms related to an increased proliferation, as described in this study, or inhibition of apoptosis, as described by others [143], may promote the enhanced growth of tumors in African-American men. Epidermal growth factor receptor (EGFR) signaling is anti-apoptotic and promotes cell proliferation by mechanisms including MAPK, PI3K/Akt and nuclear factor kappa-β activation. Furthermore, this oncogene and candidate drug target is critical in the progression to an androgen-independent disease and therefore has been widely studied in respect to advanced prostate cancer [144]. Notably, EGFR was found to be more commonly expressed in tumors from African-American patients when compared with European-American patients [145]. These studies suggest that EGFR perhaps contributes to apoptosis inhibition and enhanced proliferation in prostate tumors from African-American patients as well as possibly increasing the odds of disease recurrence.

Epigenetic DNA alterations including DNA methylation changes are present in a greater fraction of prostate tumors than any other of the known genetic defects [146]. Changes in DNA methylation are amongst the earliest somatic changes that can be detected in cancerous lesions of the prostate and occur at the stage of proliferative inflammatory atrophy, a precursor lesion in human prostate carcinogenesis. Increased DNA methylation can lead to numerous downstream effects including inactivation of tumor suppressor genes like *PTEN,* or a loss of protection against reactive chemical species. The latter typically occurs in response to hypermethylation of *GSTP1*, which is observed in 90% of prostate cancers [146]. Given the importance of epigenetic DNA alterations in contributing to prostate cancer, the occurrence of promoter DNA hypermethylation in tumors from African-American men has been evaluated. In three studies, a pattern emerged consistent with increased DNA hypermethylation in African-American tumors when compared with similar stage tumors from European-American men [147-149]. Currently, it is unknown what mechanisms may lead to race/ethnic differences in DNA hypermethylation, and whether this pattern is induced by environmental exposures, intrinsic differences in tumor biology, or a combination of the two.

Several research groups have investigated the occurrence of harmful chromosomal alterations in prostate tumors, comparing tumors from African-Americans with tumors from European-Americans. Two studies analyzed tumors from the two patient groups using the same platform [150, 151], whereas one study compared their analysis of African-American tumors with previously published data from European-Americans [152]. The first study analyzing 16 tumors from each patient group did not observe significant differences in chromosomal aberrations between the patients [150]. The second study reported several differences consistent with distinct genomic aberrations in African-American tumors at chromosomal sites that encode oncogenes (e.g., *ETV1*, *MYC*) and tumor suppressor genes (e.g., *PTEN*, *RB1*). The third study evaluated genomic alterations in both a discovery and validation cohort to describe chromosomal aberrations whose occurrence is most distinct between African-American and European-American tumors. The data revealed common genomic alterations specific to African-American tumors that may specifically target immune response genes [151]. Other studies reported population differences in the occurrence of key mutational events in prostate cancer. Recurrent genomic rearrangements like *TMPRSS2:ERG* are signature mutations of many human prostate tumors and have important biological and clinical implications [153]. Several studies examined the frequency of these recurrent gene fusions in European, European-American, Chinese, Japanese, and African-American prostate cancer patients. While common in European and European-American patients (in about 50% of patients), these mutations were less frequent in African-American patients, and rather uncommon in patients from Asia (2% to 15%) [154-157]. This pattern is not restricted to *TMPRSS2:ERG* rearrangements, but rather evidence supports that inactivation of the PTEN tumor suppressor may follow the same trend [154, 157]. The observations raise the possibility that prostate tumors in Asian populations arise from pathogenic mechanisms that are different from Western populations, which could affect disease aggressiveness. Striking differences may also exist between African-Americans and European-Americans within the United States. Thus, prostate tumor biology may display an unexpected global heterogeneity between men of European, African and Asian descent.

To date, many gene expression profiling studies have explored tumor biological differences in prostate cancer disparity research. One such study identified a novel prostate-specific gene, *PCGEM1*, with cell growth promoting function [158, 159]. *PCGEM1* was mapped to chromosome 2q32 and was shown to be over-expressed in prostate tumors. Further analyses revealed that this gene acts as a non-coding RNA and is most highly expressed in prostate tumors of African-American patients [159]. While further research is needed, this candidate gene may have important oncogenic functions in prostate cancer biology, particularly in respect to the African-American population. Other studies have used genome-wide gene expression profiling to investigate candidate differences in tumor biology between African-American and European-American prostate cancer patients [160-162]. When analyzing both prostate tumors and cultured prostate cancer epithelial cells from African-American and European-American patients, two independent studies found consistent gene expression differences between the two patient groups [157, 160] including several known metastasis-promoting genes (i.e., *AMFR*, *CXCR4*, and *MMP9)*, to be higher expressed in tumors from African-Americans. Furthermore, significant differences in tumor immunobiology and inflammation pathways between African-American and European-American patients, including a distinct interferon γ signature, have been identified [160]. This particular signature is identical to previously identified interferon-related gene signatures that are induced by both tumor-stroma interactions and a resistance to DNA damage phenotype [163-166]. Interestingly, the interferon-related signatures in these studies were all associated with metastasis and poor disease outcome. Furthermore, while additional research is required, it is possible that the presence of the interferon-related signature in prostate tumors may also suppress T cell cytotoxicity directed against cancer cells by mechanisms that may include up-regulation of the interferon-response gene, indolamine-2,3-dioxygenase [167]. It is believed that tumor-induced mechanisms, environmental stimuli, and ancestry-driven intrinsic factors may be candidate exposures that induce the interferon-related gene signature in prostate tumors (Figure 1).

In summary, a combination of genetic, tumor biological and environmental factors seemingly play a role in prostate cancer incidence and progression. As shown here, numerous studies have examined the relative contribution of these factors in the excess burden of prostate cancer among African-American men.

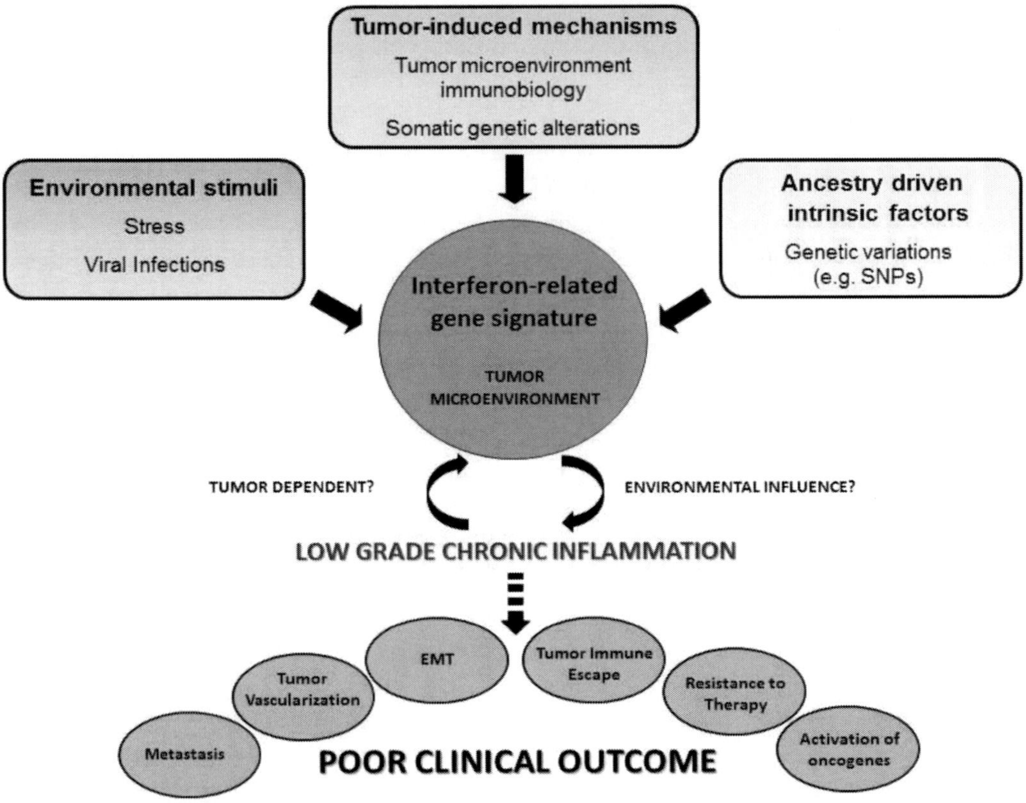

Figure 1. Potential origin and downstream effects of the interferon-related gene signature identified to be higher expressed in the tumors of African American breast and prostate cancer patients. Tumor-induced mechanisms, environmental stimuli, and/or ancestry-driven intrinsic factors are candidate exposures that may induce the interferon-related gene signature in tumor biology. It is hypothesized that this signature is maintained in an environment of low level, chronic inflammation that may arise from environmental exposures or differences in tumor subtype between the two patient groups. This signature may lead to poor clinical outcomes by promoting mechanisms including EMT, increased metastatic potential, suppression of T cell toxicity, resistance to therapy, increased tissue remodeling and tumor vascularization, and the development of tumor immune escape mechanisms. EMT = epithelial to mesenchymal transition.

Breast Cancer Disparities amongst African-American Women

Breast cancer incidence rates are about 3-fold higher in more developed than less developed countries [2]. Estrogen receptor (ER)-negative tumors represent 20–30% of all breast cancers, with a higher proportion occurring in younger women and women of African ancestry [168]. In the U.S., European-American and African-American women have a significantly higher risk to develop breast cancer than Asian/Pacific Islander, Hispanic/Latino, or American Indian/Alaska Natives [169]. Cancer epidemiology studies identified age, reproductive history, and lifestyle as major risk factors for breast cancer [170, 171]. Thus, disparities in these modifiable factors and differences in age distribution are obvious causes of the large global differences in breast cancer [1, 2].

Causes for breast cancer are multifactorial depending on the sub-type of the tumor, the timing of diagnosis, or population group effected. For example, the causes of pre-menopausal and post-menopausal diseases are different [170, 171]. There are also differences in the risk factor profile for ER-positive tumors when compared with ER-negative tumors and for African-American women compared with European-American women [172-175]. Furthermore, in general the factors that raise a woman's lifetime exposure to estrogen, like the use of hormone replacement therapy (HRT), increase the risk for the disease [170]. The decline in HRT use, triggered by studies showing that HRT increases breast cancer risk, has been credited for contributing to the marked disease decline in the U.S. in recent years [176].

Breast cancer is a heterogeneous disease and its biology is closely related to the hormone receptor status of the tumor. Landmark gene expression profiling studies discovered that breast tumors can be classified into subtypes with distinct gene expression profiles [177, 178]. Molecular signatures characterize three luminal subtypes among the estrogen receptor (ER)-positive tumors and three subtypes (basal-like, HER2-positive, and normal-like) among the ER-negative tumors. Of all subtypes, ER-negative tumors tend to produce the most aggressive disease, with the triple-negative (negative for ER, HER2, and progesterone receptor expression) being the worst in terms of treatment options and prognosis [179, 180]. It was recognized in 2006 and confirmed by other reports that the prevalence of the different breast cancer subtypes varies among U.S. population groups with the greatest difference observed between women of African and European ancestry [180-183]. Multiple investigations in West and Central Africa have provided further corroboration that women of African ancestry tend to develop early-onset, high grade, and ER-negative tumors more frequently than women of European ancestry [184-188]. These findings raise the possibility of a causative relationship between West African ancestry and an increased susceptibility for developing early onset and ER-negative breast tumors [189, 190]. Early onset, ER-negative tumors also develop more frequently in Asian Indian and Pakistani women and in women from other parts of Asia, although not as prevalent as in West Africa [2, 191].

Cancer epidemiology studies show that rates of ER-negative breast tumors in a population can change over time, suggesting environmental and reproductive factors have an influence on disease phenotypes [192]. For example, associations between age at menarche or breast-feeding with basal-like breast cancer have been observed [182, 193]. Furthermore, another study suggests the development of basal-like breast cancer in African American women is associated with breastfeeding and adiposity [182]. According to an estimate by the authors, about two-thirds of basal-like breast cancer in pre-menopausal African-American women could be prevented by promoting breastfeeding and reducing abdominal adiposity [182]. Notably however, other studies did not observe a relationship between body mass index and basal-like/triple-negative breast cancers [175, 193].

Socioeconomic status (SES) is associated with breast cancer mortality. Thus, several studies assessed if SES influences tumor characteristics such as tumor ER status. It was found that income and education are associated with tumor ER status and populations with a low SES are more likely to develop an ER-negative disease than populations with a high SES [174, 194-196]. This association may explain some of the excess risk among U.S. Hispanic and African-American women to develop an ER-negative disease yet the magnitude of the association between SES and tumor ER status seems to fall short of explaining most of the observed differences in tumor ER status between women of African and European ancestry. Furthermore, it remains uncertain whether the association of SES with tumor ER status is related to an increased risk of developing an ER-negative disease in populations with a low SES, or an increased risk of developing an ER-positive disease in populations with a high SES, or both. Generally, more affluent populations have a higher risk of breast cancer and an increased prevalence of ER-positive tumors [2, 174], with some variation across race/ethnic strata [197].

African-American women face a lower risk of being diagnosed with breast cancer, yet they are at increased risk of dying from the disease when compared with women from other U.S. population groups [198-201]. An analysis of survival disparities by disease stage using SEER data showed that the survival difference is larger for stage II/III disease than stage 0/I disease, with African-American having the worst outcomes, followed by Hispanic/Latino, European-American, and lastly, Asian American/Pacific Islander patients [201]. This survival disparity continues to exist independent of age, screening status, tumor characteristics, grade, ER status, therapy, co-morbidities and demographics. The African-American to European-American disparity in age-standardized breast cancer mortality emerged in the beginning of the 1980s and since then has widened [202]. The greatest absolute difference in the mortality hazard between the two patient groups occurs in the first years after diagnosis [202, 203], which may reflect differences in treatment, differences in response, or both. Differences in treatment exist and persist, with many African-American women receiving inadequate or delayed therapy [81, 204, 205]. However, when compared with European-American women, a poorer survival of African-American women has also been observed in equal access health care systems and clinical trials [71, 206, 207]. It has been hypothesized that the increased prevalence of early onset, ER-negative tumors and basal-like/triple-negative tumors in women of African ancestry is largely responsible for the excessive mortality among African-American women. Although this argument is very plausible, one has to be cautious because the breast cancer survival disparity in U.S. is irrespective of the tumor ER status [202]. Basal-like/triple-negative tumors confer an increased risk for early disease spread and deadly brain metastases [208], but the disease is not more aggressive in African-American patients than in European-American patients, as found by one study [209].

Few studies have examined the biology of breast tumors in African-American women beyond defining gene expression patterns associated with various subtypes. Early studies concentrated almost exclusively on hormone receptor expression and disease grading in African-American patients, describing an increased proportion of tumors that are ER-negative and poorly differentiated among this patient group [210, 211]. Additional studies describe that the ER isoform profile is different between tumors from African-American and European-American patients [212], and additionally observed differences in prostaglandin E1 receptor expression, expression of cell cycle-regulatory proteins, and aberrant/mutated expression of the p53 tumor suppressor gene [32, 213-215]. Interestingly, the tumor p53 status has been found by us and others to be associated with socio-economic deprivation, suggesting an environmental influence related to SES in the development of p53 mutations [216-218]. It was further noticed by one group that ER-negative breast tumors in young African-American women tend to exhibit promoter DNA hypermethylation [219], consistent with observations of DNA hypermethylation in prostate tumors of African-American men [149]. However, another recent study could not detect systematic differences in the tumor DNA methylation patterns between African-American and European-American patients [220].

Lastly, findings from our group and others suggest differences in the tumor microenvironment that relate to tumor vascularization and macrophage infiltration [32, 167]

(Figure 1). One study analyzing biological features of breast cancers in Nigerian women found high number of tumor infiltrating macrophages within the tumor microenvironment in these patients [221], confirming previous findings in African-American patients [32]. These differences between women with African ancestry and European-American breast cancer patients may arise from a low grade inflammation or from population differences in tumor-stroma interactions, but could also be related to tumor subtype differences between the two patient groups. In concordance with this work, others have shown that endothelial cells from African-Americans and European-Americans have different gene expression profiles that correlate with difference in protein function [222]. Those differences may influence tumor biology and the process of wound healing in general. In fact, it has been known for many years that some population groups, including African-Americans, are more prone to develop keloids, which are benign collagenous dermal tumors that form during the wound healing process [223-225]. It is aberrant wound healing involving altered fibroblast and endothelial functions that are thought to cause the condition. Future research may explore the causes of altered wound healing in populations that are prone to this disease and examine how these functions may possibly relate to the development of malignant tumors, especially the development of early-onset epithelial cancers of the breast and prostate.

SUMMARY AND PERSPECTIVES

Populations in the developing world and many minorities in more developed countries have not equally benefited from the advances in our understanding and treatment of cancer. There are substantial disparities in disease incidence and mortality between population groups nationally and globally that are largely explained by differences in lifestyle and diet, age distribution, carcinogen exposure, and inadequate or delayed access to health care. To date, there is sparse evidence that patients from any two population groups respond differently to cancer therapy when treated comparably for similar-stage cancers, arguing that inadequate or delayed access to health care is the most important factor in causing survival health disparities. However, minority populations remain underrepresented in clinical trials, and we therefore may not be able to appropriately assess whether the existing cancer therapies provide equal benefit to all population groups. Numerous studies have identified differences in tumor biology and epigenetic factors among diverse patient populations. These factors, singularly or in combination, may contribute to some of the existing cancer health disparities.

Although preliminary, there is evidence that population differences in genetic ancestry can lead to population differences in cancer susceptibility. Genome-wide association studies have dramatically impacted research on genetic determinants of cancer risk, but certainly more of this research needs to be done for populations in the developing world. As cancer health disparities research creates more understanding of the interrelationships between environmental exposures and genetic susceptibility, an extensive epidemiological assessment of various population groups worldwide is required to find out how our current understanding of environmental and genetic risk factors and their interactions in cancer development applies to all of these groups, or whether significant differences exist. Accomplishing these complex studies to investigate the contributing factors towards both domestic and global cancer health disparities will not be easily done. It is therefore vital that effective, international collaborations be established.

Key Facts

- Large global disparities in cancer incidence, prevalence, and clinical outcome are evident for almost all cancer sites.
- A complex interaction between environmental and genetic risk factors likely exist that result in variations of cancer susceptibility amongst diverse population groups, both domestic and global. Elucidating the interconnections of these factors may help explain cancer health disparities seen on an international scale.
- Numerous GWAS studies in recent years have identified cancer risk loci that may affect some population groups differently than others. As more and more of these cancer susceptibility loci are identified in the general population, further exploration is needed to identify the underlying biologically relevant variants to improve cancer risk modeling in populations with diverse ancestry.
- Despite some evidence supporting differences in tumor biology and pharmacogenomics to explain disparities in response to cancer therapy amongst different populations, it is likely that inadequate or delayed access to health care is the most important factor.
- Modifiable risk factors, including dietary patterns and obesity can have a substantial impact on prostate cancer susceptibility. Studies have shown significant differences in diet, obesity, and vitamin D deficiency amongst the African-American population, which may impact prostate cancer risk. While specific mechanisms remain to be fully explained, studies support that these risk factors may influence gene signatures and effect prostate tumor biology.
- To date, the best described risk locus for prostate cancer is located at 8q24 and confers an increase disease risk in many populations.
- Despite having lower risk of being diagnosed with breast cancer, African-American woman are at an increased risk of dying from the disease when compared with woman from other U.S. population groups. It is hypothesized that an increase in early onset, ER-negative and basal-like/triple negative tumors in African-American woman is largely responsible for the excess mortality.
- Both epidemiological and clinical studies have supported the hypothesis that tumor biological and immunological factors may contribute to cancer health disparities. Specifically, recent studies in both prostate and breast cancer have identified a distinct interferon gene signature that is more prominently expressed in the tumor microenvironment of African American patients and may influence disease aggressiveness (Figure 1).

References

[1] Parkin, D. M., Bray, F., Ferlay, J. & Pisani, P. Global cancer statistics, 2002. *CA: a cancer journal for clinicians* 55, 74-108 (2005).

[2] Kamangar, F., Dores, G. M. & Anderson, W. F. Patterns of cancer incidence, mortality, and prevalence across five continents: defining priorities to reduce cancer disparities in different geographic regions of the world. *Journal of clinical oncology : official journal of the American Society of Clinical Oncology* 24, 2137-2150, doi:10.1200/JCO.2005.05.2308 (2006).

[3] Jemal, A., Center, M. M., DeSantis, C. & Ward, E. M. Global patterns of cancer incidence and mortality rates and trends. *Cancer epidemiology, biomarkers & prevention : a publication of the American Association for Cancer Research, cosponsored by the American Society of Preventive Oncology* 19, 1893-1907, doi:10.1158/1055-9965.EPI-10-0437 (2010).

[4] Thun, M. J., DeLancey, J. O., Center, M. M., Jemal, A. & Ward, E. M. The global burden of cancer: priorities for prevention. *Carcinogenesis* 31, 100-110, doi:10.1093/carcin/bgp263 (2010).

[5] Siegel, R., Ma, J., Zou, Z. & Jemal, A. Cancer statistics, 2014. *CA: a cancer journal for clinicians* 64, 9-29, doi:10.3322/caac.21208 (2014).

[6] Haenszel, W. & Kurihara, M. Studies of Japanese migrants. I. Mortality from cancer and other diseases among Japanese in the United States. *Journal of the National Cancer Institute* 40, 43-68 (1968).

[7] Buell, P. Changing incidence of breast cancer in Japanese-American women. *Journal of the National Cancer Institute* 51, 1479-1483 (1973).

[8] Thomas, D. B. & Karagas, M. R. Cancer in first and second generation Americans. *Cancer research* 47, 5771-5776 (1987).

[9] Ziegler, R. G. et al. Migration patterns and breast cancer risk in Asian-American women. *Journal of the National Cancer Institute* 85, 1819-1827 (1993).

[10] Ioannidis, J. P., Ntzani, E. E. & Trikalinos, T. A. 'Racial' differences in genetic effects for complex diseases. *Nature genetics* 36, 1312-1318, doi:10.1038/ng1474 (2004).

[11] Lichtenstein, P. et al. Environmental and heritable factors in the causation of cancer--analyses of cohorts of twins from Sweden, Denmark, and Finland. *The New England journal of medicine* 343, 78-85, doi:10.1056/NEJM200007133430201 (2000).

[12] Amundadottir, L. T. et al. A common variant associated with prostate cancer in European and African populations. *Nature genetics* 38, 652-658, doi:10.1038/ng1808 (2006).

[13] Hunter, D. J. et al. A genome-wide association study identifies alleles in FGFR2 associated with risk of sporadic postmenopausal breast cancer. *Nature genetics* 39, 870-874, doi:10.1038/ng2075 (2007).

[14] Zheng, S. L. et al. Cumulative association of five genetic variants with prostate cancer. *The New England journal of medicine* 358, 910-919, doi:10.1056/NEJMoa075819 (2008).

[15] Wang, Y. et al. Common 5p15.33 and 6p21.33 variants influence lung cancer risk. *Nature genetics* 40, 1407-1409, doi:10.1038/ng.273 (2008).

[16] Wokolorczyk, D. et al. A range of cancers is associated with the rs6983267 marker on chromosome 8. *Cancer research* 68, 9982-9986, doi:10.1158/0008-5472.CAN-08-1838 (2008).

[17] Takata, R. et al. Genome-wide association study identifies five new susceptibility loci for prostate cancer in the Japanese population. *Nature genetics* 42, 751-754, doi:10.1038/ng.635 (2010).

[18] Wang, H. et al. Trans-ethnic genome-wide association study of colorectal cancer identifies a new susceptibility locus in VTI1A. *Nature communications* 5, 4613, doi:10.1038/ncomms5613 (2014).

[19] Haiman, C. A. et al. Multiple regions within 8q24 independently affect risk for prostate cancer. *Nature genetics* 39, 638-644, doi:10.1038/ng2015 (2007).

[20] Chang, B. L. et al. Validation of genome-wide prostate cancer associations in men of African descent. *Cancer epidemiology, biomarkers & prevention : a publication of the American Association for Cancer Research, cosponsored by the American Society of Preventive Oncology* 20, 23-32, doi:10.1158/1055-9965.EPI-10-0698 (2011).

[21] Han, Y. et al. Generalizability of established prostate cancer risk variants in men of African ancestry. *International journal of cancer. Journal international du cancer*, doi:10.1002/ijc.29066 (2014).

[22] Wu, C. et al. Joint analysis of three genome-wide association studies of esophageal squamous cell carcinoma in Chinese populations. *Nature genetics* 46, 1001-1006, doi:10.1038/ng.3064 (2014).

[23] Freedman, M. L. et al. Admixture mapping identifies 8q24 as a prostate cancer risk locus in African-American men. *Proceedings of the National Academy of Sciences of the United States of America* 103, 14068-14073, doi:10.1073/pnas.0605832103 (2006).

[24] Koutros, S. et al. Pesticide use modifies the association between genetic variants on chromosome 8q24 and prostate cancer. *Cancer research* 70, 9224-9233, doi:10.1158/0008-5472.CAN-10-1078 (2010).

[25] Neslund-Dudas, C. et al. Gene-environment interactions between JAZF1 and occupational and household lead exposure in prostate cancer among African American men. *Cancer causes & control: CCC* 25, 869-879, doi:10.1007/s10552-014-0387-1 (2014).

[26] Fejerman, L. et al. Genetic ancestry and risk of breast cancer among U.S. Latinas. *Cancer research* 68, 9723-9728, doi:10.1158/0008-5472.CAN-08-2039 (2008).

[27] Fejerman, L. et al. European ancestry is positively associated with breast cancer risk in Mexican women. *Cancer epidemiology, biomarkers & prevention : a publication of the American Association for Cancer Research, cosponsored by the American Society of Preventive Oncology* 19, 1074-1082, doi:10.1158/1055-9965.EPI-09-1193 (2010).

[28] Prokunina-Olsson, L. et al. A variant upstream of IFNL3 (IL28B) creating a new interferon gene IFNL4 is associated with impaired clearance of hepatitis C virus. *Nature genetics* 45, 164-171, doi:10.1038/ng.2521 (2013).

[29] Spielman, R. S. et al. Common genetic variants account for differences in gene expression among ethnic groups. *Nature genetics* 39, 226-231, doi:10.1038/ng1955 (2007).

[30] Storey, J. D. et al. Gene-expression variation within and among human populations. *American journal of human genetics* 80, 502-509, doi:10.1086/512017 (2007).

[31] Zhang, W. et al. Evaluation of genetic variation contributing to differences in gene expression between populations. *American journal of human genetics* 82, 631-640, doi:10.1016/j.ajhg.2007.12.015 (2008).

[32] Martin, D. N. et al. Differences in the tumor microenvironment between African-American and European-American breast cancer patients. *PloS one* 4, e4531, doi:10.1371/journal.pone.0004531 (2009).

[33] Van Dyke, A. L., Cote, M. L., Wenzlaff, A. S., Land, S. & Schwartz, A. G. Cytokine SNPs: Comparison of allele frequencies by race and implications for future studies. *Cytokine* 46, 236-244, doi:10.1016/j.cyto.2009.02.003 (2009).

[34] Kinseth, M. A. et al. Expression differences between African American and Caucasian prostate cancer tissue reveals that stroma is the site of aggressive changes. *International journal of cancer. Journal international du cancer* 134, 81-91, doi:10.1002/ijc.28326 (2014).

[35] Quan, L. et al. Cytokine and cytokine receptor genes of the adaptive immune response are differentially associated with breast cancer risk in American women of African and European ancestry. *International journal of cancer. Journal international du cancer* 134, 1408-1421, doi:10.1002/ijc.28458 (2014).

[36] Li, M. et al. Helicobacter pylori infection synergizes with three inflammation-related genetic variants in the GWASs to increase risk of gastric cancer in a Chinese population. *PloS one* 8, e74976, doi:10.1371/journal.pone.0074976 (2013).

[37] Rogers, E. N. et al. Toll-like receptor-associated sequence variants and prostate cancer risk among men of African descent. *Genes and immunity* 14, 347-355, doi:10.1038/gene.2013.22 (2013).

[38] Keller, M. F. et al. Trans-ethnic meta-analysis of white blood cell phenotypes. *Human molecular genetics*, doi:10.1093/hmg/ddu401 (2014).

[39] Coussens, L. M. & Werb, Z. Inflammation and cancer. *Nature* 420, 860-867, doi:10.1038/nature01322 (2002).

[40] Hussain, S. P., Hofseth, L. J. & Harris, C. C. Radical causes of cancer. *Nature reviews. Cancer* 3, 276-285, doi:10.1038/nrc1046 (2003).

[41] Hancock, A. M. et al. Colloquium paper: human adaptations to diet, subsistence, and ecoregion are due to subtle shifts in allele frequency. *Proceedings of the National Academy of Sciences of the United States of America* 107 Suppl 2, 8924-8930, doi:10.1073/pnas.0914625107 (2010).

[42] Genovese, G. et al. Association of trypanolytic ApoL1 variants with kidney disease in African Americans. *Science* 329, 841-845, doi:10.1126/science.1193032 (2010).

[43] Hobbs, M. R. et al. A new NOS2 promoter polymorphism associated with increased nitric oxide production and protection from severe malaria in Tanzanian and Kenyan children. *Lancet* 360, 1468-1475, doi:10.1016/S0140-6736(02)11474-7 (2002).

[44] Cramer, J. P. et al. iNOS promoter variants and severe malaria in Ghanaian children. *Tropical medicine & international health : TM & IH* 9, 1074-1080, doi:10.1111/j.1365-3156.2004.01312.x (2004).

[45] Tournamille, C., Colin, Y., Cartron, J. P. & Le Van Kim, C. Disruption of a GATA motif in the Duffy gene promoter abolishes erythroid gene expression in Duffy-negative individuals. *Nature genetics* 10, 224-228, doi:10.1038/ng0695-224 (1995).

[46] Wink, D. A. et al. The multifaceted roles of nitric oxide in cancer. *Carcinogenesis* 19, 711-721 (1998).

[47] Glynn, S. A. et al. Increased NOS2 predicts poor survival in estrogen receptor-negative breast cancer patients. *The Journal of clinical investigation* 120, 3843-3854, doi:10.1172/JCI42059 (2010).

[48] Bandyopadhyay, S. et al. Interaction of KAI1 on tumor cells with DARC on vascular endothelium leads to metastasis suppression. *Nature medicine* 12, 933-938, doi:10.1038/nm1444 (2006).

[49] Hadley, T. J. & Peiper, S. C. From malaria to chemokine receptor: the emerging physiologic role of the Duffy blood group antigen. *Blood* 89, 3077-3091 (1997).

[50] Murthy, V. H., Krumholz, H. M. & Gross, C. P. Participation in cancer clinical trials: race-, sex-, and age-based disparities. *JAMA : the journal of the American Medical Association* 291, 2720-2726, doi:10.1001/jama.291.22.2720 (2004).

[51] Bach, P. B. et al. Survival of blacks and whites after a cancer diagnosis. *JAMA : the journal of the American Medical Association* 287, 2106-2113 (2002).

[52] Shavers, V. L. & Brown, M. L. Racial and ethnic disparities in the receipt of cancer treatment. *Journal of the National Cancer Institute* 94, 334-357 (2002).

[53] Evans, S., Metcalfe, C., Ibrahim, F., Persad, R. & Ben-Shlomo, Y. Investigating Black-White differences in prostate cancer prognosis: A systematic review and meta-analysis. *International journal of cancer. Journal international du cancer* 123, 430-435, doi:10.1002/ijc.23500 (2008).

[54] Dignam, J. J. Efficacy of systemic adjuvant therapy for breast cancer in African-American and Caucasian women. *Journal of the National Cancer Institute. Monographs*, 36-43 (2001).

[55] Dignam, J. J. et al. Outcomes among African-Americans and Caucasians in colon cancer adjuvant therapy trials: findings from the National Surgical Adjuvant Breast and Bowel Project. *Journal of the National Cancer Institute* 91, 1933-1940 (1999).

[56] Freedland, S. J., Jalkut, M., Dorey, F., Sutter, M. E. & Aronson, W. J. Race is not an independent predictor of biochemical recurrence after radical prostatectomy in an equal access medical center. *Urology* 56, 87-91 (2000).

[57] Roach, M., 3rd et al. Racial differences in CYP3A4 genotype and survival among men treated on Radiation Therapy Oncology Group (RTOG) 9202: a phase III randomized trial. *International journal of radiation oncology, biology, physics* 69, 79-87, doi:10.1016/j.ijrobp.2007.03.008 (2007).

[58] Roach, M., 3rd et al. Race and survival of men treated for prostate cancer on radiation therapy oncology group phase III randomized trials. *The Journal of urology* 169, 245-250, doi:10.1097/01.ju.0000041412.57484.cd (2003).

[59] Dawood, S. et al. Triple receptor-negative breast cancer: the effect of race on response to primary systemic treatment and survival outcomes. *Journal of clinical oncology : official journal of the American Society of Clinical Oncology* 27, 220-226, doi:10.1200/JCO.2008.17.9952 (2009).

[60] Chu, Q. D. et al. Race/Ethnicity has no effect on outcome for breast cancer patients treated at an academic center with a public hospital. *Cancer epidemiology, biomarkers & prevention: a publication of the American Association for Cancer Research, cosponsored by the American Society of Preventive Oncology* 18, 2157-2161, doi:10.1158/1055-9965.EPI-09-0232 (2009).

[61] Elledge, R. M., Clark, G. M., Chamness, G. C. & Osborne, C. K. Tumor biologic factors and breast cancer prognosis among white, Hispanic, and black women in the United States. *Journal of the National Cancer Institute* 86, 705-712 (1994).

[62] Moul, J. W., Douglas, T. H., McCarthy, W. F. & McLeod, D. G. Black race is an adverse prognostic factor for prostate cancer recurrence following radical prostatectomy in an equal access health care setting. *The Journal of urology* 155, 1667-1673 (1996).

[63] Powell, I. J. et al. Prostate cancer biochemical recurrence stage for stage is more frequent among African-American than white men with locally advanced but not organ-confined disease. *Urology* 55, 246-251 (2000).

[64] Thompson, I. et al. Association of African-American ethnic background with survival in men with metastatic prostate cancer. *Journal of the National Cancer Institute* 93, 219-225 (2001).

[65] Thatai, L. C., Banerjee, M., Lai, Z. & Vaishampayan, U. Racial disparity in clinical course and outcome of metastatic androgen-independent prostate cancer. *Urology* 64, 738-743, doi:10.1016/j.urology.2004.05.024 (2004).

[66] Hoffman, R. M. et al. Racial and ethnic differences in advanced-stage prostate cancer: the Prostate Cancer Outcomes Study. *Journal of the National Cancer Institute* 93, 388-395 (2001).

[67] Field, T. S. et al. Disparities and survival among breast cancer patients. *Journal of the National Cancer Institute. Monographs*, 88-95, doi:10.1093/jncimonographs/lgi044 (2005).

[68] Woodward, W. A. et al. African-American race is associated with a poorer overall survival rate for breast cancer patients treated with mastectomy and doxorubicin-based chemotherapy. *Cancer* 107, 2662-2668, doi:10.1002/cncr.22281 (2006).

[69] Newman, L. A. et al. Meta-analysis of survival in African American and white American patients with breast cancer: ethnicity compared with socioeconomic status. *Journal of clinical oncology : official journal of the American Society of Clinical Oncology* 24, 1342-1349, doi:10.1200/JCO.2005.03.3472 (2006).

[70] Alexander, D. D. et al. African-American and Caucasian disparities in colorectal cancer mortality and survival by data source: an epidemiologic review. *Cancer biomarkers : section A of Disease markers* 3, 301-313 (2007).

[71] Polite, B. N. et al. Racial differences in clinical outcomes from metastatic breast cancer: a pooled analysis of CALGB 9342 and 9840--Cancer and Leukemia Group B. *Journal of clinical oncology : official journal of the American Society of Clinical Oncology* 26, 2659-2665, doi:10.1200/JCO.2007.13.9782 (2008).

[72] Albain, K. S., Unger, J. M., Crowley, J. J., Coltman, C. A., Jr. & Hershman, D. L. Racial disparities in cancer survival among randomized clinical trials patients of the Southwest Oncology Group. *Journal of the National Cancer Institute* 101, 984-992, doi:10.1093/jnci/djp175 (2009).

[73] Balmanoukian, A. et al. African American women who receive primary anthracycline- and taxane-based chemotherapy for triple-negative breast cancer suffer worse outcomes compared with white women. *Journal of clinical oncology : official journal of the American Society of Clinical Oncology* 27, e35-37; author reply e38-39, doi:10.1200/JCO.2008.21.5509 (2009).

[74] Boone, S. D. et al. Associations between CYP19A1 polymorphisms, Native American ancestry, and breast cancer risk and mortality: the Breast Cancer Health Disparities Study. *Cancer causes & control : CCC*, doi:10.1007/s10552-014-0448-5 (2014).

[75] Saif, M. W. et al. A DPYD variant (Y186C) specific to individuals of African descent in a patient with life-threatening 5-FU toxic effects: potential for an individualized medicine approach. *Mayo Clinic proceedings* 89, 131-136, doi:10.1016/j.mayocp.2013.09.008 (2014).

[76] Hamdy, S. I. et al. Allele and genotype frequencies of polymorphic cytochromes P450 (CYP2C9, CYP2C19, CYP2E1) and dihydropyrimidine dehydrogenase (DPYD) in the Egyptian population. *British journal of clinical pharmacology* 53, 596-603 (2002).

[77] Offer, S. M. et al. A DPYD variant (Y186C) in individuals of african ancestry is associated with reduced DPD enzyme activity. *Clinical pharmacology and therapeutics* 94, 158-166, doi:10.1038/clpt.2013.69 (2013).

[78] Xu, H. et al. ARID5B genetic polymorphisms contribute to racial disparities in the incidence and treatment outcome of childhood acute lymphoblastic leukemia. *Journal of clinical oncology : official journal of the American Society of Clinical Oncology* 30, 751-757, doi:10.1200/JCO.2011.38.0345 (2012).

[79] Krischer, J. P. et al. Clinical cardiotoxicity following anthracycline treatment for childhood cancer: the Pediatric Oncology Group experience. *Journal of clinical oncology : official journal of the American Society of Clinical Oncology* 15, 1544-1552 (1997).

[80] Hasan, S., Dinh, K., Lombardo, F. & Kark, J. Doxorubicin cardiotoxicity in African Americans. *Journal of the National Medical Association* 96, 196-199 (2004).

[81] Hershman, D. et al. Racial disparities in treatment and survival among women with early-stage breast cancer. *Journal of clinical oncology : official journal of the American Society of Clinical Oncology* 23, 6639-6646, doi:10.1200/JCO.2005.12.633 (2005).

[82] Fagerholm, R. et al. NAD(P)H:quinone oxidoreductase 1 NQO1*2 genotype (P187S) is a strong prognostic and predictive factor in breast cancer. *Nature genetics* 40, 844-853, doi:10.1038/ng.155 (2008).

[83] Glynn, S. A. et al. A mitochondrial target sequence polymorphism in manganese superoxide dismutase predicts inferior survival in breast cancer patients treated with cyclophosphamide. *Clinical cancer research : an official journal of the American Association for Cancer Research* 15, 4165-4173, doi:10.1158/1078-0432.CCR-09-0119 (2009).

[84] O'Donnell, P. H. & Dolan, M. E. Cancer pharmacoethnicity: ethnic differences in susceptibility to the effects of chemotherapy. *Clinical cancer research : an official journal of the American Association for Cancer Research* 15, 4806-4814, doi:10.1158/1078-0432.CCR-09-0344 (2009).

[85] Antoni, M. H. et al. The influence of bio-behavioural factors on tumour biology: pathways and mechanisms. *Nature reviews. Cancer* 6, 240-248, doi:10.1038/nrc1820 (2006).

[86] Duijts, S. F., Zeegers, M. P. & Borne, B. V. The association between stressful life events and breast cancer risk: a meta-analysis. *International journal of cancer. Journal international du cancer* 107, 1023-1029, doi:10.1002/ijc.11504 (2003).

[87] Glaser, R. & Kiecolt-Glaser, J. K. Stress-induced immune dysfunction: implications for health. *Nature reviews. Immunology* 5, 243-251, doi:10.1038/nri1571 (2005).

[88] Steptoe, A., Hamer, M. & Chida, Y. The effects of acute psychological stress on circulating inflammatory factors in humans: a review and meta-analysis. *Brain, behavior, and immunity* 21, 901-912, doi:10.1016/j.bbi.2007.03.011 (2007).

[89] Kiecolt-Glaser, J. K. et al. Chronic stress and age-related increases in the proinflammatory cytokine IL-6. *Proceedings of the National Academy of Sciences of the United States of America* 100, 9090-9095, doi:10.1073/pnas.1531903100 (2003).

[90] Ara, T. & Declerck, Y. A. Interleukin-6 in bone metastasis and cancer progression. *European journal of cancer* 46, 1223-1231, doi:10.1016/j.ejca.2010.02.026 (2010).

[91] Brawley, O. W. Prostate cancer and black men. *Seminars in urologic oncology* 16, 184-186 (1998).

[92] Jones, B. A. et al. Explaining the race difference in prostate cancer stage at diagnosis. *Cancer epidemiology, biomarkers & prevention : a publication of the American Association for Cancer Research, cosponsored by the American Society of Preventive Oncology* 17, 2825-2834, doi:10.1158/1055-9965.EPI-08-0203 (2008).

[93] Rapiti, E. et al. Impact of socioeconomic status on prostate cancer diagnosis, treatment, and prognosis. *Cancer* 115, 5556-5565, doi:10.1002/cncr.24607 (2009).

[94] Fedewa, S. A., Etzioni, R., Flanders, W. D., Jemal, A. & Ward, E. M. Association of insurance and race/ethnicity with disease severity among men diagnosed with prostate cancer, National Cancer Database 2004-2006. *Cancer epidemiology, biomarkers & prevention : a publication of the American Association for Cancer Research, cosponsored by the American Society of Preventive Oncology* 19, 2437-2444, doi:10.1158/1055-9965.EPI-10-0299 (2010).

[95] Tiguert, R. et al. Racial differences and prognostic significance of tumor location in radical prostatectomy specimens. *The Prostate* 37, 230-235 (1998).

[96] Tewari, A. et al. Racial differences in serum prostate-specific antigen (PSA) doubling time, histopathological variables and long-term PSA recurrence between African-American and white American men undergoing radical prostatectomy for clinically localized prostate cancer. *BJU international* 96, 29-33, doi:10.1111/j.1464-410X.2005.05561.x (2005).

[97] Powell, I. J., Bock, C. H., Ruterbusch, J. J. & Sakr, W. Evidence supports a faster growth rate and/or earlier transformation to clinically significant prostate cancer in black than in white American men, and influences racial progression and mortality disparity. *The Journal of urology* 183, 1792-1796, doi:10.1016/j.juro.2010.01.015 (2010).

[98] Shibata, A. & Whittemore, A. S. Genetic predisposition to prostate cancer: possible explanations for ethnic differences in risk. *The Prostate* 32, 65-72 (1997).

[99] Fowler, J. E., Jr. & Terrell, F. Survival in blacks and whites after treatment for localized prostate cancer. *The Journal of urology* 156, 133-136 (1996).

[100] Hsing, A. W., Tsao, L. & Devesa, S. S. International trends and patterns of prostate cancer incidence and mortality. *International journal of cancer. Journal international du cancer* 85, 60-67 (2000).

[101] Block, G. & Subar, A. F. Estimates of nutrient intake from a food frequency questionnaire: the 1987 National Health Interview Survey. *Journal of the American Dietetic Association* 92, 969-977 (1992).

[102] Swanson, C. A. et al. A comparison of diets of blacks and whites in three areas of the United States. *Nutrition and cancer* 20, 153-165, doi:10.1080/01635589309514282 (1993).

[103] Kant, A. K., Graubard, B. I. & Kumanyika, S. K. Trends in black-white differentials in dietary intakes of U.S. adults, 1971-2002. *American journal of preventive medicine* 32, 264-272, doi:10.1016/j.amepre.2006.12.011 (2007).

[104] Hayes, R. B. et al. Dietary factors and risks for prostate cancer among blacks and whites in the United States. *Cancer epidemiology, biomarkers & prevention : a publication of the American Association for Cancer Research, cosponsored by the American Society of Preventive Oncology* 8, 25-34 (1999).

[105] Rodriguez, C. et al. Meat consumption among Black and White men and risk of prostate cancer in the Cancer Prevention Study II Nutrition Cohort. *Cancer epidemiology, biomarkers & prevention : a publication of the American Association for Cancer Research, cosponsored by the American Society of Preventive Oncology* 15, 211-216, doi:10.1158/1055-9965.EPI-05-0614 (2006).

[106] Giovannucci, E. The epidemiology of vitamin D and cancer incidence and mortality: a review (United States). *Cancer causes & control : CCC* 16, 83-95, doi:10.1007/s10552-004-1661-4 (2005).

[107] Schenk, J. M. et al. Serum 25-hydroxyvitamin d concentrations and risk of prostate cancer: results from the prostate cancer prevention trial. *Cancer epidemiology, biomarkers & prevention : a publication of the American Association for Cancer Research, cosponsored by the American Society of Preventive Oncology* 23, 1484-1493, doi:10.1158/1055-9965.EPI-13-1340 (2014).

[108] Kristal, A. R. et al. Plasma vitamin d and prostate cancer risk: results from the selenium and vitamin e cancer prevention trial. *Cancer epidemiology, biomarkers & prevention : a publication of the American Association for Cancer Research, cosponsored by the American Society of Preventive Oncology* 23, 1494-1504, doi:10.1158/1055-9965.EPI-14-0115 (2014).

[109] Calle, E. E. & Thun, M. J. Obesity and cancer. *Oncogene* 23, 6365-6378, doi:10.1038/sj.onc.1207751 (2004).

[110] Spangler, E. et al. Association of obesity with tumor characteristics and treatment failure of prostate cancer in African-American and European American men. *The Journal of urology* 178, 1939-1944; discussion 1945, doi:10.1016/j.juro.2007.07.021 (2007).

[111] Cao, Y. & Ma, J. Body mass index, prostate cancer-specific mortality, and biochemical recurrence: a systematic review and meta-analysis. *Cancer prevention research* 4, 486-501, doi:10.1158/1940-6207.CAPR-10-0229 (2011).

[112] Nemesure, B., Wu, S. Y., Hennis, A., Leske, M. C. & Prostate Cancer in a Black Population Study, G. Central adiposity and Prostate Cancer in a Black Population. *Cancer epidemiology, biomarkers & prevention : a publication of the American Association for Cancer Research, cosponsored by the American Society of Preventive Oncology* 21, 851-858, doi:10.1158/1055-9965.EPI-12-0071 (2012).

[113] Su, L. J. et al. Obesity and prostate cancer aggressiveness among African and Caucasian Americans in a population-based study. *Cancer epidemiology, biomarkers & prevention : a publication of the American Association for Cancer Research,*

cosponsored by the American Society of Preventive Oncology 20, 844-853, doi:10.1158/1055-9965.EPI-10-0684 (2011).

[114] Flegal, K. M., Carroll, M. D., Ogden, C. L. & Curtin, L. R. Prevalence and trends in obesity among US adults, 1999-2008. *JAMA : the journal of the American Medical Association* 303, 235-241, doi:10.1001/jama.2009.2014 (2010).

[115] Ornish, D. et al. Changes in prostate gene expression in men undergoing an intensive nutrition and lifestyle intervention. *Proceedings of the National Academy of Sciences of the United States of America* 105, 8369-8374, doi:10.1073/pnas.0803080105 (2008).

[116] Sharad, S. et al. Prostate cancer gene expression signature of patients with high body mass index. *Prostate cancer and prostatic diseases* 14, 22-29, doi:10.1038/pcan.2010.44 (2011).

[117] Chornokur, G. et al. Variation in HNF1B and Obesity May Influence Prostate Cancer Risk in African American Men: A Pilot Study. *Prostate cancer* 2013, 384594, doi:10.1155/2013/384594 (2013).

[118] Gaston, K. E., Kim, D., Singh, S., Ford, O. H., 3rd & Mohler, J. L. Racial differences in androgen receptor protein expression in men with clinically localized prostate cancer. *The Journal of urology* 170, 990-993, doi:10.1097/01.ju.0000079761.56154.e5 (2003).

[119] Ross, R. et al. Serum testosterone levels in healthy young black and white men. *Journal of the National Cancer Institute* 76, 45-48 (1986).

[120] Ellis, L. & Nyborg, H. Racial/ethnic variations in male testosterone levels: a probable contributor to group differences in health. *Steroids* 57, 72-75 (1992).

[121] Calistro Alvarado, L. Population differences in the testosterone levels of young men are associated with prostate cancer disparities in older men. *American journal of human biology : the official journal of the Human Biology Council* 22, 449-455, doi:10.1002/ajhb.21016 (2010).

[122] Endogenous, H. et al. Endogenous sex hormones and prostate cancer: a collaborative analysis of 18 prospective studies. *Journal of the National Cancer Institute* 100, 170-183, doi:10.1093/jnci/djm323 (2008).

[123] Mohler, J. L. et al. Racial differences in prostate androgen levels in men with clinically localized prostate cancer. *The Journal of urology* 171, 2277-2280 (2004).

[124] Marks, L. S., Hess, D. L., Dorey, F. J. & Macairan, M. L. Prostatic tissue testosterone and dihydrotestosterone in African-American and white men. *Urology* 68, 337-341, doi:10.1016/j.urology.2006.03.013 (2006).

[125] Mavropoulos, J. C. et al. Do racial differences in prostate size explain higher serum prostate-specific antigen concentrations among black men? *Urology* 69, 1138-1142, doi:10.1016/j.urology.2007.01.102 (2007).

[126] Rohrmann, S. et al. Serum estrogen, but not testosterone, levels differ between black and white men in a nationally representative sample of Americans. *The Journal of clinical endocrinology and metabolism* 92, 2519-2525, doi:10.1210/jc.2007-0028 (2007).

[127] Orwoll, E. S. et al. Evidence for geographical and racial variation in serum sex steroid levels in older men. *The Journal of clinical endocrinology and metabolism* 95, E151-160, doi:10.1210/jc.2009-2435 (2010).

[128] Tam, N. N., Leav, I. & Ho, S. M. Sex hormones induce direct epithelial and inflammation-mediated oxidative/nitrosative stress that favors prostatic carcinogenesis

in the noble rat. *The American journal of pathology* 171, 1334-1341, doi:10.2353/ajpath.2007.070199 (2007).
[129] Eastham, J. A. et al. Clinical characteristics and biopsy specimen features in African-American and white men without prostate cancer. *Journal of the National Cancer Institute* 90, 756-760 (1998).
[130] Dennis, L. K., Lynch, C. F. & Torner, J. C. Epidemiologic association between prostatitis and prostate cancer. *Urology* 60, 78-83 (2002).
[131] De Marzo, A. M. et al. Inflammation in prostate carcinogenesis. *Nature reviews. Cancer* 7, 256-269, doi:10.1038/nrc2090 (2007).
[132] Penney, K. L. et al. Genome-wide association study of prostate cancer mortality. *Cancer epidemiology, biomarkers & prevention : a publication of the American Association for Cancer Research, cosponsored by the American Society of Preventive Oncology* 19, 2869-2876, doi:10.1158/1055-9965.EPI-10-0601 (2010).
[133] Hooker, S. et al. Replication of prostate cancer risk loci on 8q24, 11q13, 17q12, 19q33, and Xp11 in African Americans. *The Prostate* 70, 270-275, doi:10.1002/pros.21061 (2010).
[134] Zheng, S. L. et al. A comprehensive association study for genes in inflammation pathway provides support for their roles in prostate cancer risk in the CAPS study. *The Prostate* 66, 1556-1564, doi:10.1002/pros.20496 (2006).
[135] Sun, J., Turner, A., Xu, J., Gronberg, H. & Isaacs, W. Genetic variability in inflammation pathways and prostate cancer risk. *Urologic oncology* 25, 250-259, doi:10.1016/j.urolonc.2006.10.001 (2007).
[136] Fernandez, P., de Beer, P. M., van der Merwe, L. & Heyns, C. F. COX-2 promoter polymorphisms and the association with prostate cancer risk in South African men. *Carcinogenesis* 29, 2347-2350, doi:10.1093/carcin/bgn245 (2008).
[137] Zabaleta, J. et al. Interactions of cytokine gene polymorphisms in prostate cancer risk. *Carcinogenesis* 29, 573-578, doi:10.1093/carcin/bgm277 (2008).
[138] Panguluri, R. C. et al. COX-2 gene promoter haplotypes and prostate cancer risk. *Carcinogenesis* 25, 961-966, doi:10.1093/carcin/bgh100 (2004).
[139] Shook, S. J. et al. Association of RNASEL variants with prostate cancer risk in Hispanic Caucasians and African Americans. *Clinical cancer research : an official journal of the American Association for Cancer Research* 13, 5959-5964, doi:10.1158/1078-0432.CCR-07-0702 (2007).
[140] Rennert, H. et al. Association of susceptibility alleles in ELAC2/HPC2, RNASEL/HPC1, and MSR1 with prostate cancer severity in European American and African American men. *Cancer epidemiology, biomarkers & prevention : a publication of the American Association for Cancer Research, cosponsored by the American Society of Preventive Oncology* 14, 949-957, doi:10.1158/1055-9965.EPI-04-0637 (2005).
[141] Robbins, C. M. et al. Association of HPC2/ELAC2 and RNASEL non-synonymous variants with prostate cancer risk in African American familial and sporadic cases. *The Prostate* 68, 1790-1797, doi:10.1002/pros.20841 (2008).
[142] Powell, I. J. Epidemiology and pathophysiology of prostate cancer in African-American men. *The Journal of urology* 177, 444-449, doi:10.1016/j.juro.2006.09.024 (2007).

[143] Guo, Y., Sigman, D. B., Borkowski, A. & Kyprianou, N. Racial differences in prostate cancer growth: apoptosis and cell proliferation in Caucasian and African-American patients. *The Prostate* 42, 130-136 (2000).

[144] Mimeault, M. & Batra, S. K. Recent advances on multiple tumorigenic cascades involved in prostatic cancer progression and targeting therapies. *Carcinogenesis* 27, 1-22, doi:10.1093/carcin/bgi229 (2006).

[145] Shuch, B. et al. Racial disparity of epidermal growth factor receptor expression in prostate cancer. *Journal of clinical oncology : official journal of the American Society of Clinical Oncology* 22, 4725-4729, doi:10.1200/JCO.2004.06.134 (2004).

[146] Nelson, W. G., De Marzo, A. M. & Yegnasubramanian, S. Epigenetic alterations in human prostate cancers. *Endocrinology* 150, 3991-4002, doi:10.1210/en.2009-0573 (2009).

[147] Woodson, K., Hayes, R., Wideroff, L., Villaruz, L. & Tangrea, J. Hypermethylation of GSTP1, CD44, and E-cadherin genes in prostate cancer among US Blacks and Whites. *The Prostate* 55, 199-205, doi:10.1002/pros.10236 (2003).

[148] Enokida, H. et al. Ethnic group-related differences in CpG hypermethylation of the GSTP1 gene promoter among African-American, Caucasian and Asian patients with prostate cancer. *International journal of cancer. Journal international du cancer* 116, 174-181, doi:10.1002/ijc.21017 (2005).

[149] Kwabi-Addo, B. et al. Identification of differentially methylated genes in normal prostate tissues from African American and Caucasian men. *Clinical cancer research : an official journal of the American Association for Cancer Research* 16, 3539-3547, doi:10.1158/1078-0432.CCR-09-3342 (2010).

[150] Cher, M. L. et al. A similar pattern of chromosomal alterations in prostate cancers from African-Americans and Caucasian Americans. *Clinical cancer research : an official journal of the American Association for Cancer Research* 4, 1273-1278 (1998).

[151] Rose, A. E. et al. Copy number and gene expression differences between African American and Caucasian American prostate cancer. *Journal of translational medicine* 8, 70, doi:10.1186/1479-5876-8-70 (2010).

[152] Castro, P. et al. Genomic profiling of prostate cancers from African American men. *Neoplasia* 11, 305-312 (2009).

[153] Kumar-Sinha, C., Tomlins, S. A. & Chinnaiyan, A. M. Recurrent gene fusions in prostate cancer. *Nature reviews. Cancer* 8, 497-511, doi:10.1038/nrc2402 (2008).

[154] Mao, X. et al. Distinct genomic alterations in prostate cancers in Chinese and Western populations suggest alternative pathways of prostate carcinogenesis. *Cancer research* 70, 5207-5212, doi:10.1158/0008-5472.CAN-09-4074 (2010).

[155] Magi-Galluzzi, C. et al. TMPRSS2-ERG gene fusion prevalence and class are significantly different in prostate cancer of Caucasian, African-American and Japanese patients. *The Prostate* 71, 489-497, doi:10.1002/pros.21265 (2011).

[156] Rawal, S. et al. Low Frequency of the ERG Oncogene Alterations in Prostate Cancer Patients from India. *Journal of Cancer* 4, 468-472, doi:10.7150/jca.6568 (2013).

[157] Khani, F. et al. Evidence for Molecular Differences in Prostate Cancer between African American and Caucasian Men. *Clinical cancer research : an official journal of the American Association for Cancer Research* 20, 4925-4934, doi:10.1158/1078-0432.CCR-13-2265 (2014).

[158] Srikantan, V. et al. PCGEM1, a prostate-specific gene, is overexpressed in prostate cancer. *Proceedings of the National Academy of Sciences of the United States of America* 97, 12216-12221, doi:10.1073/pnas.97.22.12216 (2000).

[159] Petrovics, G. et al. Elevated expression of PCGEM1, a prostate-specific gene with cell growth-promoting function, is associated with high-risk prostate cancer patients. *Oncogene* 23, 605-611, doi:10.1038/sj.onc.1207069 (2004).

[160] Wallace, T. A. et al. Tumor immunobiological differences in prostate cancer between African-American and European-American men. *Cancer research* 68, 927-936, doi:10.1158/0008-5472.CAN-07-2608 (2008).

[161] Reams, R. R. et al. Microarray comparison of prostate tumor gene expression in African-American and Caucasian American males: a pilot project study. *Infectious agents and cancer* 4 Suppl 1, S3, doi:10.1186/1750-9378-4-S1-S3 (2009).

[162] Timofeeva, O. A. et al. Enhanced expression of SOS1 is detected in prostate cancer epithelial cells from African-American men. *International journal of oncology* 35, 751-760 (2009).

[163] Einav, U. et al. Gene expression analysis reveals a strong signature of an interferon-induced pathway in childhood lymphoblastic leukemia as well as in breast and ovarian cancer. *Oncogene* 24, 6367-6375, doi:10.1038/sj.onc.1208797 (2005).

[164] Buess, M. et al. Characterization of heterotypic interaction effects in vitro to deconvolute global gene expression profiles in cancer. *Genome biology* 8, R191, doi:10.1186/gb-2007-8-9-r191 (2007).

[165] Weichselbaum, R. R. et al. An interferon-related gene signature for DNA damage resistance is a predictive marker for chemotherapy and radiation for breast cancer. *Proceedings of the National Academy of Sciences of the United States of America* 105, 18490-18495, doi:10.1073/pnas.0809242105 (2008).

[166] Khodarev, N. N. et al. STAT1 pathway mediates amplification of metastatic potential and resistance to therapy. *PloS one* 4, e5821, doi:10.1371/journal.pone.0005821 (2009).

[167] Wallace, T. A., Martin, D. N. & Ambs, S. Interactions among genes, tumor biology and the environment in cancer health disparities: examining the evidence on a national and global scale. *Carcinogenesis* 32, 1107-1121, doi:10.1093/carcin/bgr066 (2011).

[168] Chu, K. C. & Anderson, W. F. Rates for breast cancer characteristics by estrogen and progesterone receptor status in the major racial/ethnic groups. *Breast cancer research and treatment* 74, 199-211 (2002).

[169] Ghafoor, A. et al. Trends in breast cancer by race and ethnicity. *CA: a cancer journal for clinicians* 53, 342-355 (2003).

[170] Key, T. J., Verkasalo, P. K. & Banks, E. Epidemiology of breast cancer. *The Lancet. Oncology* 2, 133-140, doi:10.1016/S1470-2045(00)00254-0 (2001).

[171] Michels, K. B., Mohllajee, A. P., Roset-Bahmanyar, E., Beehler, G. P. & Moysich, K. B. Diet and breast cancer: a review of the prospective observational studies. *Cancer* 109, 2712-2749, doi:10.1002/cncr.22654 (2007).

[172] Bernstein, L., Teal, C. R., Joslyn, S. & Wilson, J. Ethnicity-related variation in breast cancer risk factors. *Cancer* 97, 222-229, doi:10.1002/cncr.11014 (2003).

[173] Colditz, G. A., Rosner, B. A., Chen, W. Y., Holmes, M. D. & Hankinson, S. E. Risk factors for breast cancer according to estrogen and progesterone receptor status. *Journal of the National Cancer Institute* 96, 218-228 (2004).

[174] Vona-Davis, L. & Rose, D. P. The influence of socioeconomic disparities on breast cancer tumor biology and prognosis: a review. *Journal of women's health* 18, 883-893, doi:10.1089/jwh.2008.1127 (2009).

[175] Yang, X. R. et al. Associations of breast cancer risk factors with tumor subtypes: a pooled analysis from the Breast Cancer Association Consortium studies. *Journal of the National Cancer Institute* 103, 250-263, doi:10.1093/jnci/djq526 (2011).

[176] Jemal, A., Ward, E. & Thun, M. J. Recent trends in breast cancer incidence rates by age and tumor characteristics among U.S. women. *Breast cancer research : BCR* 9, R28, doi:10.1186/bcr1672 (2007).

[177] Perou, C. M. et al. Molecular portraits of human breast tumours. *Nature* 406, 747-752, doi:10.1038/35021093 (2000).

[178] Sorlie, T. et al. Gene expression patterns of breast carcinomas distinguish tumor subclasses with clinical implications. *Proceedings of the National Academy of Sciences of the United States of America* 98, 10869-10874, doi:10.1073/pnas.191367098 (2001).

[179] Nielsen, T. O. et al. Immunohistochemical and clinical characterization of the basal-like subtype of invasive breast carcinoma. *Clinical cancer research : an official journal of the American Association for Cancer Research* 10, 5367-5374, doi:10.1158/1078-0432.CCR-04-0220 (2004).

[180] Carey, L. A. et al. Race, breast cancer subtypes, and survival in the Carolina Breast Cancer Study. *JAMA : the journal of the American Medical Association* 295, 2492-2502, doi:10.1001/jama.295.21.2492 (2006).

[181] Ihemelandu, C. U. et al. Molecular breast cancer subtypes in premenopausal and postmenopausal African-American women: age-specific prevalence and survival. *The Journal of surgical research* 143, 109-118, doi:10.1016/j.jss.2007.03.085 (2007).

[182] Millikan, R. C. et al. Epidemiology of basal-like breast cancer. *Breast cancer research and treatment* 109, 123-139, doi:10.1007/s10549-007-9632-6 (2008).

[183] Cunningham, J. E., Montero, A. J., Garrett-Mayer, E., Berkel, H. J. & Ely, B. Racial differences in the incidence of breast cancer subtypes defined by combined histologic grade and hormone receptor status. *Cancer causes & control : CCC* 21, 399-409, doi:10.1007/s10552-009-9472-2 (2010).

[184] Ijaduola, T. G. & Smith, E. B. Pattern of breast cancer among white-American, African-American, and nonimmigrant west-African women. *Journal of the National Medical Association* 90, 547-551 (1998).

[185] Ikpatt, O. F. & Ndoma-Egba, R. Oestrogen and progesterone receptors in Nigerian breast cancer: relationship to tumour histopathology and survival of patients. *The Central African journal of medicine* 49, 122-126 (2003).

[186] Gukas, I. D. et al. Clinicopathological features and molecular markers of breast cancer in Jos, Nigeria. *West African journal of medicine* 24, 209-213 (2005).

[187] Bird, P. A., Hill, A. G. & Houssami, N. Poor hormone receptor expression in East African breast cancer: evidence of a biologically different disease? *Annals of surgical oncology* 15, 1983-1988, doi:10.1245/s10434-008-9900-7 (2008).

[188] Awadelkarim, K. D. et al. Pathological, clinical and prognostic characteristics of breast cancer in Central Sudan versus Northern Italy: implications for breast cancer in Africa. *Histopathology* 52, 445-456, doi:10.1111/j.1365-2559.2008.02966.x (2008).

[189] Bowen, R. L., Duffy, S. W., Ryan, D. A., Hart, I. R. & Jones, J. L. Early onset of breast cancer in a group of British black women. *British journal of cancer* 98, 277-281, doi:10.1038/sj.bjc.6604174 (2008).
[190] Hennis, A. J. et al. Breast cancer incidence and mortality in a Caribbean population: comparisons with African-Americans. *International journal of cancer. Journal international du cancer* 124, 429-433, doi:10.1002/ijc.23889 (2009).
[191] Kakarala, M., Rozek, L., Cote, M., Liyanage, S. & Brenner, D. E. Breast cancer histology and receptor status characterization in Asian Indian and Pakistani women in the U.S.--a SEER analysis. *BMC cancer* 10, 191, doi:10.1186/1471-2407-10-191 (2010).
[192] Hausauer, A. K., Keegan, T. H., Chang, E. T. & Clarke, C. A. Recent breast cancer trends among Asian/Pacific Islander, Hispanic, and African-American women in the US: changes by tumor subtype. *Breast cancer research : BCR* 9, R90, doi:10.1186/bcr1839 (2007).
[193] Yang, X. R. et al. Differences in risk factors for breast cancer molecular subtypes in a population-based study. *Cancer epidemiology, biomarkers & prevention : a publication of the American Association for Cancer Research, cosponsored by the American Society of Preventive Oncology* 16, 439-443, doi:10.1158/1055-9965.EPI-06-0806 (2007).
[194] Gordon, N. H. Association of education and income with estrogen receptor status in primary breast cancer. *American journal of epidemiology* 142, 796-803 (1995).
[195] Thomson, C. S. et al. Prognostic factors in women with breast cancer: distribution by socioeconomic status and effect on differences in survival. *Journal of epidemiology and community health* 55, 308-315 (2001).
[196] Bauer, K. R., Brown, M., Cress, R. D., Parise, C. A. & Caggiano, V. Descriptive analysis of estrogen receptor (ER)-negative, progesterone receptor (PR)-negative, and HER2-negative invasive breast cancer, the so-called triple-negative phenotype: a population-based study from the California cancer Registry. *Cancer* 109, 1721-1728, doi:10.1002/cncr.22618 (2007).
[197] Vainshtein, J. Disparities in breast cancer incidence across racial/ethnic strata and socioeconomic status: a systematic review. *Journal of the National Medical Association* 100, 833-839 (2008).
[198] Eley, J. W. et al. Racial differences in survival from breast cancer. Results of the National Cancer Institute Black/White Cancer Survival Study. *JAMA : the journal of the American Medical Association* 272, 947-954 (1994).
[199] Newman, L. A. et al. African-American ethnicity, socioeconomic status, and breast cancer survival: a meta-analysis of 14 studies involving over 10,000 African-American and 40,000 White American patients with carcinoma of the breast. *Cancer* 94, 2844-2854, doi:10.1002/cncr.10575 (2002).
[200] Chlebowski, R. T. et al. Ethnicity and breast cancer: factors influencing differences in incidence and outcome. *Journal of the National Cancer Institute* 97, 439-448, doi:10.1093/jnci/dji064 (2005).
[201] Curtis, E., Quale, C., Haggstrom, D. & Smith-Bindman, R. Racial and ethnic differences in breast cancer survival: how much is explained by screening, tumor severity, biology, treatment, comorbidities, and demographics? *Cancer* 112, 171-180, doi:10.1002/cncr.23131 (2008).

[202] Menashe, I., Anderson, W. F., Jatoi, I. & Rosenberg, P. S. Underlying causes of the black-white racial disparity in breast cancer mortality: a population-based analysis. *Journal of the National Cancer Institute* 101, 993-1000, doi:10.1093/jnci/djp176 (2009).

[203] Demicheli, R. et al. Racial disparities in breast cancer outcome: insights into host-tumor interactions. *Cancer* 110, 1880-1888, doi:10.1002/cncr.22998 (2007).

[204] Shavers, V. L., Harlan, L. C. & Stevens, J. L. Racial/ethnic variation in clinical presentation, treatment, and survival among breast cancer patients under age 35. *Cancer* 97, 134-147, doi:10.1002/cncr.11051 (2003).

[205] van Ravesteyn, N. T. et al. Race-specific impact of natural history, mammography screening, and adjuvant treatment on breast cancer mortality rates in the United States. *Cancer epidemiology, biomarkers & prevention : a publication of the American Association for Cancer Research, cosponsored by the American Society of Preventive Oncology* 20, 112-122, doi:10.1158/1055-9965.EPI-10-0944 (2011).

[206] Wojcik, B. E., Spinks, M. K. & Optenberg, S. A. Breast carcinoma survival analysis for African American and white women in an equal-access health care system. *Cancer* 82, 1310-1318 (1998).

[207] Jatoi, I., Becher, H. & Leake, C. R. Widening disparity in survival between white and African-American patients with breast carcinoma treated in the U. S. Department of Defense Healthcare system. *Cancer* 98, 894-899, doi:10.1002/cncr.11604 (2003).

[208] Anders, C. K. et al. The prognostic contribution of clinical breast cancer subtype, age, and race among patients with breast cancer brain metastases. *Cancer* 117, 1602-1611, doi:10.1002/cncr.25746 (2011).

[209] O'Brien, K. M. et al. Intrinsic breast tumor subtypes, race, and long-term survival in the Carolina Breast Cancer Study. *Clinical cancer research : an official journal of the American Association for Cancer Research* 16, 6100-6110, doi:10.1158/1078-0432.CCR-10-1533 (2010).

[210] Rose, D. P. & Royak-Schaler, R. Tumor biology and prognosis in black breast cancer patients: a review. *Cancer detection and prevention* 25, 16-31 (2001).

[211] Joslyn, S. A. Hormone receptors in breast cancer: racial differences in distribution and survival. *Breast cancer research and treatment* 73, 45-59 (2002).

[212] Poola, I., Clarke, R., DeWitty, R. & Leffall, L. D. Functionally active estrogen receptor isoform profiles in the breast tumors of African American women are different from the profiles in breast tumors of Caucasian women. *Cancer* 94, 615-623 (2002).

[213] Ma, X. et al. Prostaglandin E receptor EP1 suppresses breast cancer metastasis and is linked to survival differences and cancer disparities. *Molecular cancer research : MCR* 8, 1310-1318, doi:10.1158/1541-7786.MCR-10-0003 (2010).

[214] Porter, P. L. et al. Racial differences in the expression of cell cycle-regulatory proteins in breast carcinoma. *Cancer* 100, 2533-2542, doi:10.1002/cncr.20279 (2004).

[215] Jones, B. A. et al. African-American/White differences in breast carcinoma: p53 alterations and other tumor characteristics. *Cancer* 101, 1293-1301, doi:10.1002/cncr.20500 (2004).

[216] Baker, L. et al. p53 mutation, deprivation and poor prognosis in primary breast cancer. *British journal of cancer* 102, 719-726, doi:10.1038/sj.bjc.6605540 (2010).

[217] Greenblatt, M. S., Bennett, W. P., Hollstein, M. & Harris, C. C. Mutations in the p53 tumor suppressor gene: clues to cancer etiology and molecular pathogenesis. *Cancer research* 54, 4855-4878 (1994).
[218] Starks, A. M. et al. Household income is associated with the p53 mutation frequency in human breast tumors. *PloS one* 8, e57361, doi:10.1371/journal.pone.0057361 (2013).
[219] Mehrotra, J. et al. Estrogen receptor/progesterone receptor-negative breast cancers of young African-American women have a higher frequency of methylation of multiple genes than those of Caucasian women. *Clinical cancer research : an official journal of the American Association for Cancer Research* 10, 2052-2057 (2004).
[220] Ambrosone, C. B. et al. Genome-wide methylation patterns provide insight into differences in breast tumor biology between American women of African and European ancestry. *Oncotarget* 5, 237-248 (2014).
[221] Adisa, C. A. et al. Biology of breast cancer in Nigerian women: a pilot study. *Annals of African medicine* 11, 169-175, doi:10.4103/1596-3519.96880 (2012).
[222] Wei, P. et al. Differential endothelial cell gene expression by African Americans versus Caucasian Americans: a possible contribution to health disparity in vascular disease and cancer. *BMC medicine* 9, 2, doi:10.1186/1741-7015-9-2 (2011).
[223] Butler, P. D., Longaker, M. T. & Yang, G. P. Current progress in keloid research and treatment. *Journal of the American College of Surgeons* 206, 731-741, doi:10.1016/j.jamcollsurg.2007.12.001 (2008).
[224] Smith, J. C., Boone, B. E., Opalenik, S. R., Williams, S. M. & Russell, S. B. Gene profiling of keloid fibroblasts shows altered expression in multiple fibrosis-associated pathways. *The Journal of investigative dermatology* 128, 1298-1310, doi:10.1038/sj.jid.5701149 (2008).
[225] Russell, S. B. et al. Epigenetically altered wound healing in keloid fibroblasts. *The Journal of investigative dermatology* 130, 2489-2496, doi:10.1038/jid.2010.162 (2010).

In: Gene-Environment Interactions and Human Diseases
Editor: Lu Qi
ISBN: 978-1-63483-680-7

Chapter 8

THE MISSING HERITABILITY IN CORONARY ARTERY DISEASE

Pierre A. Zalloua and Daniel E. Platt
Lebanese American University, Beirut, Lebanon

ABSTRACT

All epidemiological studies have consistently and repeatedly demonstrated that the total heritability of Coronary Artery Disease (CAD) in all studied populations varies between 40-60%. However, collective results from all the large successful genomic studies conducted to date have not revealed the substantial causation or major susceptibilities to CAD. It is hypothesized that CAD can develop as a result of the combined additive contribution of a number of relatively frequent genetic variants (polygenic) that are directly or indirectly involved in any of a series of biochemical pathways involved in CAD pathogenesis. Considering the large number of CAD pathogenic pathways, and their unpredictability in disease causing in a particular individual, common variants will invariably show weak association with CAD in a given population genomic study. Genotype imputation, and Whole Exome Sequencing approaches are available to extend SNP explorations into some of these candidate pathways. It is now also possible to combine the various sources of associations integrating genomic transcriptomic, and metabolomic connections with disease phenotypes to further identify CAD determinants.

1. INTRODUCTION

Coronary Artery Disease (CAD) and its main complication, myocardial infarction (MI), continue to be the leading causes of mortality in men and women of different geographic backgrounds [1]. Several risk factors are associated with an increased risk of CAD, notably elevated LDL-cholesterol values, arterial hypertension, smoking, diabetes mellitus, sedentary lifestyle, and obesity [2]. Controlling traditional risk factors has led to effective preventive treatments of CAD. However, up to 20% of patients with MI do not carry any of the known risk factors [3].

CAD is predicted to be the leading cause of death in the developing world by 2020, with projected estimates of 7.8 million annual deaths. These data emphasize the potential impact of significant reductions in incidence and severity CAD to patient care costs, morbidity, and mortality [1-3].

It has become accepted that CAD and similar complex disease etiologies possess both genetic and environmental components that interact to determine disease manifestation, including severity and early/late onset. For example, two individuals with similar risk factors can still exhibit huge disparities in the percent occlusion and number of arteries affected and age of onset of the disease. Under the same environmental conditions, some individuals seem to have inherent "defense" mechanisms while others are predisposed to develop CAD, which can most likely be attributed to genotypic differences manifest in compensatory mutations.

All epidemiological studies have consistently and repeatedly demonstrated that the total heritability of CAD in studied populations varies between 40-60% [4-6]. However, collective results from all successful large genomic studies conducted to date have not revealed the substantial causation or major susceptibilities to CAD. Despite the enormous progress in genomics studies that has been made recently, the identified susceptibility genomic variants from the largest CAD consortia map to 46 loci and account for about 10% of CAD heritability and most scientists are struggling to uncover susceptibility alleles that can explain the remaining "missing heritability" [7].

The pathological elements of CAD have been shown to have complex etiology and pathogenesis that dilute an individual's relative risk of developing these diseases given exposure [8]. The genetic input is complex, as it involves combinations of multiple genes contributing to susceptibility or resistance to CAD risk factors. Moreover CAD is polygenic, involving complex gene-gene and gene-environment interactions; these interactions confer to CAD its multi-factorial phenotypes. The identification of genetic variants of CAD susceptibility from genomic studies is based on odds-ratios (OR) estimates that assume that any given locus predicting disease can be accurately identified as a disease-causing variant. The validity of this assumption depends on the precision of the clinical information used to classify each case and control. Furthermore, locus-specific heritability depends on accurate measurements of CAD prevalence, which in turn has substantial population variability due to differences in life style changes, exposures, sex and age. In addition, odds-ratios predicting disease due to exposure are identical to those indicating an exposure for a given disease. This implies that CAD will show a weakened association with any one pathway, since the association will not be any stronger than implied by the chances that any one of the many pathways would be implicated by the presence of CAD in any given subject. Therefore, the genetics of risk factors characterizing each individual pathway must be isolated and validated in the context of those risk factors and of CAD. While additional CAD susceptibility alleles will be uncovered with the continuously evolving field of high throughput genomics, various novel approaches to study designs as well as result interpretations are needed to unravel further the "missing" CAD heritability.

Another cause for losing penetrant SNPs in GWAS for any given confidence level, is that the probability is not small that variations in the frequencies of alleles simply due to sampling randomness would produce a highly significant odds ratio. For exploratory searches, there is no way to distinguish these false positives from real signal without bringing in other information to focus attention on specific genes or potential candidate regulatory regions. The approach most commonly taken is to select a confidence level so severe that no false positives would be expected among a large number of GW SNPs. The result is that a large number of false negatives are discarded [9]. The impact of whole sequence analysis or imputation is to increase the effective number of SNPs. Exploration of gene-gene analysis has the raw potential of increasing the possible number of candidates from a number of SNPs $N \approx 10^5$ for standard GWAS to N^2 for gene-gene interactions. These increases vastly exacerbate the problem unless the effective number of hypotheses are reduced, or the disease phenotype description is itself made multifactorial and complex by increasing the number of traits describing the phenotype. For example, the chances of seeing seven or more out of 10 traits defining a phenotype state (or pathway hypothesis, or other association specified by a complex of traits) supported by a single SNP at the 95% confidence level is $8.198 \times 10\text{-}^{8}$ [10].

2. Underlying CAD Complexity: A Web of Interactions

2.1. Study Populations and the Choice of Cases and Controls

A large number of genomic studies that have been conducted to date, on a sizable number of samples, had to rely primarily on cases and controls that have been recruited under varying criteria of inclusions and exclusion. Different studies have used different diagnostic criteria (MI, angiographic results, stale angina) and were combined in large consortia to increase the power of analysis. In CAD, the heritability estimate is diluted due to the clinically/phenotypically heterogeneous nature of the trait and additional variability due to study design will only be adding a significant level of complication to the association analysis. CAD phenotype varies from one case series to another, where coronary angiography documented stenosis and Myocardial Infarction (MI) are consistently interchanged and confounded to designate the same trait. Indeed, it is well established that MI often results from narrowing of the coronary arteries as a result of atherosclerosis, however MI may occur without atherosclerosis and atherosclerosis does not necessarily lead to MI. The variant rs4977574 located within the 9p21 locus, one of the most significant and replicated susceptibility alleles for CAD, has not been universally associated with MI across all of the populations studied. In fact, most of the studies to date that have based their CAD phenotype on coronary angiography results, have shown a significant association of the 9p21 variant with stenosis and a lack of association with MI. It is to be noted, however, that none of these studies were prospective longitudinal studies where follow up on CAD patients would be critical to validate any association or the lack of association [11-16].

Furthermore, variability in study enrolment, especially the use of different inclusion and exclusion criteria for the control group, such as age, sex, life style patterns and dietary habits, all have the potential to impact susceptibility and risk estimates. Longitudinal prospective studies from the general population, with adequate numbers, unambiguous angiographic visualization of CAD phenotypic characterization, and follow up availability can minimize phenotypic variability, document disease evolution, and exclude disease progression among controls, and therefore are ideal to conduct association studies. Unfortunately such studies are logistically very demanding and require substantial funding over long time periods to be adequately conducted.

2.2. SNP Chips and Ascertainment Bias

It is hypothesized that CAD can develop as a result of the combined additive contribution of a number of relatively frequent genetic variants (polygenic) that are directly or indirectly involved in any of a series of biochemical pathways involved in CAD pathogenesis. Under the same additive hypothesis model, some environmental factors compound the genetic effect and the combination of factors may also lead to CAD development. An alternative hypothesis stipulates that one (monogenic) or a few (oligogenic) rare genetic variants with high impact on the regulation of the same set of CAD biochemical pathways can lead to CAD. The impact on CAD development of these rare genetic variants can be modulated (attenuated or exaggerated) by their interactions with other genetic and environmental triggers.

Considering the large number of CAD pathogenic pathways, and their unpredictability in causing disease in a particular individual, common variants will invariably show weak association with CAD in a given population GWAS. A weak association can also result with high contribution rare variants due to their variable penetrance or absence in that particular population under investigation.

The above considerations suggest that in order to fully characterize CAD risk factors, identifying each pathological pathway with its various individual genetic components and their modulators (genes and environmental triggers) must be fully elucidated. Consequently, the kinds of significant insights into the underlying disease pathogenesis that these pathways and apparently weaker SNPs can provide will come from unraveling the genetic determinants of CAD's risk factors on each of these major pathways,, and reveal novel therapeutic targets.

2.2.1. Determining the Causative SNPs

Gene chips are biased towards specific populations as the chip SNPs that are used on the Illumina and Affymetrix platforms (the only high throughput available chips to date) are almost exclusively built from the HapMap project SNPs that originally included only the following groups of individuals: Yoruba in Ibadan, Nigeria (YRI), Japanese in Tokyo, Japan (JPT), Han Chinese in Beijing, China (CHB) and CEPH (Utah residents with ancestry from northern and western Europe, CEU). The SNPs used on the chips have typical minimal allele frequencies of 5% (MAF:0.05) or higher and vary significantly among these populations. Alleles with MAF lower than 5% were largely absent from the early genome-wide SNP arrays, which had been designed to type common variants in GWAS. Furthermore, some SNPs may be rare in some populations but not others, leading to uneven power in recognizing penetrant SNPs. These frequency variations that limit the effectiveness of Gene Chip studies across global populations setting can be reduced by the use of genotype imputation.

In GWAS, the association of a SNP at a given locus with a phenotype and its effective impact are frequently based on odds ratio estimates. These locus heritability estimates assume that the given SNP accurately tags the disease-causing variant under a particular disease model. In these studies, a low significance value per-SNP ($p < 5x10^{-7}$) is considered to hold false positive rates to a relatively low level. In addition, it has become common practice that for a SNP to be considered significantly associated with a trait, this association has to be replicated in multiple populations. Further, GWAS with a relatively small number of samples are typically underpowered to detect small genetic effects at a stringent genome-wide significance level. However, since most large GWAS as well as meta-analyses are conducted on a relatively heterogeneous CAD phenotype with less stringent inclusion criteria for controls that are clinically unscreened, there is a potential bias in overestimating the association. The 'winner's curse' [17, 18], where extreme values are likely to include effects of random sampling variation, can inflate false positives or, alternatively, use of clinically unscreened control data [19] can lead to misclassifications, diluting signal and promoting false negatives.

As previously stated, the large number of SNPs tested in standard GWAS implies that a large number of SNPs would be expected to test significantly due to random sampling variations alone. This is avoided by choosing significance levels high enough to exclude false positives from any of all of those SNPs tested due to random sampling variation. The result is large numbers of false negatives. All of this is exacerbated by more complex analyses, such as gene-gene interactions, or more complex gene-environment interaction studies where tests for environment-driven epigenetic change impacts are also being included. Consideration of other information or relationships among processes would effectively increase the number of tests being performed on each SNP and lower the GWAS threshold required compared to that of a single phenotype test per SNP. Therefore, designs considering multiple associations characterizing disease may increase power per SNP. The reasoning would be that SNPs that are relevant to a disease process, marked by distinct phenotype characteristics *A* and *B* of a disease, would likely correlate with both, whereas a SNP that would have correlated with one or the other by chance sampling variation would be unlikely to have correlated with both.

Results from the 1000 genome project have opened the door to the development of several methods to impute the genotype of a given SNP not present on the chip based on the Linkage Disequilibrium structure of the genome [20, 21]. Populations represented in the 1000 genome will have more accurate imputed results than those who are not represented. Although computationally demanding, genotype imputations methods have become a standard component of any GWAS as have meta-analysis tool kits [22, 23]. Genotype imputation can significantly reduce the genomic coverage disparities between chips and substantially improve the power of the study [24].

2.2.2. Rare variants and Whole Genome Sequencing

The HapMap and the 1000 Genomes Projects (1000GP) provided a catalogue of genetic variations in humans of great informative value, allowing SNPs to be identified across the genome. Using this catalogue, a number of GWAS have identified several loci marked by hundreds of these SNPs with a moderate increased disease risk [25]. These GWAS-identified loci may differ across traits, be attributable to linkage with nearby multiple risk variants within the locus, or combinations of both common variants with weak effects and rare variants with large effects [26-28]. However these studies have systematically failed to identify any rare variants that have a great impact on CAD development and thus potentially explain some of the missing genetic contribution to CAD. Whole genome and exome sequencing have recently been shown to be effective in the identification of clinically relevant rare genomic variants. Whole genome sequencing (WGS) witnessed a rapid evolution in recent years. Exome sequencing technologies have already accelerated genetic studies of Mendelian disorders, and there is great interest in extending them to complex traits such as CAD [29]. With the rapid development of sequencing techniques, it is now possible to sequence an entire genome in little more than a week. High-throughput sequencing platforms are generating massive amounts of genetic variation data for diverse genomes and these are reported in several databases such as the 1000 Genomes Project, dbSNP and HapMap Project.

Several statistical software packages are now available for the analysis of the high volume of genetic variant data generated (e. g., Burrows-Wheeler Alignment tool (BWA) [20], SAMtools variant calling program [21], SequenceVariantAnalyzer (SVA) [22], Ingenuity Pathway Analysis (IPA) and ANNOVAR [23].

Rare variants with functional impact that have been discovered are usually clustered within genes, their promoters and adjacent sequences, and, for CAD, are frequently associated with functional changes with deleterious effect leading to CAD [30-33]. Several analyses using HapMap data have demonstrated the extreme allelic heterogeneity and underscored the detrimental effect of rare variants in complex disease [26-28]. These characteristics, rarity and heterogeneity, of these variants make their identification elusive in GWAS. In order to identify these SNPs and distinguish them from read errors, multiple reads must be performed. Deep sequencing provides verification since read overlaps provide multiple independent support of the presence of a rare variant. Once observed, even if perfectly penetrant disease association of 100% is observed in a sample size of 5000 with MAF of 0.1% yielding around 5 such SNPs, a balanced case-control study will have produced that 100% penetrance by chance with a probability of $0.5^5 = 0.03125$ – hardly genome-wide significant. A perfectly penetrant but rare allele (MAF 0.1%) would require a sample size of around 18,000 in a balanced case-control study to observe genome wide significance. Restriction to exomes greatly reduces the size of a study required to observe exome-wide significance. Whole genome sequencing cost suggests application of particular rare-variant hunting strategies, but these bring with them inherent bias [34]. If the rare SNPs are significant under analysis focused for specific mechanisms or genes, then validation still must be achieved through functional analysis, which is why the literature on rare alleles is so rich in functional analysis. Therefore, applying whole genome technologies and computational biology, along with detailed clinical and demographic information, together with functional analysis, on the same population used in GWASs will undoubtedly lead to the identification of additional CAD susceptibility variants with robust disease impact. While rare variants are likely to be identified and characterized in this method, the actual fraction of rare variants ultimately discovered that may be producing disease in a population is likely to remain relatively low[34], accounting for a fair fraction of unexplained variance.

Further characterization of GWAS-identified loci of interest in complex diseases by direct analysis, such as sequencing, has been shown in several studies to yield promising results. Among the variants that were found to be associated with CAD, rheumatoid arthritis, diabetes and other complex disorders, rare variants with direct functional consequence (partial- or total-loss-of-function variations), have been significantly overrepresented [35]. The most recent Type 2 Diabetes Mellitus (T2DM) GWAS studies that were conducted on diabetics have resulted in the identification of several functional SNPs involved in regulating glucose-stimulated insulin secretion fasting glucose concentration [22, 36-40]. A large scale sequencing analysis focusing on exonic regions of the genome, conducted on more than 7000 Europeans, identified several functional rare variants with a combined T2DM OR of 5.67 [41]. These recent studies when combined with careful biological evaluation of the newly identified rare variant with high impact, have demonstrated that careful analyses at the sequencing a level can effectively unravel the genetic architecture of complex diseases such as CAD.

2.2.3. The Network Approach

SNPs identified by GWAS provide little connection between individual SNP associations and the biological pathways they mediate [42]. SNPs connect to phenotypic biological impact changes in protein structure through non-synonymous coding (impact on folding), protein activity (changes in the active site, hydrophobicity, charge, polarity, polarizability, steric interactions, allosteric interactions, and others), synonymous coding (regulatory expression of protein), and changes in non-coding regions (changes in transcription factor binding sites or transcription factors themselves), and even epigenetic changes (impacts on methylation sites) [43]. Changes in catalytic activity will affect cascades and feedback loops impacting the expression responses of these mutated genes. Techniques to measure these biological phenomena on a genome wide basis have emerged over time, increasing the ability to connect these variants to phenotype. Mutability may be a phenotype [44], with epigenetic methylation-based silencing being blocked by mutation of CpG sites, as well as possibly promoting that mutation [45]. This mutability may be either positive [46] (perhaps due to plastic response to environmental stresses) or negative [47]. It is now possible to combine the various sources of associations integrating genomic transcriptomic, and metabolomic connections with disease phenotypes [48].

One of the problems with the explosion of information is that, not only is there more of it at the raw level, but much of that information connects through interactions with each other. Even at low level, this results in combinatorial explosion. This implies many more candidates which stresses even highly parallel systems to analyze, and greatly increases the exclusion of false negatives in seeking to identify true positives. Biological systems are not simply random; they are organized into networks of interactions, such as transcriptional regulation networks [49]. This greatly limits the hypothesis space for interactions. Further, associations connecting impacts of variants on network components more directly connects to the biology of pathogenesis. However, these networks may also involve peculiar relationships where non-coding SNPs in one gene may affect expression levels of another gene which may be located at a significant distance from the controlling SNPs. For example, associations were established between intronic α-ketogluterate-dependent dioxygenase FTO SNPs and obesity. Effectively, the FTO SNPs in question were actively implicated in obesity because of their direct impact on the regulation of Iroquois homeobox protein IRX3, a gene with direct impact on obesity [50]. An alternative way to limit candidate linkages is to apply gene ontologies and known transcriptional regulators to identify networks and candidates for gene-gene interactions [51-53].

2.3. The Elusive Epigenetics Components

Another assumption explaining some of the missing heritability in CAD is that gene expression alterations may be consequent to epigenetic modifications, many of which cannot be directly detected by a GWAS or WGS approach [8].

Epigenetic modifications are increasingly being recognized as a key factor in the pathogenesis of complex disorders, and cardiovascular diseases including CAD are now being approached from an epigenetic perspective [54]. Several studies have indeed already shown the role of epigenetic mechanisms in critical pathways related to CAD development in human and animal models [55,56]. The same is true about atherosclerosis, which is one of the main known cardiovascular pathways.

Epigenetic regulation is not only involved in the response to pathological challenges that affect the vascular system, but is also a major player during the development and maintenance of normal blood vessel physiology [57]. Throughout an organism's lifespan, epigenetic factors are responsible for providing the vascular system with high flexibility and the potential to regenerate and to generate new structures in response to sudden physiological changes [58]. Accordingly, abnormal regulation that takes place in this context has been shown to result in atherosclerosis, restenosis, myocardial infarctions as well as CAD. Indeed, within the atherosclerotic lesion, several epigenetic signatures are now known to regulate vascular and other tissue-gene expression by modulation the transcription machinery.

The field of epigenetics can be broadly categorized into three areas: DNA based modifications, including cytosine methylations and cytosine hydroxymethylations, post-translational modifications of histone proteins, and RNA-based mechanisms that operate in the nucleus. All of these alter gene expression levels by modulating accessibility of the transcription machinery to the DNA, rather than altering the DNA sequence itself.

2.3.1. DNA Methylation

The principal epigenetic tag found in mammalian cells is DNA methylation. Methylation of CpG islands results in transcriptional gene silencing, and serves to control the expression of some tissue-specific genes during the development and differentiation processes [59]. Removal of methylation marks are yet to be fully understood as they seem to be either replication- dependent or independent.

Several associations have been established between DNA methylation and cardiovascular conditions including atherosclerosis and CAD. A notable example is the folic acid deficiency and the associated elevated homocysteine serum levels. Hyperhomocystinemia (HHcy), similar to hyperlipidemia, hypertension and smoking, is a well established common risk factor for cardiovascular disease in the general population. HHcy has been reported to be linked with a marked decrease in global CpG methylation levels in leukocytes as well as gene-specific methylation of promoters leading to their aberrant expression [60]. Higher plasma levels of homocysteine correlated with increased concentrations of its intracellular precursor S-adenosylhomocysteine [61], also known to function as a strong inhibitor of DNMTs [62]. These studies and others conducted in animal models [63] provide supporting evidence that the observed DNA hypomethylation may be responsible in part for the vascular complications associated with HHcy. Specifically, it is becoming evident that homocysteine and adenosylhomocysteine can inhibit DNMT activity and can have complex effects on DNA methylation targets and their transcriptional potential. This could partially explain the study by Sharma and colleagues, in which confounding results were reported: In CAD patients, a significant positive correlation of global DNA methylation with plasma homocysteine levels was found. Furthermore, within a higher range of serum homocysteine levels, global DNA methylation, obtained by measuring difference in mean incorporate ion of [^{3}H]dCTP between digestions with *Msp*I and *Hpa*II, was significantly higher in CAD patients than in controls [64].

Few of the specific genes whose expression is altered by DNA methylation have been reported so far, and results remain variable depending on the cell types that are analyzed. For instance, in atherosclerosis and CAD, the expression of estrogen receptor genes (*ESR1 and*

ESR2), which play a protective role in cardiovascular health, are found to be repressed by promoter hypermethylation in human coronary atherosclerotic plaques compared to proximal normal vascular tissue [65, 66]. Another important gene found to be epigenetically regulated at the promoter region is *coagulation factor VII*. The plasma concentrations of the protein are used to assess the risk of developing CAD. When comparing methylation levels of plasma factor VII in CAD versus CAD-free patients, Friso and colleagues reported that in CAD, hypomethylation at the promoter of the gene is shown to result in higher plasma concentration of the protein and higher risks of CAD. Furthermore, higher methylation levels resulting in lower plasma factor VII was a dominant feature within the CAD-free group [67]. Finally, the *ApoE* gene promoter was previously shown to have varying degrees of methylation depending on plasma homocysteine levels [68]. However, when tested in peripheral blood lymhpocytes by Sharma and colleagues, the group reported similar methylation levels between CAD patients and controls [64]. Other genes found to have differential methylation patterns in patients with HHcy include Cyclin A. Demethylation of its promoter was shown to impair histone acetylation mechanisms, another epigenetic alteration, histone modification, described in the following section, lead to increased acetylation of histones H3 and H4.

These results stress the need for elucidation and characterization of the precise genes that are targeted by the DNA methylation machinery, with emphasis on the affected tissues, potentially providing an explanation for the precise mechanisms that govern DNMT activity. Another important aspect of DNA methylation is its unequal distribution in the genome, with hypo and hyper-methylated regions coexisting. Thus global DNA methylation levels do not provide a very accurate picture of what is happening at the gene level, further emphasizing the need for a tissue-specific analysis of the epigenome. So far, most studies have been focused on global methylation rates, or on methylation at specific genes with tissue specificity. Methods, such as single-cell genome-wide bisulfite sequencing [69] are emerging that can measure epigenetic heterogeneity.

2.3.2. Histone Modifications

The second major epigenetic area targets the N-terminal ends of histones; specifically, post-translational modifications of these tails modulate histone-DNA interactions, and significantly influence chromatin structure and thereby affect the accessibility of transcriptional elements to certain genes. The most well characterized modifications include acetylations and methylations at lysine and arginine residues as well as phosphorylation on serine residues. In contrast with DNA methylation which usually correlates with gene repression, little is known about the role of histone methylation which results in either transcriptional activation or inactivation, depending on the residue modified and the degree of methylation. Histone deacetyltransferases (HDACs) remove acetyl groups from histone lysine residues thus resulting in increased interaction between DNA and histones and reduced accessibility of the transcription machinery. In contrast, histone transferases (HATs) increase histone acetylation and correlate with increased gene transcription. It has been shown that certain factors such as the MeCP2 (methyl CpG-binding protein 2) preferentially bind to methyl-CpG and contribute to transcriptional repression by recruiting HDACs, which illustrates how epigenetic modification contribute to the transcriptional on or off state of target genes [70].

Lysine methylation patterns seem to be more complex in that some methylation sites are associated with active chromatin, whereas others are repressive. Lysine residues have also been shown to be mono-methylated (me), dimethylated (me2) or trimethylated (me3). Well described silencing marks include the H3K27me3 and the H3K9me signatures, whereas H3K4me3 and H3K36me3 are associated with gene activation [71].

In vascular disease, the effects of HDACs on arteriosclerosis development in animal models have been observed after HDAC inhibition using specific inhibitors. For instance, Choi and colleagues reported findings suggesting that histone acetylation plays a role in atherogenesis by using the specific HDAC inhibitor Trichostatin A TSA. In atherosclerosis-prone mice, treatment with TSA aggravated atherosclerosis without affecting plasma lipid profiles, by modulating the expression of oxidative low-density lipoprotein (oxLDL) receptors and other proatherogenic genes. It also induced the formation of fatty streak lesions suggesting that histone acetylations signatures are protective [72].

While most prior research on epigenetics of heart failure has focused on HATs and HDACs, the genes regulated by these histone modifications or how they impact cardiac hypertrophy are not yet well understood. Genome-wide profiles of epigenetic changes in hypertrophic myocytes chromatin state maps between normal and diseased cells have been obtained by chromatin immunoprecipitation coupled with massively parallel DNA sequencing (CHiP-seq) [73]. A total of 726 genes were found to be upregulated, while 383 were found to be downregulated, and specifically identified and distinguished by subsets of antibodies against H3K9ac, H3K27ac, H3K4me3, H3K9me2, H3K9me3, and H3K27me3. Studies and methods for genome-wide epigenetic changes are becoming more common [74, 75].

2.3.3. RNA-Based Mechanisms

The most recently described type of epigenetic regulation involves RNA-based mechanisms. These include long noncoding RNAs (lncRNAs) as well as microRNAs. LncRNAs consist of transcripts longer than 200 nucleotides which do not code for proteins, with thousands already shown to play an important role in epigenetic regulation [76]. LncRNAs are highly tissue specific and regulate transcription through chromatin structure remodelling, by associating with other activating or repressive chromatin remodelling complexes [77,78].

One such lncRNA concerns chromosome 9p21 which is the most replicated marker of CAD. We and others have shown by GWAS that multiple polymorphisms on this region are tightly associated with CAD. However, the mechanism linking this risk locus to CAD remains unclear [79,80]. A recent report by Zhuang and colleagues now suggests that a link exists between the epigenetic changes observed in $p15^{INK4b}$ methylation and the expression of the lncRNA *ANRIL* providing an explanation for the involvement of chromosome 9p21 in CAD development [81]. Antisense noncoding RNA in the INK4 locus (ANRIL) mediates epigenetic regulation by recruiting chromatin-modifying polycomb repressive complex 1 and 2 to silence target genes [82]. Interestingly, ANRIL is found to be highly expressed in tissues and cell types affected by atherosclerosis, mainly in atherosclerotic plaques, with levels of expression directly correlated with the severity of atherosclerosis [83].

Small non coding RNAs also play a role in chromatin-based silencing. These small non coding RNAs called microRNAs (miRs) are first produced as primary miRNAs, and possess a hairpin shaped structure of around 70 bases long with a 5' cap and poly-adenylated tail. Identification by DGCR8 is followed by cleaving by Drosha of this structure gives rise to precursor miRNA which is then transported out of the nucleus by exportin-5. This molecule is then cleaved by DICER into short double stranded RNA fragments of about 20-25 nucleotides long. An enzymatic complex unwinds the double strands and carries the mature single stranded miRNA to its target mRNA that will be degraded [84]. MiRNAs are present in several copies and can negatively affect post-transcriptional events, and mediate translational suppression.

In CAD patients, several miRs including miR-126, miR-17 and miR-155, have been shown to have reduced plasma levels, whereas others such as miR-133 and mir-208, are found to be markedly upregulated [85]. To identify the heart as a potential source for miRs released into the circulation during cardiac events, De Rosa and colleagues analysed concentration gradients of key miRs across the coronary circulation. Notably, miR-126, miR-499 and miR-133 were found to be released from the heart following myocardial injury. Interestingly, the levels of expression of the latter 2 miRs are directly correlated with the extent of myocardial injury. In line with previous results, circulating levels of miR-126 were found to decrease and are lost following transcoronary passage. These data show a differential regulation of circulating miRs during cardiac events and provides insights into their role as CAD prognostic biomarkers [86]. Other miRs have been suggested as potential preventive biomarkers. MiR-624 and miR-340 reside within patient's platelets, and have been described to be upregulated in patients with premature CAD and a family history of cardiovascular disease. Mechanistically, these miRs have been proposed to be involved in vessel occlusion, however, whether they are a cause or a consequence of the disease remains to be elucidated [87].

Whereas the profiling of circulating miRs has been proven to be crucial in the assessment of cardiovascular disease, a study by Menghini et al., highlights the importance of studying local injured environments such as the atherosclerotic plaques [88]. In the study, miR217 is described as a novel prognostic biomarker, whose increased levels within human atheromas leads to the disintegration of the endothelium. Endothelial dysfunction is demonstrated to be an early step in the development of atherosclerosis, contributing to the formation, progression, and complication of atherosclerotic plaques [89]. Interestingly, the function of miR217 in accelerating vascular senescence is found to be mediated by downregulation of the NAD-dependent HDAC silent information regulator 1 (SirT1) [88]. The deacetylase activity of SirT1 controls a variety of major transcription factors such as p53, Forkhead box O (FoxO) and others [90]. Interestingly, within the lesions, overexpression of miR217 resulted in marked FoxO1 acetylation and subsequent inactivation, without effect on p53 acetylation status, pointing towards a targeted modulation of angiogenic rather than apoptotic SirT1 targets [88, 91].

These examples all emphasize how finely tuned the epigenetic machinery is; in which all the different players are tightly coordinated to provide feedback loops in order to control the expression of specific target genes. It is, therefore, imperative to go beyond a GWAS and WGS approaches, which do not allow for the detection of gene expression alterations that are consequent to epigenetic modifications, especially since GWAS have found that approximately 43% of polymorphisms are located in intergenic regions of the genome, potentially in regions expressing non coding RNAs [92]. It is also evident from these studies, that CAD epigenetics vary widely depending on the tissue or cells that are being investigated. Focusing on atherosclerotic plaques will allow for a better understanding of the pathogenesis of atherosclerosis and the subsequent CAD development as well as for the possible definition of unique biomarkers for the disease.

Identification of candidate miRNAs has proved complex, and the list currently associated with CAD is very short. In contrast, data mining techniques applied to the identification of segments that would likely form reverse-complement bonding impacting regulation, along with the characteristic hairpins of the right sizes, have yielded tens of thousands of candidates with identified matching segments in hundreds of thousands of control regions, and tend to align well by ontology [93, 94]. Further, these have been associated with a number of cancers and auto-immune issues [95], and networks [96]. The list of miRNAs associated with expression changes and networks along disease pathways can provide a significant opportunity to characterize candidate segments that mining has already enumerated. Imputation approaches are available to extend SNP explorations into some of these candidate regions, explore SNP relative SNP densities in these regions, and other facets that may lead to correlations with CAD and its risk factors.

2.4. Gene-Environment Interactions

2.4.1. Describing Gene–Environment Interactions

CAD is perhaps most famous for its well defined cluster of known risk factors. These include hypertension (HT), high LDL, low HDL, high triglycerides, type II diabetes (T2D), obesity (often defined in terms of BMI), diet and smoking. This cluster (except for smoking and diet) are medical conditions in their own right, and cluster together in a syndrome known as metabolic disorder. They tend to be highly correlated. T2D couples with the hyperlipidemia group through triglyceride metabolism, for example. The insensitivity to diet, and insulin resistant response to sugar and diet even after having brought these under control argues for epigenetic changes. Metagenomic impacts are noted in terms of gut bacteria, T2D and obesity, as well as oral bacteria and CAD.

Interactions among risk factors are worth noting. Smoking impacts how LDL affects endothelial cells [97-99] with impacts on genetic activity[100]. A simple logistic regression will tend to show increases in smoking associate with decreases in HDL. However, inclusion of exercise plus interaction terms between exercise and smoking show that the effective association between smoking and HDL is primarily mediated through exercise. Since the primary known modality for increasing HDL (reducing a risk factor) is exercise, this is not unexpected.

Homocysteine is a transitional peptide intermediary in the production of homocysteine. Excess homocysteine is typically removed from the system since excess homocysteine produces a well defined pathway of damage, first involving arterial microvasculature, then production of scar tissue, then more substantial plaques in the epithelial layers, leading to well defined atherosclerosis. This specific pathway is well documented, and used in animal model studies to induce atherosclerosis to observe a standard progression of the disease. Disease progression occurs over a period of a decade or so. Since this includes formation of scar tissues and other irreversible changes, it is not clear wither homocysteine reduction improves CAD outlooks for patients. Manifestation of early damage tends to be reflected in hypertension, which is often treated with diuretics. This can actually promote homocysteine levels, actually exacerbating CAD development. One suggestion is that diuretics carry away water soluble vitamins, including B vitamins implicated in removal of excess homocysteine. Analysis of homocysteine production and its association with CAD, and their GWAS must include whether hypertension was diagnosed, whether diuretics was a part of the treatment, and their interaction term.

From the standpoint of logic, the standard inclusion of risk factors as "adjustment variables" in logistic regressions identifying SNP associations with disease reflect an intention to factor out the largest and most well defined physiological pathways leading to CAD in order to identify leading genetic variates directly implicated in CAD. These SNPs will tend to be the highly penetrant SNPs not specific to any particular pathway as reflected in the risk factors. This suggests that a major source of information may be missing. SNP associations mediated upwards by adjustment with risk factors may reflect compensatory responses, while SNP associations mediated downwards may reflect pathway-dependent SNPs that impact CAD through specific pathways.

2.4.2. Endophenotypes

Heritable characteristics clearly connected to genetics that associate with an illness, such as CAD, in the population, that may be observed with and without illness, and segregates in families according to illness, are likely to appear in complex diseases. Metabolic syndrome variables, such as hypertension, obesity, and T2D appear to fit this definition, serving as endophenotypes for CAD. These phenotypes are all highly correlated to each other, show distinct genetic associations under GWAS, and tend to associate with CAD. This could be used as a basis to characterize and identify distinct pathways or clusters of pathways leading to CAD, as well as reducing the number of gene-gene combinations that may be relevant to the pathogenesis of CAD. Further, they more directly relate to regulatory networks. Such combinatorial reductions may serve to increase the statistical power available from smaller sample sets. Further, complexes of multiple phenotype characterizations expected to correlate in the presence of a pathogenic process, that would otherwise be uncorrelated in random sample variations, would be expected to show significant increases in statistical power when considered in combination. That level of improvement (7 of 10 traits at 95% confidence would yield a combined confidence of 8.198×10^{-8}) would greatly reduce the number of discarded false negatives at genome wide significance levels.

2.4.3. EWAS

An analysis that parallels GWAS in structure, but including, instead, a large number of environmental factors has been christened Environment Wide Association Studies, or EWAS. Such studies have found multiple environmental associations with a number of diseases, including some remarkable associations involving T2D [101]. In a very real sense, environmental factor associations with GWAS studies of CAD have been common since the beginning, yet primarily as a way to try to isolate those SNPs directly implicated in CAD, as if environmental and impacts of other conditions such as metabolic syndrome were contaminants of genetic CAD mechanisms, rather than as being markers of pathways important in the production of CAD. Within the last decade, there has been a huge explosion of information regarding regulatory mechanisms. A number of these mechanisms allow for some plasticity in the response of tissues to environment, which may also be key to the weaknesses leading to CAD – they could be simply normal tissues responding through normal pathways to very adverse environmental effects. The interaction of these environmental impacts across genetic variations in the population may show much more detail in how normal pathways break. An understanding of the genetic basis of these interactions that forms some of the first steps in understanding the network view of GWAS would be to include genetic and environmental interactions in association studies – a GEWAS.

3. Summary and Perspectives

The pathological elements of CAD have been shown to have complex etiology and pathogenesis. As pointed out earlier, odds ratios for disease given exposure is identical to odds ratio of exposure given disease. This implies that mutations promoting individual pathways will be diluted by the presence of multiple pathways [8], the diversity of which confers to CAD its multi-factorial phenotypes. Corollary to this, is that any highly penetrant SNP likely to be discovered in a basic GWAS is unlikely to be associated with any of those specific pathways. As has been made clear in the development presented here, there are a number of possible mechanisms by which mutations can impact the activity of proteins, a number of specific ones have been elucidated in CAD and its associated risk factors. Yet, both the selection of genome-wide significant SNPs and adjustment by metabolic disorder risk factors tend to exclude SNPs that might point to pathway-specific mutations.

Genome-wide significance selects snps showing sufficient significance to be unlikely to have been observed by chance among all SNPs tested, at the expense of rejecting a large number of false negatives. At the same time, gene expression, interactions among gene-environment hypotheses, and gene-gene interactions, along with expression data and other phenotype data being considered, greatly exacerbate the problem of large hypothesis counts. Yet, the likely etiology involves organization of processes into regulatory networks with a limited number of pathways to disease manifestation. Use of these organizations to reduce the number of SNPs involved in hypotheses, can reduce the number of hypotheses being tested, boosting power. Also, the organization of hypotheses into multi-factor complexes of phenotype markers can drastically increase the power of the tests for each SNP.

The wealth of information actually makes it more difficult to account for the missing variability explained by measured SNPs since it becomes more difficult to account for which SNPs are actually carrying valid information. Yet, the amount of the variation accounted for may not be the right measure of how much understanding of the underlying biological mechanisms involved in the etiology of CAD. If that is understood from bottom to top, the functional impact of any given SNP could be deduced. It is clear that the elucidation of the pathways involved is taking center stage.

Key Facts

I. CAD is a multifactorial disease with both genetic and behavioral components that tends to run in families. It tends to manifest multiple subtly different phenotypes reflecting multiple possible pathways. Since the odds ratio predicting disease given exposure is identical to the odds ratio of an exposure given the disease, if any of those pathways is sufficient to cause CAD, and they are independent, the odds ratio for that pathway must be diluted if it is just one of many pathways leading to CAD in the population. SNPs implicated by a specific pathway are likely to be relatively weak. Corollary to that is that highly penetrant SNPs are likely active regardless of pathway. Further, the general notion of applying an "adjustment" by including environmental and risk factor variates in the logistic regression to isolate the impact of individual SNPs on disease essentially exclude pathways marked by those risk factors. For CAD, the primary risk factors often included in that "adjustment" are metabolic syndrome diagnoses. Essentially some of the primary candidates of genetic pathways get explicitly factored out in this modality of analysis. So pathway information is hit two ways in standard GWAS analysis of CAD.

II. The large number of hypotheses being tested in GWAS lead to the possibility of large numbers of false positives simply by chance (variations among random samples will show significant odds ratios even if there is no underlying interactions between disease and exposures) unless p-value thresholds are set very strictly. This leads to large numbers of false negatives. Efforts to identify interactions between SNPs, gene expression levels and SNPs, environmental changes and expression levels, and other features that explore the correlation of combinations of features will drastically expand that number of hypotheses, resulting in much stricter thresholds and a much higher percentage of false negative rejections. The solutions to this are to restrict the space of hypotheses considered, and/or to consider more complex phenotype structures. For example, the chances of seeing seven or more out of 10 traits supported by a single SNP at the 95% confidence level is 8.198×10^{-8}. There is some overlap between these two in that network or ontology membership could be viewed as a matchable trait.

III. The most striking feature of most of the emerging areas being explored, such as epigenetic effects is the emergence of genome wide methods. In some cases, such as miRNA, the proliferation of candidates that likely impact regulatory events and for which mappings among genes has already been performed, anticipated the availability of information that could have taken advantage of those efforts. Even Environmental Wide Association Studies (EWAS) are being explored, where a much broader range of environment variables are mined for their associations with disease in much the way that GWAS has been done. The preliminary work involving regulatory changes given environmental exposures immediately suggests the combination of these on a broad basis, Gene-Environment Wide Association Studies (GEWAS) is a logical next step.

IV. Since the larger part of variability in accounting for CAD phenotypic variation has not been accounted for, a central goal of this review was to explore the possible sources of most of the variation. GWAS gives just a bit of accounting, which represents how much can be grasped without accounting for deeper pathways and mechanisms. Within the last decade, there has been a huge explosion of information regarding regulatory mechanisms. A number of them allow for some plasticity in the response of tissues to environment, which may also be key to the weaknesses leading to CAD – they could be simply normal tissues responding through normal pathways to very adverse environmental effects. The interaction of these environmental impacts across genetic variations in the population may show much more detail in how normal pathways break. Alternatively, GWAS discards a large part of its potential accounting among false positives since none of these can rise above threshold to exclude the possibility from those due to random sampling variations.

V. Is that the right question? In a sense, accounting for all the variability implies that all the information needed to put together the puzzle pieces have been assembled. Without that accounting, you don't know if you have enough pieces to understand the real question: "Why?" Of these, some of the ontological and network classifications that are being pieced together are really core to such explanations – and they focus the combinatorial hypotheses to much more manageable numbers all at the same time.

References

[1] Lopez, AD; Mathers, CD; Ezzati, M; Jamison, DT; Murray, CJ. Global and regional burden of disease and risk factors, 2001: systematic analysis of population health data. *Lancet*, 367, 1747-1757, doi:S0140-6736(06)68770-9 [pii] 10.1016/S0140-6736(06) 68770-9 (2006).

[2] Yusuf, S; et al. Effect of potentially modifiable risk factors associated with myocardial infarction in 52 countries (the INTERHEART study): case-control study. *Lancet*, 364, 937-952, doi:10.1016/S0140-6736(04)17018-9S0140-6736(04)17018-9 [pii] (2004).

[3] Lloyd-Jones, DM; et al. Parental Cardiovascular Disease as a Risk Factor for Cardiovascular Disease in Middle-aged Adults. *JAMA: The Journal of the American Medical Association*, 291, 2204-2211, doi:10.1001/jama.291.18.2204 (2004).

[4] Fischer, M; et al. Distinct heritable patterns of angiographic coronary artery disease in families with myocardial infarction. *Circulation*, 111, 855-862, doi:10.1161/ 01.CIR.0000155611.41961.BB (2005).

[5] Chan, L; Boerwinkle, E. Gene-environment interactions and gene therapy in atherosclerosis. *Cardiol Rev.*, 2, 130-137 (1994).

[6] Zdravkovic, S; et al. Heritability of death from coronary heart disease: a 36-year follow-up of 20 966 Swedish twins. *Journal of internal medicine*, 252, 247-254 (2002).

[7] CARDIoGRAMplusC4D Consortium et al. Large-scale association analysis identifies new risk loci for coronary artery disease. *Nature genetics*, 45, 25-33, doi:10.1038/ng.2480 (2013).

[8] Wilson, PW. Diabetes mellitus and coronary heart disease. *Am J Kidney Dis*, 32, S89-100 (1998).

[9] Casas, JP; Cooper, J; Miller, GJ; Hingorani, AD; Humphries, SE. Investigating the genetic determinants of cardiovascular disease using candidate genes and meta-analysis of association studies. *Annals of human genetics*, 70, 145-169, doi:10.1111/j.1469-1809.2005.00241.x (2006).

[10] Zalloua, PA; et al. Identifying genetic traces of historical expansions: Phoenician footprints in the Mediterranean. *Am J Hum Genet*, 83, 633-642, doi:10.1016/ j.ajhg.2008.10.012 (2008).

[11] Horne, BD; Carlquist, JF; Muhlestein, JB; Bair, TL; Anderson, JL. Association of variation in the chromosome 9p21 locus with myocardial infarction versus chronic

coronary artery disease. *Circulation. Cardiovascular genetics*, 1, 85-92, doi:10.1161/ CIRCGENETICS. 108.793158 (2008).

[12] Ardissino, D; et al. Influence of 9p21.3 genetic variants on clinical and angiographic outcomes in early-onset myocardial infarction. *Journal of the American College of Cardiology*, 58, 426-434, doi:10.1016/ j.jacc.2010.11.075 (2011).

[13] Helgadottir, A; et al. The same sequence variant on 9p21 associates with myocardial infarction, abdominal aortic aneurysm and intracranial aneurysm. *Nature genetics*, 40, 217-224, doi:10.1038/ng.72 (2008).

[14] Gschwendtner, A; et al. Sequence variants on chromosome, 9p21.3 confer risk for atherosclerotic stroke. *Annals of neurology*, 65, 531-539, doi:10.1002/ana.21590 (2009).

[15] Matarin, M; et al. Whole genome analyses suggest ischemic stroke and heart disease share an association with polymorphisms on chromosome, 9p21. *Stroke; a journal of cerebral circulation*, 39, 1586-1589, doi:10.1161/STROKEAHA.107.502963 (2008).

[16] Patel, RS; et al. The chromosome 9p21 risk locus is associated with angiographic severity and progression of coronary artery disease. *European heart journal*, 31, 3017-3023, doi:10.1093/eurheartj/ehq272 (2010).

[17] Bush, WS; Moore, JH. in *PLoS Comput Biol* Vol. 8(12) *Translational Bioinformatics* Ch. Chapter 11, e1002822 (PLoS, 2011).

[18] Bland, JM; Altman, DG. Regression towards the mean. *Bmj*, 308, 1499 (1994).

[19] Moskvina, V; Holmans, P; Schmidt, KM; Craddock, N. Design of case-controls studies with unscreened controls. *Annals of human genetics*, 69, 566-576, doi:10.1111/j.1529-8817.2005.00175.x (2005).

[20] Marchini, J; Howie, B; Myers, S; McVean, G; Donnelly, P. A new multipoint method for genome-wide association studies by imputation of genotypes. *Nature genetics*, 39, 906-913, doi:10.1038/ng2088 (2007).

[21] Scheet, P; Stephens, M. A fast and flexible statistical model for large-scale population genotype data: applications to inferring missing genotypes and haplotypic phase. *Am J Hum Genet*, 78, 629-644, doi:10.1086/502802 (2006).

[22] Zeggini, E; et al. Meta-analysis of genome-wide association data and large-scale replication identifies additional susceptibility loci for type, 2 diabetes. *Nature genetics*, 40, 638-645, doi:10.1038/ng.120 (2008).

[23] Barrett, JC; et al. Genome-wide association defines more than, 30 distinct susceptibility loci for Crohn's disease. *Nature genetics*, 40, 955-962, doi:10.1038/ng.175 (2008).

[24] Spencer, CC; Su, Z; Donnelly, P; Marchini, J. Designing genome-wide association studies: sample size, power, imputation, and the choice of genotyping chip. *PLoS genetics*, 5, e1000477, doi:10.1371/ journal.pgen.1000477 (2009).

[25] Deloukas, P; et al. Large-scale association analysis identifies new risk loci for coronary artery disease. *Nature genetics*, 45, 25-33, doi:10.1038/ ng.2480 (2012).

[26] Haiman, CA; et al. Multiple regions within, 8q24 independently affect risk for prostate cancer. *Nature genetics*, 39, 638-644, doi:10.1038/ ng2015 (2007).

[27] Manolio, TA; et al. Finding the missing heritability of complex diseases. *Nature*, 461, 747-753, doi:10.1038/nature08494 (2009).

[28] McCarroll, SA; et al. Deletion polymorphism upstream of IRGM associated with altered IRGM expression and Crohn's disease. *Nature genetics*, 40, 1107-1112, doi:10.1038/ng.215 (2008).

[29] Kiezun, A; et al. Exome sequencing and the genetic basis of complex traits. *Nature genetics*, 44, 623-630, doi:10.1038/ng.2303 (2012).
[30] Frazer, KA; Murray, SS; Schork, NJ; Topol, EJ. Human genetic variation and its contribution to complex traits. *Nature reviews. Genetics*, 10, 241-251, doi:nrg2554 [pii]10.1038/nrg2554 (2009).
[31] Gorlov, IP; Gorlova, OY; Sunyaev, SR; Spitz, MR; Amos, CI. Shifting paradigm of association studies: value of rare single-nucleotide polymorphisms. *Am J Hum Genet*, 82, 100-112, doi:S0002-9297(07)00012-2 [pii]10.1016/j.ajhg.2007.09.006 (2008).
[32] Kryukov, GV; Pennacchio, LA; Sunyaev, SR. Most rare missense alleles are deleterious in humans: implications for complex disease and association studies. *Am J Hum Genet*, 80, 727-739, doi:S0002-9297(07)61104-5 [pii]10.1086/513473 (2007).
[33] Pritchard, JK. Are rare variants responsible for susceptibility to complex diseases? *Am J Hum Genet*, 69, 124-137, doi:S0002-9297(07)61452-9 [pii]10.1086/321272 (2001).
[34] Li, B; Leal, SM. Discovery of rare variants via sequencing: implications for the design of complex trait association studies. *PLoS genetics*, 5, e1000481, doi:10.1371/ journal.pgen.1000481 (2009).
[35] Eyre, S; et al. Examining the overlap between genome-wide rare variant association signals and linkage peaks in rheumatoid arthritis. *Arthritis and rheumatism*, 63, 1522-1526, doi:10.1002/art.30315 (2011).
[36] Zeggini, E; et al. Replication of genome-wide association signals in UK samples reveals risk loci for type 2 diabetes. *Science*, 316, 1336-1341, doi:10.1126/ science.1142364 (2007).
[37] Diabetes Genetics Initiative of Broad Institute of, H; et al. Genome-wide association analysis identifies loci for type 2 diabetes and triglyceride levels. *Science*, 316, 1331-1336, doi:10.1126/science.1142358 (2007).
[38] Bouatia-Naji, N; et al. A variant near MTNR1B is associated with increased fasting plasma glucose levels and type 2 diabetes risk. *Nature genetics*, 41, 89-94, doi:10.1038/ng.277 (2009).
[39] Sladek, R; et al. A genome-wide association study identifies novel risk loci for type, 2 diabetes. *Nature*, 445, 881-885, doi:10.1038/nature05616 (2007).
[40] Scott, LJ; et al. A genome-wide association study of type 2 diabetes in Finns detects multiple susceptibility variants. *Science*, 316, 1341-1345, doi:10.1126/science.1142382 (2007).
[41] Bonnefond, A; et al. Rare MTNR1B variants impairing melatonin receptor 1B function contribute to type 2 diabetes. *Nature genetics*, 44, 297-301, doi:10.1038/ng.1053 (2012).
[42] Abraham, G; Bhalala, OG; de Bakker, PI; Ripatti, S; Inouye, M. Towards a molecular systems model of coronary artery disease. *Current cardiology reports*, 16, 488, doi:10.1007/s11886-014-0488-1 (2014).
[43] Kerkel, K; et al. Genomic surveys by methylation-sensitive SNP analysis identify sequence-dependent allele-specific DNA methylation. *Nature genetics*, 40, 904-908, doi:10.1038/ng.174 (2008).
[44] King, DG; Kashi, Y. Indirect selection for mutability. *Heredity*, 99, 123-124, doi:10.1038/sj.hdy.6800998 (2007).
[45] Scarano, E; Iaccarino, M; Grippo, P; Parisi, E. The heterogeneity of thymine methyl group origin in DNA pyrimidine isostichs of developing sea urchin embryos.

Proceedings of the National Academy of Sciences of the United States of America, 57, 1394-1400 (1967).

[46] Hernando-Herraez, I; et al. Dynamics of DNA methylation in recent human and great ape evolution. *PLoS genetics*, 9, e1003763, doi:10.1371/journal.pgen.1003763 (2013).

[47] Medvedeva, YA; et al. Effects of cytosine methylation on transcription factor binding sites. *BMC genomics*, 15, 119, doi:10.1186/1471-2164-15-119 (2014).

[48] Inouye, M; et al. Novel Loci for metabolic networks and multi-tissue expression studies reveal genes for atherosclerosis. *PLoS genetics*, 8, e1002907, doi:10.1371/journal.pgen.1002907 (2012).

[49] Alon, U. *An Introduction to Systems Biology: Design Principals of Biological Circuits*. (Chapman & Hall/CRC, 2007).

[50] Smemo, S; et al. Obesity-associated variants within FTO form long-range functional connections with IRX3. *Nature*, 507, 371-375, doi:10.1038/nature13138 (2014).

[51] Nair, J; Ghatge, M; Kakkar, VV; Shanker, J. Network analysis of inflammatory genes and their transcriptional regulators in coronary artery disease. *PloS one*, 9, e94328, doi:10.1371/journal.pone.0094328 (2014).

[52] Nair, J; Shanker, J; Jambunathan, S; Arvind, P; Kakkar, VV. Expression analysis of leukotriene-inflammatory gene interaction network in patients with coronary artery disease. *Journal of atherosclerosis and thrombosis*, 21, 329-345 (2014).

[53] Zhang, Y; et al. Network-based analysis reveals distinct association patterns in a semantic MEDLINE-based drug-disease-gene network. *Journal of biomedical semantics*, 5, 33, doi:10.1186/2041-1480-5-33 (2014).

[54] Stenvinkel, P; et al. Impact of inflammation on epigenetic DNA methylation - a novel risk factor for cardiovascular disease? *Journal of internal medicine*, 261, 488-499, doi:10.1111/j.1365-2796.2007.01777.x (2007).

[55] Baccarelli, A; Rienstra, M; Benjamin, EJ. Cardiovascular epigenetics: basic concepts and results from animal and human studies. *Circulation. Cardiovascular genetics*, 3, 567-573, doi:10.1161/CIRCGENETICS. 110.958744 (2010).

[56] Creemers, EE; Tijsen, AJ; Pinto, YM. Circulating microRNAs: novel biomarkers and extracellular communicators in cardiovascular disease? *Circulation research*, 110, 483-495, doi:10.1161/CIRCRESAHA. 111.247452 (2012).

[57] Schleithoff, C; Voelter-Mahlknecht, S; Dahmke, IN; Mahlknecht, U. On the epigenetics of vascular regulation and disease. *Clinical epigenetics*, 4, 7, doi:10.1186/1868-7083-4-7 (2012).

[58] McDonald, OG; Owens, GK. Programming smooth muscle plasticity with chromatin dynamics. *Circulation research*, 100, 1428-1441, doi:10.1161/01.RES.0000266448.30370.a0 (2007).

[59] Weber, M; et al. Distribution, silencing potential and evolutionary impact of promoter DNA methylation in the human genome. *Nature genetics*, 39, 457-466, doi:10.1038/ng1990 (2007).

[60] Jiang, M; et al. Hypermethylation of hepatic glucokinase and L-type pyruvate kinase promoters in high-fat diet-induced obese rats. *Endocrinology*, 152, 1284-1289, doi:10.1210/en.2010-1162 (2011).

[61] Castro, R; et al. Increased homocysteine and S-adenosylhomocysteine concentrations and DNA hypomethylation in vascular disease. *Clinical chemistry*, 49, 1292-1296 (2003).

[62] Yi, P; et al. Increase in plasma homocysteine associated with parallel increases in plasma S-adenosylhomocysteine and lymphocyte DNA hypomethylation. *The Journal of biological chemistry*, 275, 29318-29323, doi:10.1074/jbc.M002725200 (2000).

[63] Castro, R; Rivera, I; Blom, HJ; Jakobs, C; Tavares de Almeida, I. Homocysteine metabolism, hyperhomocysteinaemia and vascular disease: an overview. *Journal of inherited metabolic disease*, 29, 3-20, doi:10.1007/s10545-006-0106-5 (2006).

[64] Sharma, P; et al. Detection of altered global DNA methylation in coronary artery disease patients. *DNA and cell biology*, 27, 357-365, doi:10.1089/dna.2007.0694 (2008).

[65] Post, WS; et al. Methylation of the estrogen receptor gene is associated with aging and atherosclerosis in the cardiovascular system. *Cardiovascular research*, 43, 985-991 (1999).

[66] Kim, J; et al. Epigenetic changes in estrogen receptor beta gene in atherosclerotic cardiovascular tissues and in-vitro vascular senescence. *Biochimica et biophysica acta*, 1772, 72-80, doi:10.1016/j.bbadis. 2006.10.004 (2007).

[67] Friso, S; et al. Promoter methylation in coagulation F7 gene influences plasma FVII concentrations and relates to coronary artery disease. *Journal of medical genetics*, 49, 192-199, doi:10.1136/jmedgenet-2011-100195 (2012).

[68] Yideng, J; et al. Homocysteine-mediated expression of SAHH, DNMTs, MBD2, and DNA hypomethylation potential pathogenic mechanism in VSMCs. *DNA and cell biology*, 26, 603-611, doi:10.1089/dna. 2007.0584 (2007).

[69] Smallwood, SA; et al. Single-cell genome-wide bisulfite sequencing for assessing epigenetic heterogeneity. *Nature methods*, 11, 817-820, doi:10.1038/nmeth.3035 (2014).

[70] Goldberg, AD; Allis, CD; Bernstein, E. Epigenetics: a landscape takes shape. *Cell*, 128, 635-638, doi:10.1016/j.cell.2007.02.006 (2007).

[71] Volkel, P; Angrand, PO. The control of histone lysine methylation in epigenetic regulation. *Biochimie*, 89, 1-20, doi:10.1016/j.biochi. 2006.07.009 (2007).

[72] Choi, JH; et al. Trichostatin A exacerbates atherosclerosis in low density lipoprotein receptor-deficient mice. *Arteriosclerosis, thrombosis, and vascular biology*, 25, 2404-2409, doi:10.1161/01.ATV. 0000184758. 07257.88 (2005).

[73] Papait, R; et al. Genome-wide analysis of histone marks identifying an epigenetic signature of promoters and enhancers underlying cardiac hypertrophy. *Proceedings of the National Academy of Sciences of the United States of America*, 110, 20164-20169, doi:10.1073/pnas. 1315155110 (2013).

[74] Dayeh, T; et al. Genome-wide DNA methylation analysis of human pancreatic islets from type, 2 diabetic and non-diabetic donors identifies candidate genes that influence insulin secretion. *PLoS genetics*, 10, e1004160, doi:10.1371/journal.pgen.1004160 (2014).

[75] Bibikova, M; et al. High density DNA methylation array with single CpG site resolution. *Genomics*, 98, 288-295, doi:10.1016/j.ygeno. 2011.07.007 (2011).

[76] Guttman, M; et al. Chromatin signature reveals over a thousand highly conserved large non-coding RNAs in mammals. *Nature*, 458, 223-227, doi:10.1038/nature07672 (2009).

[77] Malecova, B; Morris, KV. Transcriptional gene silencing through epigenetic changes mediated by non-coding RNAs. *Current opinion in molecular therapeutics*, 12, 214-222 (2010).

[78] Saxena, A; Carninci, P. Long non-coding RNA modifies chromatin: epigenetic silencing by long non-coding RNAs. *BioEssays : news and reviews in molecular, cellular and developmental biology*, 33, 830-839, doi:10.1002/bies.201100084 (2011).

[79] Congrains, A; et al. Genetic variants at the, 9p21 locus contribute to atherosclerosis through modulation of ANRIL and CDKN2A/B. *Atherosclerosis*, 220, 449-455, doi:10.1016/j.atherosclerosis. 2011.11. 017 (2012).

[80] McPherson, R; et al. A common allele on chromosome, 9 associated with coronary heart disease. *Science*, 316, 1488-1491, doi:10.1126/ science.1142447 (2007).

[81] Zhuang, J; et al. Methylation of p15INK4b and expression of ANRIL on chromosome, 9p21 are associated with coronary artery disease. *PloS one*, 7, e47193, doi:10.1371/journal.pone.0047193 (2012).

[82] Aguilo, F; Zhou, MM; Walsh, MJ. Long noncoding RNA, polycomb, and the ghosts haunting INK4b-ARF-INK4a expression. *Cancer research*, 71, 5365-5369, doi:10.1158/0008-5472.CAN-10-4379 (2011).

[83] Holdt, LM; et al. ANRIL expression is associated with atherosclerosis risk at chromosome, 9p21. *Arteriosclerosis, thrombosis, and vascular biology*, 30, 620-627, doi:10.1161/ATVBAHA.109.196832 (2010).

[84] Carthew, RW; Sontheimer, EJ. Origins and Mechanisms of miRNAs and siRNAs. *Cell*, 136, 642-655, doi:10.1016/j.cell.2009.01.035 (2009).

[85] Fichtlscherer, S; et al. Circulating microRNAs in patients with coronary artery disease. *Circulation research*, 107, 677-684, doi:10.1161/ CIRCRESAHA.109.215566 (2010).

[86] De Rosa, S; et al. Transcoronary concentration gradients of circulating microRNAs. *Circulation*, 124, 1936-1944, doi:10.1161/ CIRCULATIONAHA. 111.037572 (2011).

[87] Sondermeijer, BM; et al. Platelets in patients with premature coronary artery disease exhibit upregulation of miRNA340* and miRNA624*. *PloS one*, 6, e25946, doi:10.1371/journal.pone.0025946 (2011).

[88] Menghini, R; et al. MicroRNA, 217 modulates endothelial cell senescence via silent information regulator, 1. *Circulation*, 120, 1524-1532, doi:10.1161/CIRCULATION AHA.109.864629 (2009).

[89] Davignon, J; Ganz, P. Role of endothelial dysfunction in atherosclerosis. *Circulation*, 109, III27-32, doi:10.1161/01.CIR.0000131515.03336.f8 (2004).

[90] Potente, M; Dimmeler, S. Emerging roles of SIRT1 in vascular endothelial homeostasis. *Cell cycle*, 7, 2117-2122 (2008).

[91] Potente, M; et al. Involvement of Foxo transcription factors in angiogenesis and postnatal neovascularization. *The Journal of clinical investigation*, 115, 2382-2392, doi:10.1172/JCI23126 (2005).

[92] Hindorff, LA; et al. Potential etiologic and functional implications of genome-wide association loci for human diseases and traits. *Proceedings of the National Academy of Sciences of the United States of America*, 106, 9362-9367, doi:10.1073/pnas. 0903103106 (2009).

[93] Rigoutsos, I; et al. Short blocks from the noncoding parts of the human genome have instances within nearly all known genes and relate to biological processes. *Proceedings*

of the National Academy of Sciences of the United States of America, 103, 6605-6610, doi:10.1073/ pnas.0601688103 (2006).

[94] Tsirigos, A; Rigoutsos, I. Human and mouse introns are linked to the same processes and functions through each genome's most frequent non-conserved motifs. *Nucleic acids research*, 36, 3484-3493, doi:10.1093/ nar/gkn155 (2008).

[95] Glinsky, GV. Human genome connectivity code links disease-associated SNPs, microRNAs and pyknons. *Cell cycle*, 8, 925-930 (2009).

[96] Lai, X; et al. Computational analysis of target hub gene repression regulated by multiple and cooperative miRNAs. *Nucleic acids research*, 40, 8818-8834, doi:10.1093/nar/gks657 (2012).

[97] Harats, D; et al. Cigarette smoking renders LDL susceptible to peroxidative modification and enhanced metabolism by macrophages. *Atherosclerosis*, 79, 245-252 (1989).

[98] Pech-Amsellem, MA; et al. Enhanced modifications of low-density lipoproteins (LDL) by endothelial cells from smokers: a possible mechanism of smoking-related atherosclerosis. *Cardiovascular research*, 31, 975-983 (1996).

[99] Vruwink, KG; Gershwin, ME; Sachet, P; Halpern, G; Davis, PA. Modification of human LDL by in vitro incubation with cigarette smoke or copper ions: implications for allergies, asthma and atherosclerosis. *Journal of investigational allergology & clinical immunology*, 6, 294-300 (1996).

[100] Steffen, Y; et al. Cigarette smoke and LDL cooperate in reducing nitric oxide bioavailability in endothelial cells via effects on both eNOS and NADPH oxidase. *Nitric oxide : biology and chemistry / official journal of the Nitric Oxide Society*, 27, 176-184, doi:10.1016/j.niox.2012.06.006 (2012).

[101] Patel, CJ; Bhattacharya, J; Butte, AJ. An Environment-Wide Association Study (EWAS) on type, 2 diabetes mellitus. *PloS one*, 5, e10746, doi:10.1371/ journal.pone.0010746 (2010).

In: Gene-Environment Interactions and Human Diseases
Editor: Lu Qi
ISBN: 978-1-63483-680-7

Chapter 9

RECENT ADVANCES IN THE EXPLORATION OF GENE-ENVIRONMENT INTERACTIONS IN PARKINSON'S DISEASE

John P. Wise, Jr. and Jason R. Cannon*
School of Health Sciences, Purdue University, West Lafayette, IN, US

ABSTRACT

Parkinson's disease (PD) is a very complex progressive neurodegenerative disorder that affects millions of people around the world, but has few known causes and no known cure. In most cases, the onset of the disease is likely due to a combination of genetic susceptibility and exposure to environmental risk factors – though these interactions are only just starting to be identified. In this review we summarize what is known about the genetic and environmental risk factors separately, what interactions have been studied so far, the most prominent pathogenic mechanisms underlying PD, and experimental tools that are used to study these interactions and mechanisms. Though the current understanding of gene-environment interactions is limited in scope and depth, there is much value in what has been discovered and the future direction of this area of research has started down an exciting and promising path towards understanding PD etiology.

1. PD OVERVIEW

In 1817, Dr. James Parkinson described a condition of involuntary resting tremor, rigid and stooped posture, a short shuffling gait, and weakness of the limbs, which he called 'shaking palsy' [1]. This condition, now referred to as Parkinson's disease (PD), has since been characterized pathologically as the progressive loss of dopaminergic (DA) neurons in the substantia nigra pars compacta (SNpc), and formation of intracellular cytoplasmic inclusions in surviving DA neuronal cell bodies (Lewy bodies) and processes (Lewy

* Corresponding author: Jason R. Cannon, Ph.D. Assistant Professor of Health Sciences and Toxicology, Purdue University, HAMP-1271, 550 Stadium Mall Dr.West Lafayette, IN 47907,Tel: (765) 494-0794 Fax: (765) 496-1377, E-mail: cannonjr@purdue.edu.

neurites), which contain α-synuclein, other proteins, ubiquitin, and cellular breakdown products [2-6]. Today, PD is the second most common neurodegenerative disease, affecting more than 5 million people globally, with prevalence significantly increasing among those aged 65 or older [7]. Diagnostic tests are utilized by neurologists, but PD cannot be absolutely verified until post-mortem analysis. The most reliable diagnostic tests require specific criteria for motor signs and symptoms (e.g., rigidity, bradykinesia, L-DOPA responsiveness), which only manifest after ~60% loss of DA neurons in the substantia nigra pars compacta [8-10]. Unfortunately, due to this late-stage diagnosis and the complexity of the disease pathogenesis and progression, a cure remains elusive and current therapies are only able to treat symptoms for a limited time.

For the majority of cases, determining causative factors for PD onset has been nearly impossible – only about 10% of all cases have known genetic links. The remaining 90% of cases are suspected to be caused by environmental exposures to various toxicants in combination with genetic predisposition. The drug derivative 1-methyl-4-phenyl-1,2,3,6-tetrahydropyridine (MPTP) provided the first specific clear evidence that neurotoxicants could reproduce major features of PD and served as a launch pad for an expansive amount of PD mechanistic research, demonstrating that a chemical could cross the blood brain barrier (BBB), enter dopaminergic cells, and target the mitochondrial electron transport chain [11-14]. The chemical structure of MPTP has been used to identify other potential PD toxicants with similar structures and to test genetic variants that may result in heightened sensitivity to DA toxicants.

Experimental evaluation of PD neuropathology has revealed (1) while other brain regions may be affected, selective loss of nigral dopamine neurons typically occurs, (2) the distribution of Lewy bodies and Lewy neurites gradually progresses from the dorsal motor vagal nucleus to the cortex, and (3) enteric nervous system pathology may precede brain pathology. This progression has been classified into six different stages with clinical symptoms appearing between stages 3 and 4. By stage 3 the Lewy neurites have reached the SNpc but it is still macroscopically intact and there are no signs of neuronal loss. Other regions that become involved in pathology at this stage include the cholinergic tegmental pedunculopontine nucleus, the oral raphe nuclei, the cholinergic magnocellular nuclei of the basal forebrain, and the hypothalamic tuberomamillary nucleus. Stage 4 shows the first signs of pathology in the cerebral cortex (more specifically the temporal mesocortex) and in the second sector of Ammon's horn [2, 15]. All brain regions affected by these protein inclusions share two common properties; the neuronal projections are poorly myelinated or completely unmyelinated, and the neurons all have an unusually large number of projections [15, 16]. For example, Matsuda et al. [16] predict ~75,000 striatal neurons are influenced by a single dopaminergic neuron in the substantia nigra. These two properties result in a heightened metabolic demand for normal cellular function and neurotransmission, which causes increased stress on the electron transport chain of the mitochondria and increased production of reactive oxygen species (ROS), leading to oxidative stress. Oxidative stress is a major component in PD progression and likely an important factor in the earliest stages. The electron transport chain normally produces ROS during cellular respiration primarily at mitochondrial complex I, with contribution from complex III, the adverse effects of which are tightly controlled by mitochondrial reductive enzymes (e.g., manganese superoxide dismutase). In addition, dopamine metabolism produces ROS and can auto-oxidize, rendering DA cells more vulnerable to additional oxidative insults [17]. Oxidative stress occurs when the amount of ROS produced exceeds the capacity that reductive enzymes can neutralize. Given the high metabolic demand on neurons with large, unmyelinated projections, it is plausible that this population of cells is especially vulnerable to toxicants that induce ROS and/or target mitochondrial respiration – both of which have been demonstrated epidemiologically and mechanistically to be likely risk factors for idiopathic PD.

Numerous other mechanisms have been linked to PD pathogenesis, but it is difficult to determine which are causes or byproducts of the disease. At least three overarching pathogenic pathways have been repeatedly implicated in genetically inherited and sporadic cases: mitochondrial dysfunction, oxidative stress, and impaired protein clearance pathways. Oxidative stress, andresulting oxidative damage have been implicated in PD models and clinical PD due to the high metabolic demand and unique cellular environment of dopaminergic cells. In addition, many of the other mechanisms involved in PD pathogenesis are exacerbated by and increase the amount of oxidative stress present within the cell, undoubtedly overwhelming the reductive capacity of antioxidants and neutralizing enzymes. The cause for mitochondrial dysfunction has been linked to multiple genetic mutations (e.g., Parkin, PINK1, and DJ-1) and environmental toxicants (e.g., MPTP, rotenone, paraquat). As with many other neurodegenerative diseases, PD is characterized by the formation of protein aggregates in neuronal bodies and neurites. These aggregates likely initially form to sequester improperly folded proteins tagged for disassembly and amino acid recycling. However, the two major pathways for aggregate clearance, the ubiquitin proteasome system and autophagy, have been shown to malfunction and causing aggregates to persist, expand, and spread.

2. EPIDEMIOLOGY

Conclusions from epidemiological studies investigating PD incidence have been inconclusive due to inconsistent methodology, non-standardized diagnosis, and difficulty diagnosing PD until post-mortem [18, 19]. The method with the least variation reported across studies was conducted in Europe and found overall incidence rates between 410 and 529 per 100,000 person-years [20-22]. Age has been repeatedly determined to be the strongest risk factor for PD diagnosis with incidence increasing from an average of 2.5 per 100,000 at age 50 to greater than 100 per 100,000 by age 80 [23]. Other risk factors include genetic links and exposure to various environmental toxicants.

2.A. Genetic Epidemiology

Genetic epidemiology studies have demonstrated that 12 genes can be linked to heritable causes of PD with at least 20 additional loci; these monogenic forms of PD are listed in Table 1 [19, 24, 25]. The first of these to be identified were autosomal dominant mutations in the SNCA gene, which generates the α-synuclein protein [26]. The normal function of the α-synuclein protein is not fully understood, but it is suggested to be involved in vesicle turnover, synaptic release, synapse development, synaptic plasticity, and intracellular trafficking within the endoplasmic reticulum/Golgi network [27-32]. Patients with SNCA-linked PD are often diagnosed at earlier ages, have a more rapidly progressing disease, exhibit less tremor, sometimes have severe dementia, and demonstrate the typical midbrain dopaminergic neuronal loss and accumulation of Lewy bodies [33]. Three SNCA point mutations have been reported to cause PD – A53T, A30P, and E46K – but this is extremely rare [34]. Two additional mutations have been reported more recently – G51D, H50Q [35, 36]. The most common of these, A53T, was found in 12 Mediterranean families and a Korean family, suggesting two different common ancestors with the initial mutation [33, 37]. SNCA gene expansions are more common and are due to the duplication or triplication of the locus, which also could be caused by mutations in the promoter region [38-41].

Table 1. Genetic mutations associated with Parkinson's disease

Gene	Biological Function	Mutations (Amino Acid Change)	Age of Onset	Epidemiology	References
SNCA/α-synuclein	Involved in synaptic vesicle formation	G209A (A53T), duplication/triplication, G88C (A30P), G188A (E46K), T150G (H50Q), G152A (G51D)	<50	A53T – 1 in 1,921 (1 in 237 with family history) (Berg et al., 2005)	[24, 26, 41, 58, 187]
LRRK2 (PARK8)	Protein kinase	G6099A (G2019S), (N1437H), (R1441C), (R1441G), (R1441H), (Y1699C), (I2020T)	>50	10% autosomal dominant, 4% sporadic	[24, 45, 58, 188]
Parkin	E3 ligase (component of UPS), also involve in mitophagy	>100 mutations (see Grünewald and Klein, 2013)	<50	~50% cases w/ autosomal recessive early onset, 10-20% non-familial cases w/ early onset, 0.4-0.7% of non-familial cases	[58, 189, 190]
PARK3	Unknown	None found	Average 59	Unknown	[19, 190]
UCH-L1	Deubiquitinating enzyme, involved in UPS	C277G (I93M)	49, 51	2 cases documented	[51, 58]
PINK1	Mitochondrial kinase, involved in mitophagy	>40 point mutations (see Tan, 2013)	<50	1-8% early-onset and/or familial PD	[58, 65, 191]
DJ-1	Involved in oxidative stress response, likely serves a protective role	>10 point mutations, T497C (L166P); (see Tan, 2013)	<50	<2% of early-onset or recessive parkinsonism	[58, 65, 192-194]
ATP13A2	Lysosomal ATPase	>5 point mutations 3057delC (1019GframesshiftX1021), 1306+5G→A, 1632_1653dup22	Juvenile	Unknown	[58, 60]

Gene	Biological Function	Mutations (Amino Acid Change)	Age of Onset	Epidemiology	References
VPS35	Subunit of the retromer complex, involved in retrograde transport between endosomes and trans-Golgi network	G1858A (D620N)	>50	<1/500 to 1/100	[24, 25, 195, 196]
EIF4G1	Core scaffold for translation initiation complex that regulates translation of mRNAs of mitochondrial, cell survival and growth genes	(R1502H), C1505T (A502V), G2056T (G686C), A3490C (S1164R), C3589T (R1197W)	>50 (mean= 64)	0.2% for R1502H, 0.02% for A502V	[25, 197]
PLA2G6	Catalytic enzyme involved in inflammatory regulation and apoptosis	(R741Q), (R747W), (Q452X), (R635Q), (R632W), (D331Y)	Juvenile	Unknown	[24, 198, 199]
FBXO7	Involved in ubiquitin labeling/proteasom e pathway	C1132G (R378G), C1492T (R498X), C65T (T22M), C1441T (R481C)	Juvenile	Unknown	[24, 199, 200]
OMI/ HTRA2	Mitochondrial-targeted serine protease	G399S, A141S, R404W	>50	Unknown	[41, 45, 201-203]

The most common form of autosomal dominant monogenic PD is located on chromosome 12 at the PARK8 locus and encodes the LRRK2 protein, a very large multi-domain protein with over 50 variants identified, but its function is not fully understood [42]. The most common LRRK2 mutation is a missense mutation in the LRRK2 protein (G2019S) that has been reported to occur in 2-5% of familial PD cases and 1-2% of sporadic PD cases [43-46]. This G2019S mutation has variable penetrance depending on ethnicity and increasing penetrance with age. The most prevalent occurrence for the G2019S mutation is found in populations from North Africa (30-40%) and Ashkenazi Jews (10-30%), whereas it is very rare in Asia, South Africa, and some European countries [45]. Two different reports about the age-specific penetrance have shown slightly different results, most likely due to variation in study design and/or sample size. Kachergus et al. [47] reported an increase in G2019S prevalence from 17% at age 50 to 85% by age 70 in a family based study; whereas Healy et al. [48] determined prevalence to be 28% at 59, 51% at 69, and 74% by 79 years, regardless of gender. While these studies vary slightly, it is still evident that age is a prominent factor. Patients with LRRK2 mutations exhibit Lewy body pathology, increased LRRK2 kinase activity, nigral degeneration, and the G2019S mutation has been proposed to increase SNCA expression [49, 50].

One case study has documented a point mutation (I93M) in the ubiquitin-carboxy-terminal hydrolase L1 (UCH-L1) gene in a pair of siblings with autosomal dominant PD, though the transmitting parent was unaffected. UCH-L1 is a deubiquitinating enzyme involved in the ubiquitin proteasome system, responsible for removal of ubiquitin monomers from the carboxy-terminus of proteins. The clinical phenotype observed in these two individuals was similar to typical PD with onset at ages 49 and 51; however, there is no pathological or observational data available [51]. Despite the lack of pathological data, mouse models with the I93M mutation have demonstrated a loss of DA neurons in the SNpc and of dopamine in the striatum [52]. These last three genes mentioned, SNCA, LRRK2, and UCH-L1, are the only known autosomal dominant causes of monogenic PD, many more have been found for autosomal recessive PD.

Four genes have been conclusively associated with early-onset autosomal recessive PD (PINK1, Parkin, DJ-1, and ATP13A2) [45]. All four of these seem to be involved in clearance of pathogenic protein aggregates and/or dysfunctional mitochondria. Parkin is an E3 ligase that targets α-synuclein among other proteins, with over 100 documented mutations of this gene associated with PD [45, 53]. It is suspected that Parkin accounts for up to 50% of patients with recessive inheritance and early onset (<45 years of age) and about 18% of early onset cases that are apparently sporadic [54]. PINK1 encodes a mitochondrial kinase that localizes to the outer mitochondrial membrane and is believed to protect neurons from stress-induced mitochondrial dysfunction, and has been demonstrated to interact with Parkin under PD pathological conditions [55-57]. Associations of PINK1 mutations in patients with PD are rare (1-2% of early-onset cases), with over 40 point mutations or nucleotide substitutions identified [45, 58]. PINK1-induced PD exhibits DA neuron loss in the SNpc accompanied with astrocytic gliosis and microgliosis, with Lewy bodies in the remaining DA neurons in the SNpc and in many other brain regions [59]. The endogenous function of DJ-1 is not completely known, though it is most likely involved with sensing and protecting against oxidative stress within cells and may act as an antioxidant protein or a scavenger [19]. Mutations in DJ-1 are also rare, accounting for less than 1% of early onset PD cases, and have at least 10 identified point mutations and large deletions [45, 58]. Finally, ATP13A2 protein is a lysosomal ATPase with an unknown target but is upregulated in brains of patients with idiopathic PD [60]. Only 5 point mutations have been reported thus far, most of which are associated with Kufor-Rakeb syndrome – an autosomal recessive, early-onset neurological disorder that is characterized by akinetic-rigid parkinsonism with response to L-DOPA, pyramidal tract dysfunction, supranuclear gaze paresis, and dementia [45, 58, 61].

While monogenic forms of PD represent only about 10% of all PD cases, it is suspected that genetic variation can increase susceptibility to dysfunctional cellular mechanisms or increase vulnerability to toxicant insult, which may account for a large portion of the sporadic cases. Due to the difficulty in identifying and characterizing susceptible genes, only a few have been described thus far (see reference reviews [19, 45]). A prime example of the susceptibility factor is the MDR1 gene which produced P-glycoprotein (P-gp), which contributes to the functional permeability of the BBB and prevents penetration of toxic xenobiotics and metabolites into the brain. Two allelic single nucleotide polymorphisms with decrease P-gp expression have been associated with PD risk – C3435T and G2677(A/T) – and were found to be most present in early-onset and late-onset PD patients compared to controls [62, 63]. Of particular importance it should be pointed out that these MDR1 polymorphisms do not seem to be causative for PD [64-66]. Rather, these polymorphisms are only associated with PD in conjunction with chronic exposure to pesticides [62, 67-69]. However, unlike the other two studies, Kiyohara et al. [68] reported a lack of association between pesticide exposure and C3435T in a Japanese population – this difference may be due to ethnicity in study groups (the other two being European). Furthermore, experimental studies with MDR1 knockout mice exhibited almost 100-fold increased accumulation of the pesticide ivermectin [70]. For a description of more susceptibility genes involved in gene-environment interactions for PD, refer to Table 2.

2.B. Epidemiology for Environmental Exposures Linked to PD

Although modern genetics has been a powerful tool for identifying key causative factors for PD, it provides answers for a very limited number of PD cases. For decades epidemiologists have been trying to find answers as to what environmental factors might induce or increase the risk of PD. The results have pointed to increased risks due to occupation (mostly welding and pesticide mixing or application), rural living, age, well-water drinking, and farming [10]. Classes of compounds repeatedly linked to PD include pesticides, solvents, metals, and others [71]. Pesticides, particularly those with chemical structures similar to known DA toxicants, such as MPTP, have received significant attention as putative risk factors. Recently, a detailed epidemiological study identified both paraquat and rotenone as PD risk factors from a large number of pesticides [72]. Both of these compounds have been utilized to model PD in laboratory animals, and have received heightened regulatory status in developed nations in recent years [72, 73]. Due to unknown etiological origins of ~90% of sporadic PD cases, there is an expansive amount of literature that attempts to determine associations between chemical exposures and incidences of disease. Associations between PD incidence and exposure to pesticides, metals, and solvents have been on the forefront of epidemiological study for many decades.

Many epidemiological studies have reported a convincing association for pesticides increasing risk of PD (see reference reviews [10, 74]), and these studies combined with case studies have added to the growing list of PD-risk factors, including rotenone, maneb, trichloroethylene, and organophosphorus pesticides [72, 75-79]. However, the scientific rigor utilized in such studies ranges widely, notably in sample size and applied statistical stringency. Regulations for pesticides have generally increased since many of the older studies have been published or since patients were exposed. Additionally, there have been significant methodological improvements, particularly related to tracking exposures in recent years. Interestingly, most studies and meta-analyses indicate association with herbicides and insecticides but not fungicides [74, 80, 81]. van der Mark et al. [81] suggest this is due to the high correlation coefficient between insecticide and herbicide use, which they calculated to be 0.79 between the study-specific risk ratios of each. In most cases, the association between PD and pesticide exposure occurs in occupations where the individual is mixing or applying the pesticide (e.g., farming practices), though there have been studies demonstrating less direct exposure via consumption of contaminated well water, ambient air in rural areas and areas where pesticides are frequently used, or exposure to contaminated clothing (see reference review [82]).

Table 2. Gene-Environment interactions for PD

Gene	Key function(s)	Polymorphisms	Toxicant Exposure	Potential Mechanism(s)	Reported Odds Ratio with Exposure to Toxicant (95% CI)	Number of Subjects (mean age ± standard deviation, range in years)	References
PON1	Hydrolyzes organophosphates	L54M	Pesticides	Altered toxicant metabolism and neurotransmission of acetylcholine	No significant link	125 with PD(69.06. 51 to 86), 162 controls (68.0, 54 to 84)	[204]
		E191R				92 with PD (NR), 122 controls (NR)	[205]
		At position 55	Diazinon, chlorpyrifos, parathion	Increased susceptibility to toxicity from organophosphates due to decreased PON1 levels	2.24 (1.12-4.48), 2.61 (1.25-5.44), no association	351 with PD (NR), 363 controls (NR)	[206]
PON2	Unknown	At codon 311	Pesticides	Unknown	No significant link	92 with PD (NR), 122 controls (NR)	[205]
SOD2	MnSOD dismutates superoxide radicals to become superoxide anions	V>A (T>C)	Pesticides	Accumulation of free radicals and increased ROS generation	2.49 (1.18-5.26), 4.09 when combined with NQO1 (1.34-10.64)	153 with PD (71.66±8.25, NR); 155 (70.07±6.31, NR)	[207]
NQO1	Catalyzes two-electron reduction of various toxic quinoid compounds to reduced states	P>S (C609T)	Pesticides	Additional ROS generation due to reduced ability to catalyze quinine metabolites during redox cycling	2.42 (1.16-4.76), 4.09 when combined with SOD2 (1.34-10.64)	153 with PD (71.66±8.25, NR); 155 (70.07±6.31, NR)	[207]

Gene	Key function(s)	Polymorphisms	Toxicant Exposure	Potential Mechanism(s)	Reported Odds Ratio with Exposure to Toxicant (95% CI)	Number of Subjects (mean age ± standard deviation, range in years)	References
DAT	Dopamine transporter	AA diplotype in 5' region, 9/9 3' VNTR (max= 4 susceptibility genes)	Paraquat and maneb	Increased expression may lead to increased toxicant accumulation; altered expression may alter DA neurotransmission, increasing susceptibility to toxic insults	5.66 (1.73-18.53) with 2+ risk alleles	261 with PD (66.7±9.9, NR), 376 controls (69.1±8.6, NR)	[208]
					2.99 (0.88-10.21) with one DAT susceptibility allele, 4.53 (1.70-12.09) with 2+ susceptibility alleles	324 with PD (NR), 334 controls (NR)	[209]
CYP2D6	Xenobiotic metabolism	G/A transition at intron 3-exon 4 junction (classified as poor metabolizers)	Pesticides	Decreased metabolism of toxic xenobiotics	4.74 (1.29-17.45)	190 with PD (NR), 419 controls (NR);	[210]
					2.09 (0.80-5.43) occasional exposure, 8.41 (1.01-69.76) regular exposure	393 with PD (67±10, NR), 389 controls (64±11, NR), 3.27 (1.21-8.80) for carriers with regular exposure to pesticides	[211]
			Solvents		14.47 (1.16-185.23)	100 with PD (66.6±9.7, NR), 200 controls (64.2±9.2, NR)	[212]

Table 2. (Continued)

Gene	Key function(s)	Polymorphisms	Toxicant Exposure	Potential Mechanism(s)	Reported Odds Ratio with Exposure to Toxicant (95% CI)	Number of Subjects (mean age ± standard deviation, range in years)	References
ALDH2	Aldehyde dehydrogenase, detoxifying enzyme	Clade 2 (see ref for description)	Pesticides	Inhibition of ALDH2 prevents detox of DA metabolites and increases ROS	2.21 (1.01-4.82) when exposed to 1 or 2 pesticides, 5.30 (1.42-19.8) when exposed to 3+ pesticides	353 with PD (68.3, 34-88), 518 controls (67.0, 35-94)	[213]
MDR1/ ABCB1	Expressed in BBB, produces P-glycoprotein, which plays crucial role protecting CNS against toxicants	3435T	Pesticides	Increased toxicant entry into the brain, access for compounds that are normally excluded	4.744 (1.009-22.306)	415 with PD (56.2±12.3, NR), 184 controls (50.9±13.1, NR)	[69]
					2.9 (1.2-7.1) for heterozygous, 1.3 (0.4-4.4) for homozygous, 4.9 (1.7-14.6) for both	59 with PD exposed to pesticides (58.9±11.1, NR), 48 with PD not exposed to pesticides (57.1±11.2, NR), 103 controls (74.1±5.9, 65-93)	[62]
		C3435T, G2677(A,T)			7.2 (2.1-24.8), 7.9 (2.2-28.9)	118 with PD (NR), 286 controls (NR)	[67]
		C3435T	Smoking		1.62 (0.84-3.10)	238 with PD (68.5, 67.4-69.6), 371	[68]
			Alcohol consumption Pesticides		1.83 (1.07-3.15) 1.16 (0.52-2.38)	controls (66.6, 65.7-67.5)	

Gene	Key function(s)	Polymorphisms	Toxicant Exposure	Potential Mechanism(s)	Reported Odds Ratio with Exposure to Toxicant (95% CI)	Number of Subjects (mean age ± standard deviation, range in years)	References
MAO-B	Catabolizes dopamine, activates MPTP	Intron 13, G or A allele	Smoking	Unknown	0.24 (0.10-0.55) for G allele and ever smoked, 1.44 (0.45-4.68) for A allele and ever smoked	98 with PD (NR, 37-88), 146 controls (NR, 44-84)	[214]
NOS1	Regulates cellular levels of nitric oxide	Multiple investigated	Pesticides	Increased nitric oxide progression leading to neuronal loss	No significance in exposure and risk-allele carriers	163 with PD (NR), 178 controls (NR)	[215]
NOS2A	Regulates cellular levels of nitric oxide	Multiple investigated CT + TT allele in nNOS-29 and -18	Smoking	Increased nitric oxide progression leading to neuronal loss; loss of protection by smoking	No significance in exposure and risk-allele carriers 1.13 (0.62-2.06) in nNOS-29 ex-smokers, 1.06 (0.42-2.64) in nNOS-29 current smokers, 0.84 (0.47-1.51) in nNOS-18 ex-smokers, 0.79 (0.32-1.99) in nNOS-18 current smokers	163 with PD (NR), 178 controls (NR) 209 cases (68 (37-75), 488 controls (68, 36-79)	[215] [216]
APOE	Cholesterol distribution for neuronal growth and repair	ε2,3,4	Caffeine	Unknown	0.339 (0.198-0.580) for ε2 carriers and	932 with PD (67.3, 32.4-96.7), 664	[217]
SNCA	Codes for α-synuclein	REP1 allelic length (257 confers reduced PD risk, 261 confers increased PD risk)	Smoking	Unknown	0.576 (0.396-0.838) for 257 and largest cumulative smoking, 0.268 (0.115-0.574) for 261 and largest accumulative smoking	932 with PD (67.3, 32.4-96.7), 664 controls (66.6, 21.2-93.8)	[217]

Table 2. (Continued)

Gene	Key function(s)	Polymorphisms	Toxicant Exposure	Potential Mechanism(s)	Reported Odds Ratio with Exposure to Toxicant (95% CI)	Number of Subjects (mean age ± standard deviation, range in years)	References
SKP1	Component of complex involved in target identification for ubiquitination and cell cycle regulation, also in presynaptic differentiation	rs2284314 T	Pesticides	Likely due to UPS inhibition	7.57 (3.14-18.3)	360 with PD (NR), 953 controls (NR)	[218]
BST1	NAD-metabolizing ectoenzyme and a glycosylphosphatidylinositol-anchored glycoprotein, produces cyclic ADP-ribose that releases Ca^{2+} from intracellular stores leading to Ca^{2+} influx	rs11724635 A/C or C/C	Well water consumption Smoking	Unknown, potential role in Ca^{2+} flux	1.453 (1.052-2.010), 1.623 (1.137-2.324) 0.523 (0.276-0.976) for former smokers, 0.859 (0.555-1.330) for current smokers	468 with PD (66.02±9.46, NR), 487 controls (67.16±9.33, NR)	[219]

NR= not reported

Occupational exposure to only a few metals has been observed so far as a risk factor for PD; manganese (Mn), iron (Fe), lead (Pb), and mercury (Hg). Though both Fe and Mn are essential elements for human health, abnormally high levels have been observed in brains of PD patients with strong evidence to suggest a pathogenic role for these increased levels [83, 84]. It has been long known that Fe plays a pivotal role in PD pathogenesis due to the basal ganglia having higher concentrations than other brain regions, Fe's tendency to cause oxidative stress through Fenton reactions, and potentially form strong binding to neuromelanin (see reference reviews [10, 71]). Despite how much is understood about Fe's role in PD pathogenesis, there has yet to be an epidemiological study to provide a clear, causative link to Fe exposure. Dietary exposure has been a hypothesized route, but is unlikely due to the tight regulation of dietary Fe uptake. Mn, however, has been determined to cause a movement disorder, termed manganism that is clinically similar to PD but pathologically distinct. Mn exposure is primarily associated with occupations involving mining and refining Mn ore, and synthesis of the pesticide maneb [85]. Thus, while high dose Mn causes a distinct neurological disorder, links between long-term, lower dose exposure and PD remains controversial. There has been minimal to modest epidemiological support for association of Pb or Hg exposures with PD incidence, though there has been some experimental data to indicate reduced striatal dopamine and increased oxidative stress after exposure to either of these metals (see reference review [10]). Thus, while high dose Mn causes a distinct neurological disorder, links between long-term, lower dose exposure and PD remain controversial.

Solvents are a very broad class of chemicals that are often lipophilic and can easily enter the central or peripheral nervous systems. Despite their widespread use and long history of being a suspected PD risk factor, there is very little mechanistic data available. Two solvents in particular have received the majority of attention for PD-solvent association studies; trichloroethylene (TCE) and perchloroethylene (PCE). TCE is an industrial solvent used for dry-cleaning and degreasing, as an additive in common household cleaning products, and was used as an anesthetic, grain fumigant, and caffeine extractant until 1977 [10, 71]. A few case studies and twin studies have demonstrated significant links between occupational exposure to TCE and onset of PD [86-88]. Further investigation into the effects of TCE exposure in rats have demonstrated selective DA neuronal loss in the nigrostriatal system (though DA levels in the striatum appear unaffected), inhibition of mitochondrial complex I, microglial activation, and intracellular accumulation of α-synuclein [89]. TCE is known to be metabolized by the cytochrome P450 isoform CYP2E1 to trichloroethylene oxide intermediate, which rearranges into chloral and likely reacts with tryptamine to form 1-trichloromethyl-1,2,3,4-tetrahydro-β-carboline (TaClo). TaClo-induced cell death included caspase 3-mediated apoptosis, increased ROS production, and inhibition of mitochondrial respiration, and appears to be more potent to mitochondrial complex I than MPP^+ at similar concentrations [90-92]. PCE has been the primary dry-cleaning solvent since the 1960s and has induced similar clinical conditions as TCE after occupational exposure; neurological symptoms, delayed reaction times, selective loss of dopaminergic neurons of the SNpc, and likely an increased risk of PD [10, 93]. Clinical case reports have been reported for other solvents, including methanol, n-hexane, toluene, and carbon disulfide – though these are predominantly rare and isolated cases [93, 94].

3. Mechanisms and Models of Parkinson's Disease

Oxidative stress, mitochondrial dysfunction, and impaired autophagic and proteasomal clearance of protein aggregates have been repeatedly identified as broad pathogenic pathways important in PD. Recently, toxicants, PD genes, and gene-environment interactions have been mechanistically investigated for roles in these pivotal pathways.

The clearest example of a toxicant as a direct cause of a PD-like syndrome occurred in the early 1980s, when a handful of young persons were exposed to 1-methyl-4-phenyl-1,2,3-tetrahydropyridine (MPTP), an unintended byproduct of designer drug synthesis. Exposure to MPTP resulted in severe and rapid onset of a Parkinsonian syndrome that was clinically indistinguishable from PD [11, 13, 14]. These key (albeit unfortunate) incidents led to a boom in research investigating how MPTP induces PD and a search for other chemicals that are structurally similar which may induce similar effects. This MPTP model for PD has provided thorough insight into many key pathogenic features. The results of this research produced a very detailed explanation for how MPTP induced PD. First, MPTP crosses the BBB due to high lipophilicity, then in glial cells, becomes oxidized by monoamine oxidase (MAO) B and oxidized again into its ultimate metabolite, MPP^+ [95, 96]. MPP^+ is the toxic metabolite of MPTP that enters DA cells of the SNpc via the dopamine transporter, where it concentrates in the mitochondria and disrupts complex I of the electron transport chain [12, 97]. Mitochondrial complex I is essential for ATP synthesis and is the primary source of endogenous cellular oxidative stress; hence, disruption of its function results in decreased ATP production and increased release of superoxide anions. Within the mitochondria, superoxide anions are converted to hydrogen peroxide by manganese superoxide dismutase (MnSOD) catalysis, and leads to massively increased ROS. This is likely the major oxidative stress pathway in the MPTP model.

3.A. Oxidative Stress

There is an expansive collection of literature regarding the role for oxidative stress in PD, much more than can be discussed in the scope of this chapter (for excellent reviews, see references [98-100]). Roles for and effects of oxidative stress have been reported in all the major experimental models of PD, and markers of oxidative stress are repeatedly observed in the postmortem brains of PD patients. While much data suggests that oxidative stress is an early pathogenic event in most PD models, it is often unclear if oxidative stress causes the onset of clinical PD or if it is a result of some other pathogenic event. Regardless, several important factors have been established about the pertinence of oxidative stress in PD and the susceptibility of DA neurons. While rotenone also inhibits mitochondrial complex I and induces oxidative stress similar to PD, a key difference further supports the importance of oxidative stress in PD pathogenesis. Rotenone accumulates in all cells due to lipophilicity, yet nigral dopamine neurons are selectively lesioned, providing evidence on selective sensitivity of this cell population to oxidative stress [101, 102]. Paraquat, also linked to PD, reportedly causes lesions in nigral dopamine neurons *in vivo* and *in vitro* and is believed to undergo redox cycling to generate ROS [103-108]. In addition to heightened oxidative burden stemming from energy demands and dopamine metabolism, primate brains have high levels of neuromelanin which displays a high affinity for iron and induces iron accumulation in the SNpc [109]. Iron levels have been reported to be elevated in the SN in brains of PD patients [110-113]. The iron that is accumulated is typically in the ferrous form, which readily conducts Fenton reactions and converts the hydrogen peroxide formed by MnSOD to the highly reactive hydroxyl radical and causes oxidative damage of various cellular components [114]. Another critical aspect for oxidative stress in PD is that normal dopamine metabolism generates ROS, and dopamine itself is oxidative and can auto-oxidize, especially when it is incidentally released into the cytosol, which has been reported to occur due to synaptic vesicle trafficking complications in PD [115-117]. It is believed that DA neurons express such high levels of neuromelanin as a means of neuroprotection due to dopamine's tendency to produce ROS [99]. Finally, oxidative stress in PD can also be induced and/or released by reactive microglia (see reference reviews [118, 119]).

3.B. Mitochondrial Dysfunction

Oxidative stress and mitochondrial dysfuntion appear to be very closely related in PD pathogenesis, as illustrated by DA neurotoxicants that produce oxidative stress through inhibiting mitochondrial respiration. Postmortem analysis of PD brains has indicated significant reduction in mitochondrial complex I activity, though the causes and consequences of this in PD are poorly understood [120-123]. This decrease in complex I activity has been observed not only in the SNpc of PD patients, but also in their platelets and skeletal muscle – though the abnormality in skeletal muscle has been more difficult to detect reliably [124-127]. Rotenone is an insecticide and piscicide that has been used extensively for pest control, and is a mitochondrial toxicant that has a high affinity for mitochondrial complex I [102]. Inhibition of complex I activity results in increased free radical production decreased ATP production, which can affect other mechanisms in PD such as the proteasomal system. For example, an *in vitro* study with rotenone demonstrated a 20% loss of ATP, increased ROS, and reduction in proteasomal activity [128]. Following chronic exposure to rotenone in experimental models there is selective loss of DA neurons in SNpc, formation of α-synuclein positive protein inclusions similar to Lewy bodies, and loss of TH immunoreactive terminals in striatum. Furthermore, rotenone can cause mitochondrial depolarization, which leads to the opening of the mitochondrial permeability transition pore and caspase 3 activation, thus initiating the apoptotic pathway [102, 129-132]. Although systemic infusion of rotenone induces mitochondrial dysfunction throughout the brain, neuronal loss appears to be specific to dopaminergic regions of the SNpc and the olfactory bulb. These regions also have the highest levels of soluble protein carbonyls after rotenone exposure, indicating that the selectivity for neurodegeneration in these regions is likely due to an oxidative stress mechanism, presumably combining the ROS production from mitochondrial dysfunction and dopamine metabolism. Observations from rotenone models closely replicate the oxidative damage to lipids, DNA, and protein that is observed in postmortem PD brains [102]. The data collected from MPTP and rotenone experiments and the observations in postmortem brains of patients with PD suggest that mitochondrial complex I disruption is likely a good indicator that chemicals with similar mitochondrial targets are a PD risk factor. More recently, epidemiology studies have indicated trichloroethylene as a PD risk factor and experimental studies have provided some evidence for a similar mechanism of mitochondrial dysfunction [86].

Furthermore, six of the genes associated with PD involve clearly affect mitochondria – α-synuclein, parkin, PINK1, DJ-1, LRRK2, and OMI/HTRA2. Alterations to these proteins (either polymorphisms or expression levels) have demonstrated decreased mitochondrial protection against oxidative stress, which cycles back to increasing mitochondrial dysfunction. The roles for α-synuclein and LRRK2 in mitochondrial dysfunction are less clear than some of the other proteins. Nonetheless, the overexpression of α-synuclein (wild type or A53T) has been shown to associate with the inner mitochondrial membrane, inhibit mitochondrial complex I, induce oxidative stress, increase mitophagy, and likely causes aberrant mitochondrial morphology [133-137]. About 10% of the LRRK2 protein content in mammalian brains is found associated with the outer mitochondrial membrane, though it is not yet clear how it may affect mitochondrial integrity [138]. Sanders et al. [139] have provided some insight, demonstrating LRRK2 induced mitochondrial DNA damage in neural cells derived from LRRK2-PD patients. Parkin, PINK1, and DJ-1 are well known to be important proteins in mitochondrial protection and clearance, and their misexpression or mutations often lead to pathogenic consequences. PINK1 deficiency confers decreased mitochondrial membrane potential (which leads to cytochrome c release and can induce apoptosis), activity of complexes I and IV, ATP production, increased ROS production and abnormal mitochondrial morphology [140].

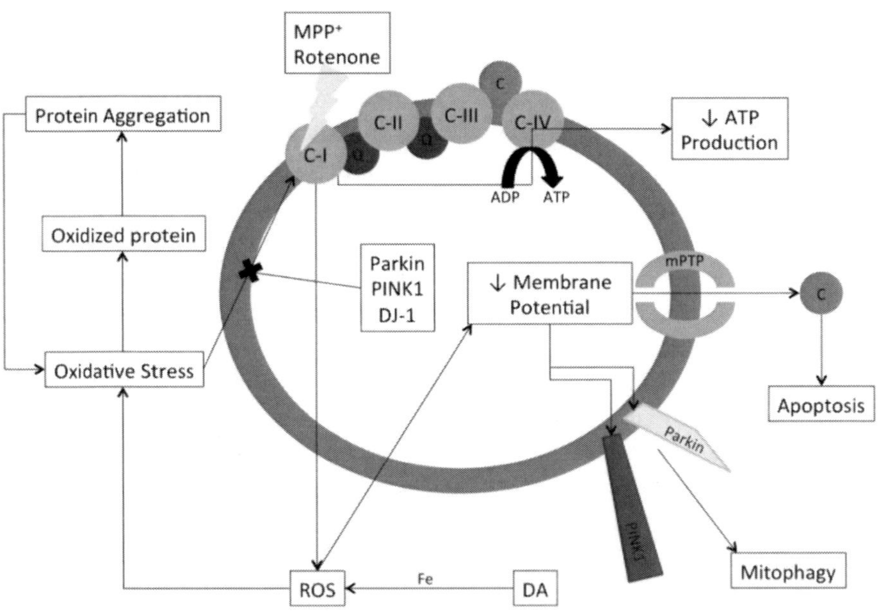

Toxicant exposure (e.g., MPP^+ or rotenone) are inducers of mitochondrial dysfunction, and may be exacerbated by various genetic mutations in mitochondrial proteins. C-I, -II, -III, -IV represent mitochondrial complexes I-IV; C represents cytochrome C; mPTP represents the mitochondrial permeability transition pore; Q represents coenzyme Q10; MPP^+ is 1-methyl-4-phenylpyridinium; PINK1 is PTEN-induced putative kinase 1; ROS is reactive oxygen species; DA is dopamine.

Figure 1. Key molecular mechanisms of mitochondrial dysfunction.

Parkin can inhibit cytochrome c release and exhibits other similar functions to PINK1, and their interaction is important for the turnover of dysfunctional or damaged mitochondria by autophagy [141]. Hence, mutation in either one can result in persistence of dysfunctional mitochondria. DJ-1 has multiple neuroprotective functions and is translocated to mitochondria upon ROS stimulation, likely to play a role in antioxidative stress mechanisms [142]. However, excessive oxidative stress can also render DJ-1 inactive. See Figure 1 for a representation of the major mechanistic molecules and proteins in mitochondrial dysfunction.

3.C. Protein Clearance by Autophagy and the Ubiquitin Proteasome System

Both autophagy and the ubiquitin proteasome system (UPS) are cellular processes responsible for clearing damaged, misfolded, and/or aggregated proteins in cells and normally serve as an effective mechanism for recycling macromolecules from their targets. Briefly, autophagy is the formation of double-membraned large vacuoles (called autophagosomes) that engulf protein aggregates and/or subcellular organelles that have been tagged for degradation. These vacuoles then fuse with lysosomes, forming autophagolysosomes, which digest the inner membrane and all of the components within [143, 144]. The UPS targets short-lived proteins that have been tagged with ubiquitin by an ubiquitin protein ligase (E3) and utilizes an ATP-dependent, non-lysosomal degradation pathway through a series of enzyme-mediated reactions in the 26S proteasome [145]. Though these two systems have very different mechanisms for clearance and normally have different targets, it has been demonstrated with various pharmacological blockers that increased autophagic activity can be a compensatory response to UPS impairment or dysfunction [146, 147]. For example, experimental studies using rotenone and MPTP have demonstrated a marked decrease in proteasome activity after exposure to either chemical [148]. The effect of this inhibited activity would be the persistence of misfolded proteins, and likely increasing formation of protein aggregates. If unaffected, the autophagy-lysosomal pathway can eliminate these aggregates. However, the autophagic machinery can also be impaired by PD relevant toxicants, and high levels of ROS have been shown to impair autophagic flux by preventing autophagosome biogenesis, fusion with lysosomes, or lysosomal dysfunction [149]. In the case of impaired autophagic flux, not only is the cell unable to clear misfolded proteins and protein aggregates, but it is unable to clear dysfunctional mitochondria and other organelles. Due to its dependency on ATP, the UPS is easily impaired by mitochondrial dysfunction. Some proteasomal targets include pro-apoptotic proteins (e.g., p53 and BCL2), and its impairment can redistribute cytochrome c into the cytosol; thus putting cells at risk for apoptosis [150, 151]. These interactions create further pathogenic feedback loop, where dysfunctional mitochondria produce increased ROS which interrupts autophagic flux and results in impaired clearance of dysfunctional mitochondria; and decreased ATP production impairs the UPS, which inhibits degradation of pro-apoptotic proteins among others. For a more detailed review on how PD-associated toxicants affect autophagy machinery see reference review [152].

Specific genetic models of PD have demonstrated that one or both of these pathways can be impaired, which likely leads to the formation and persistence of Lewy bodies and PD progression. It has been determined that the UPS system degrades α-synuclein under normal conditions, but if α-synuclein is deubiquitinated or overexpressed the autophagy pathway takes over as the primary degradation pathway [148]. However, when α-synuclein is overexpressed, it can inhibit the UPS by interacting with the regulatory cap of its 26S subunit [153]. Furthermore, overexpression or mutant α-synuclein can impair both pathways, and α-synuclein inclusions are resistant to degradation [154-156] – thus creating yet another pathogenic feedback loop similar to the one described for mitochondrial clearance. LRRK2 G2019S mutation causes neurite shortening by autophagy in neuroblastoma cells and exhibit increased number of autophagosomes [157]. Pharmacologic inhibition of LRRK2 kinase activity stimulates macroautophagy, suggesting a role for LRRK2 in autophagosome formation [158]. Conversely, mutations in LRRK2 result in impaired autophagic flux and the accumulation of LC3 and p62 (two marker proteins of autophagy) [159, 160]. Figure 2 outlines the autophagy pathway and various points where toxicants and genetic mutations interfere.

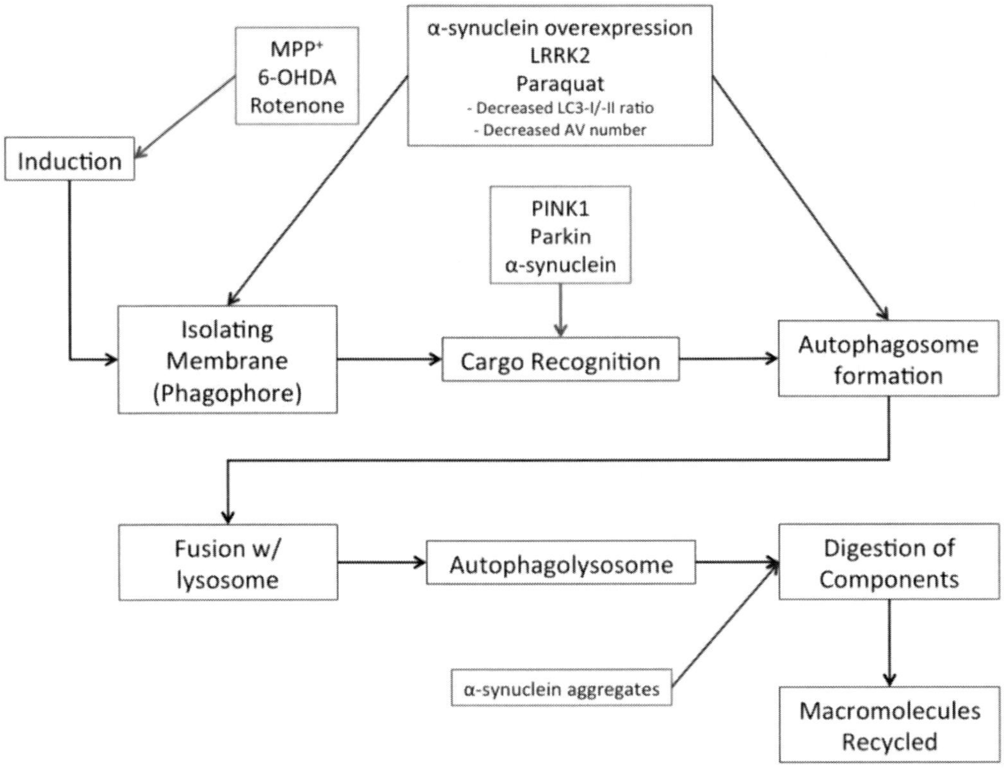

Black boxes/lines represent normal autophagic flux; red boxes/lines indicate toxicants/genes that induce autophagy, whereas blue boxes/lines represent ones that suppress autophagy. Lines indicate where the toxicants/genes are acting in the autophagic pathway. MPP^+ is 1-methyl-4-phenylpyridinium; 6-OHDA is 6-hydroxydopamine; LRRK2 is leucine-rich repeat kinase 2; PINK1 is PTEN-induced putative kinase 1; LC3 is microtubule-associated protein 1A/1B-light chain 3; AV is autophagosome.

Figure 2. Interactions of PD toxicants with the autophagic pathway.

Table 3. Established Toxicant-Induced Animal Models of PD

Toxicant Name	Justification for Use	Clinical/Pathological Features	Mechanism of Action	References
6-OHDA	Similar structure to DA; postulated to form in vivo	SNpc neuronal loss, asymmetric circular behavior due to unilateral lesion in striatum	ROS production and inactivation of macromolecules by quinone adducts	[220-224]
MPTP	Case studies and experimental studies produce accurate replication of most PD features and pathogenesis	Bradykinesia, rigidity, tremor, SNpc neuronal loss, Lewy bodies	Uptake via DA transporter, then mitochondrial complex I impairment and ROS production	[13, 14]
Paraquat	Positive epidemiological link; structural similarities to DA and DA toxicants	Bradykinesia, SNpc neuronal loss	ROS production by redox cycling	[71, 103, 108]
Rotenone	Epidemiological links; Proof of concept: systemic mitochondrial dysfunction leads to SNpc lesions	Bradykinesia, rigidity, SNpc neuronal loss, Lewy bodies	Inhibition of mitochondrial complex I, ROS generation, microglia activation	[10, 71, 101, 225]
Trichloroethylene	Case studies and experimental animal studies suggest link	Locomotor impairment, SNpc neuronal loss, buccal-lingual dyskinesia, rigidity	Unknown; inhibition of mitochondrial complex I?	[71, 226, 22]
Iron (Fe)	Accumulation observed in human PD and experimental models, unknown if cause or effect	Unknown/Not applicable	Exacerbates oxidative stress through Fenton reactions	[71, 228-232]
Dieldrin	Case-control studies suggest as risk factor, present in PD brains	In vivo experiments show decreased DA levels	Unknown; oxidative stress?	[71, 233-235]

4. iPSCs: A Cutting Edge Research Technique to Examine Gene-Environment Interactions

For decades researchers have used a variety of cell cultures, animal strains, and tissue sections to investigate the pathogenesis and mechanisms of PD. As is the case with all models for human disease, each model has significant advantages and disadvantages, and some models are superior in replicating key disease process. Animal models for human disease, though controversial, have been crucial for developing a better understanding of disease mechanisms and discovering therapeutic drugs and practices to cure or alleviate suffering. Despite their obvious importance for health studies, no animal model is an exact replica of that which occurs in humans – due in part to anatomical and physiological differences, lifespan, and limited focus of experimental studies on a limited number of disease-related factors. PD in particular has been difficult to model, as the causes and genesis of the disease are still poorly understood. Most notably, numerous therapeutic approaches found to be efficacious in PD animal models have failed in clinical trials. Thus, there is no animal model that completely recapitulates the clinical and pathological definition, but researchers are continually seeking to improve models to better represent what occurs in this disease. The immediately obvious challenge that likely will not be overcome is the difference in lifespan – most animal models only live a couple years (which also makes them useful for rapid study), whereas PD takes decades to develop and display symptoms. While the limitations faced using animal models seem daunting, the last several decades of research have produced numerous exciting advances. PD research has brought together a wide array of genetic and environmental models that have been well characterized and continue to be used to understand the intricate molecular mechanisms behind the disease pathology and progression. These models can be broadly defined as genetic-induced, toxicant-induced, or induced by a combination of genetic and toxicant exposures. Table 3 summarizes the most commonly used and well established animal models in PD research. For cell culture models, the challenge that persists is the lack of a proper human neuronal DA phenotype, as primary human neurons are mostly impossible to acquire from the midbrain for culture. The majority of primary dopaminergic neuronal cultures that have been used come from lab animals (mostly rats or mice). The human cell cultures that have been used are typically neuroblastoma cells or cell types from non-neuronal origin that can be modified to express near-dopaminergic profiles (e.g., dopamine transporters or monoamine oxidase B). To overcome some of these challenges, some groups have utilized organotypic cultures to preserve more of the natural neuronal environment and neural connections. These culture types also present methodological challenges, such as determining viability and biochemical alterations (see review for more details on limitations of cellular models [161]). However, recent advances using induced pluripotent stem cells (iPSCs) provide new hope and enthusiasm for overcoming many of these technical challenges.

Perhaps the most exciting of all the PD models that have been developed and characterized are the iPSCs – human fibroblast cells collected from patients diagnosed with monogenic PD and transformed into DA neuronal cells by de-differentiating fibroblasts into induced stem cells and re-differentiating the stem cells into neurons by defined factors, using modified methods from those first described by Takahashi et al. [162-165]. Thus far, iPSCs have been described for four forms of monogenic PD – LRRK2, SNCA, PARK2, and PINK1. One of the major concerns for this new technology is reducing variability in differentiation efficiency, as the successful induction of iPSCs is not consistent even for cells derived from the same patient. Another concern is the same as any cell culture model – the lack of the cell's natural environment, as studies utilizing iPSCs have not yet investigated the effects of microglia, neuronal synaptic connections, or an extensive arborization of dendrites. Nonetheless, the development of this technology is a major step forward in our ability to understand the cellular and molecular mechanisms for monogenic forms of PD. Already this technique has provided impressive experimental links for PD-associated genes and toxicant exposure for development of PD pathologies.

Three recent studies have looked at iPSCs from monogenic PD caused by SNCA mutations –investigating SNCA triplication or the A53T mutation [166-168]. The first two of these studies have only characterized the iPS cell lines to determine their usefulness and validity as a model for PD. Both reported the usefulness of these iPSCs in future studies of patient-specific PD, reporting pathological phenotypes including the increased accumulation of α-synuclein, but neither demonstrated formation of Lewy bodies that occurs in the brains of patients with these forms of monogenetic PD, perhaps because it takes decades for these to develop in humans. In addition, Soldner et al. [168] demonstrated the applicability of zinc finger nuclease-mediated genome editing to induce point mutations causing PD in iPSCs derived from patients without PD. This approach adds to the strength of evidence for the role of point mutations in PD and will provide exciting new insight into PD-related cellular and molecular mechanisms. A very exciting study by Ryan et al. [167] demonstrated a potential early stage contributor to PD pathogenesis after genetic and environmental insult involving the transcriptional pathway formed between myocyte enhancer factor 2C (MEF2C) and peroxisome proliferator-activated receptor-γ coactivator-1α (PGC1α). MEF2C is a transcriptional factor for many DA-neuron specific targets, and PGC1α is a master regulator known for promoting mitochondrial neuroprotection in PD models [169, 170]. Both genetic (A53T) and environmental (exposure to paraquat/maneb or rotenone) models induced posttranslational redox modification of MEF2C (i.e., sulfonation or S-nitrosylation), resulting in PGC1α inhibition and increased neuronal death [167]. They further demonstrated this redox-induced neuronal death could be counteracted by pretreatment with a nitric oxide synthase inhibitor or treatment with isoxazole, a pharmacological inducer of MEF2C transcription.

To date, several studies have used iPSCs derived from LRRK2 PD patients [50, 89, 139, 171, 172]. Multiple important PD features were observed in these studies, including accumulation of α-synuclein, impaired autophagy, upregulation of ROS-responsive genes, and vulnerability to neurotoxins. Sánchez-Danés et al. [172] reported a significantly different morphological phenotype in LRRK2(G2019S)-iPSCs when compared to control, both the length and number of neurites of DA neurons were significantly higher in control iPSCs than those from PD iPSCs. This morphological distinction was determined to be caused by deficient autophagic clearance pathway, due to improper maturation of autophagosomes into autophagolysosomes. Reinhardt et al. [171] confirmed these results in LRRK2(G2019S) iPSCs, and further reported a likely role for TAU and phospho-TAU protein in this process. This study further investigated potential changes of gene expression caused by LRRK2(G2019S), and demonstrated a statistically significant increase in *CPNE8*, *ANXA1*, *MAP7*, and *CADPS2*, and a significantly decreased expression of *UHRF2* in dopaminergic neurons from LRRK2(G2019S) iPSCs compared to gene corrected iPSCs. These gene expression dysregulations were suggested to be due in part to aberrant ERK activation, and that ERK inhibitors restored neurite outgrowth of differentiated neurons. Sanders et al. [139] is the only group so far who has used iPSCs to investigate the effects of a PD-linked gene on mitochondrial DNA damage. This group used two allelic forms of LRRK2 (G2019S and R1441C) from five different subjects (compared to three different controls) to show these mutations increase the amount of mitochondrial DNA damage, and confirmed their results using gene-editing to correct the G2019S mutation with zinc-finger nucleases to account for any genetic variability between the patient-derived cells.

A variety of *in vitro* studies have shown the interaction between PINK1 and Parkin plays an important role in mitochondrial function and clearance, though these *in vitro* findings need to be confirmed *in vivo* [55-57, 173-175]. To verify the use of iPSCs and further investigate the Parkin-PINK1 interaction, a few groups utilized iPSCs derived from patients expressing one gene or the other [56, 176-178]. Taken together, these studies have demonstrated a role for Parkin and PINK1 point mutations causing mitochondrial dysfunction, impaired mitochondrial clearance, increased oxidative stress, dysfunctional dopamine management (i.e., release and uptake), and α-synuclein accumulation (though no aggregates were reported).

Mutations in glucocerebrosidase (GBA) have been investigated using iPSCs from patients with Gaucher's disease, and have also provided some insight into PD pathogenesis. When GBA mutations are homozygous, the individual develops Gaucher's disease; heterozygous mutations, however, are susceptibility factors for idiopathic PD. Gaucher's disease is a lysosomal storage disorder that often exhibits formation of Lewy bodies [179,180]. Heterozygous GBA1 mutations were found in a large cohort of PD patients (4-7%), and the GBA protein was observed to colocalize with α-synuclein in Lewy bodies [181-183]. iPSCs developed from patients suffering from Gaucher's disease demonstrate increased α-synuclein protein in DA neurons and poor macrophagic clearance, in addition to the expected reduction in GBA protein and activity [184-186].

Though iPSC technology is still very new, it has demonstrated its potential as a PD model, and has the major advantage of utilizing primary human cells. Considering the potential this technology may hold, relatively little has been achieved to understand the cellular and molecular mechanisms behind monogenic PD and therapeutic efficacy has not yet been widely applied using iPSCs. It will be very exciting to see the development of our understanding of gene-specific diseases, variability in applicability for novel therapeutics, and the interactions of neurotoxicants and genetic susceptibility factors.

5. Future Directions

There are many potential mechanisms for gene-environment interactions inducing Parkinson's disease – most of which likely have yet to be identified. Thus, it is likely that solely investigating one or the other, as has been done for the last few decades, may not be enough to detect important information about PD. Such simplified approaches may explain the inconsistencies in environmental and genetic association studies. Obviously these studies have been, and will continue to be vital to our understanding of PD pathogenesis and progression, but we must expand our scope if we are to truly understand the underlying mechanisms and risk factors of this disease. The field has now reached a point where we can begin to combine multi-faceted experimental situations to more accurately portray real-life situations and by combining results from multiple scientific disciplines. Methods for human epidemiological studies have vastly improved such that we can robustly measure interactions between genetic polymorphisms, environmental exposures, and disease risk. We can also further verify these findings using animal and cell culture models in which we can more precisely control the roles of genes, toxicants, and environmental parameters. The development and verification of iPSCs derived from PD patients and patients with susceptibility genes for PD is an exciting advancement. Though these are very new relative to established models, early promise suggests that iPSCs have a strong probability of proving to be extremely useful in verifying susceptibility genes and could also be used to discover and study protective genes. Ultimately the goal of improved techniques is to achieve a better understanding for mechanisms of disease and to potentially discover new pathways underlying disease states. Eventually with advanced understanding of gene-environment interactions physicians might be able to conduct genetic screens of patients and warn them to avoid exposure to certain classes of chemicals.

Key Facts

i. Multiple genes have been identified as causative factors for Parkinson’s disease, though many more have been implicated as susceptibility genes and more are identified each year.
ii. Most environmental risk factors are lipophilic chemicals that have some capacity to cross the blood brain barrier; pesticides have been the most robustly identified, though some metals and solvents have also been linked.
iii. Oxidative stress plays a major role in the pathogenesis of Parkinson’s disease and likely has an important role in the earliest stages of the disease.
iv. The key mechanisms underlying Parkinson’s disease (oxidative stress, mitochondrial dysfunction, and malfunction of protein clearance pathways) create complex pathogenic feedback loops, and likely have an important role in the progression of the disease.
v. Recent advances in stem cell technology deriving induced pluripotent stem cells from patients with monogenic forms of Parkinson’s disease are a very exciting development and have promising application to better understand the mechanisms underlying causes of disease (genetic, environmental, and interactions between genes and environment).

References

[1] Parkinson, J. An essay on the shaking palsy. 1817. *The Journal of neuropsychiatry and clinical neurosciences*, 14, 223-236; discussion 222 (2002).
[2] Braak, H; et al. Staging of brain pathology related to sporadic Parkinson's disease. *Neurobiology of aging*, 24, 197-211 (2003).
[3] Fornai, F; et al. Recent knowledge on molecular components of Lewy bodies discloses future therapeutic strategies in Parkinson's disease. *Current drug targets. CNS and neurological disorders*, 2, 149-152 (2003).
[4] Forno, LS. Neuropathology of Parkinson's disease. *Journal of neuropathology and experimental neurology*, 55, 259-272 (1996).

[5] Gai, WP; et al. In situ and in vitro study of colocalization and segregation of alpha-synuclein, ubiquitin, and lipids in Lewy bodies. *Experimental neurology*, 166, 324-333, doi:10.1006/exnr.2000.7527 (2000).

[6] Spillantini, MG; et al. Alpha-synuclein in Lewy bodies. *Nature*, 388, 839-840, doi:10.1038/42166 (1997).

[7] Dorsey, ER; et al. Projected number of people with Parkinson disease in the most populous nations, 2005 through 2030. *Neurology*, 68, 384-386, doi:10.1212/01.wnl.0000247740.47667.03 (2007).

[8] Bernheimer, H; Birkmayer, W; Hornykiewicz, O; Jellinger, K; Seitelberger, F. Brain dopamine and the syndromes of Parkinson and Huntington. Clinical, morphological and neurochemical correlations. *Journal of the neurological sciences*, 20, 415-455 (1973).

[9] Gelb, DJ; Oliver, E; Gilman, S. Diagnostic criteria for Parkinson disease. *Archives of neurology*, 56, 33-39 (1999).

[10] Goldman, SM. Environmental toxins and Parkinson's disease. *Annual review of pharmacology and toxicology*, 54, 141-164, doi:10.1146/annurev-pharmtox-011613-135937 (2014).

[11] Davis, GC; et al. Chronic Parkinsonism secondary to intravenous injection of meperidine analogues. *Psychiatry research*, 1, 249-254 (1979).

[12] Javitch, JA; D'Amato, RJ; Strittmatter, SM; Snyder, SH. Parkinsonism-inducing neurotoxin, N-methyl-4-phenyl-1,2,3,6-tetrahydropyridine: uptake of the metabolite N-methyl-4-phenylpyridine by dopamine neurons explains selective toxicity. *Proceedings of the National Academy of Sciences of the United States of America*, 82, 2173-2177 (1985).

[13] Langston, JW; Ballard, P; Tetrud, JW; Irwin, I. Chronic Parkinsonism in humans due to a product of meperidine-analog synthesis. *Science*, 219, 979-980 (1983).

[14] Langston, JW; Ballard, PA; Jr. Parkinson's disease in a chemist working with 1-methyl-4-phenyl-1,2,5,6-tetrahydropyridine. *The New England journal of medicine*, 309, 310 (1983).

[15] Braak, H; Ghebremedhin, E; Rub, U; Bratzke, H; Del Tredici, K. Stages in the development of Parkinson's disease-related pathology. *Cell and tissue research*, 318, 121-134, doi:10.1007/s00441-004-0956-9 (2004).

[16] Matsuda, W; et al. Single nigrostriatal dopaminergic neurons form widely spread and highly dense axonal arborizations in the neostriatum. *The Journal of neuroscience : the official journal of the Society for Neuroscience*, 29, 444-453, doi:10.1523/JNEUROSCI.4029-08.2009 (2009).

[17] Hastings, TG; Zigmond, MJ. Identification of catechol-protein conjugates in neostriatal slices incubated with [3H]dopamine: impact of ascorbic acid and glutathione. *Journal of neurochemistry*, 63, 1126-1132 (1994).

[18] Alves, G; Forsaa, EB; Pedersen, KF; Dreetz Gjerstad, M; Larsen, JP. Epidemiology of Parkinson's disease. *Journal of neurology*, 255 Suppl 5, 18-32, doi:10.1007/s00415-008-5004-3 (2008).

[19] Wirdefeldt, K; Adami, HO; Cole, P; Trichopoulos, D; Mandel, J. Epidemiology and etiology of Parkinson's disease: a review of the evidence. *European journal of epidemiology*, 26 Suppl, 1, S1-58, doi:10.1007/s10654-011-9581-6 (2011).

[20] Baldereschi, M; et al. Parkinson's disease and parkinsonism in a longitudinal study: two-fold higher incidence in men. ILSA Working Group. Italian Longitudinal Study on Aging. *Neurology*, 55, 1358-1363 (2000).

[21] Benito-Leon, J; et al. Incidence of Parkinson disease and parkinsonism in three elderly populations of central Spain. *Neurology*, 62, 734-741 (2004).

[22] de Lau, LM; et al. Incidence of parkinsonism and Parkinson disease in a general population: the Rotterdam Study. *Neurology*, 63, 1240-1244 (2004).

[23] Van Den Eeden, SK; et al. Incidence of Parkinson's disease: variation by age, gender, and race/ethnicity. *American journal of epidemiology*, 157, 1015-1022 (2003).

[24] Bonifati, V. Genetics of Parkinson's disease--state of the art, 2013. *Parkinsonism & related disorders*, 20 Suppl, 1, S23-28, doi:10.1016/S1353-8020(13)70009-9 (2014).

[25] Puschmann, A. Monogenic Parkinson's disease and parkinsonism: clinical phenotypes and frequencies of known mutations. *Parkinsonism & related disorders*, 19, 407-415, doi:10.1016/j.parkreldis.2013.01.020 (2013).

[26] Polymeropoulos, MH; et al. Mutation in the alpha-synuclein gene identified in families with Parkinson's disease. *Science*, 276, 2045-2047 (1997).

[27] Abeliovich, A; et al. Mice lacking alpha-synuclein display functional deficits in the nigrostriatal dopamine system. *Neuron*, 25, 239-252 (2000).

[28] Chandra, S; et al. Double-knockout mice for alpha- and beta-synucleins: effect on synaptic functions. *Proceedings of the National Academy of Sciences of the United States of America*, 101, 14966-14971, doi:10.1073/pnas.0406283101 (2004).

[29] Cooper, AA; et al. Alpha-synuclein blocks ER-Golgi traffic and Rab1 rescues neuron loss in Parkinson's models. *Science*, 313, 324-328, doi:10.1126/science.1129462 (2006).

[30] Gitler, AD; et al. The Parkinson's disease protein alpha-synuclein disrupts cellular Rab homeostasis. *Proceedings of the National Academy of Sciences of the United States of America*, 105, 145-150, doi:10.1073/pnas.0710685105 (2008).

[31] Liu, S; et al. alpha-Synuclein produces a long-lasting increase in neurotransmitter release. *The EMBO journal*, 23, 4506-4516, doi:10.1038/sj.emboj.7600451 (2004).

[32] Roy, S. The paradoxical cell biology of alpha-Synucle. *Results and problems in cell differentiation*, 48, 159-172, doi:10.1007/400_2009_23 (2009).

[33] Spira, PJ; Sharpe, DM; Halliday, G; Cavanagh, J; Nicholson, GA. Clinical and pathological features of a Parkinsonian syndrome in a family with an Ala53Thr alpha-synuclein mutation. *Annals of neurology*, 49, 313-319 (2001).

[34] Berg, D; et al. Alpha-synuclein and Parkinson's disease: implications from the screening of more than 1,900 patients. *Movement disorders : official journal of the Movement Disorder Society*, 20, 1191-1194, doi:10.1002/mds.20504 (2005).

[35] Appel-Cresswell, S; et al. Alpha-synuclein p.H50Q, a novel pathogenic mutation for Parkinson's disease. *Movement disorders : official journal of the Movement Disorder Society*, 28, 811-813, doi:10.1002/mds.25421 (2013).

[36] Kiely, AP; et al. alpha-Synucleinopathy associated with G51D SNCA mutation: a link between Parkinson's disease and multiple system atrophy? *Acta neuropathologica*, 125, 753-769, doi:10.1007/s00401-013-1096-7 (2013).

[37] Ki, CS; et al. The Ala53Thr mutation in the alpha-synuclein gene in a Korean family with Parkinson disease. *Clinical genetics*, 71, 471-473, doi:10.1111/j.1399-0004.2007.00781.x (2007).

[38] Chartier-Harlin, MC; et al. Alpha-synuclein locus duplication as a cause of familial Parkinson's disease. *Lancet*, 364, 1167-1169, doi:10.1016/S0140-6736(04)17103-1 (2004).

[39] Farrer, M; et al. Comparison of kindreds with parkinsonism and alpha-synuclein genomic multiplications. *Annals of neurology*, 55, 174-179, doi:10.1002/ana.10846 (2004).

[40] Ibanez, P; et al. Causal relation between alpha-synuclein gene duplication and familial Parkinson's disease. *Lancet*, 364, 1169-1171, doi:10.1016/S0140-6736(04)17104-3 (2004).

[41] Singleton, AB; et al. alpha-Synuclein locus triplication causes Parkinson's disease. *Science*, 302, 841, doi:10.1126/science.1090278 (2003).

[42] Cookson, MR. The role of leucine-rich repeat kinase 2 (LRRK2) in Parkinson's disease. *Nature reviews. Neuroscience*, 11, 791-797, doi:10.1038/nrn2935 (2010).

[43] Di Fonzo, A; et al. A frequent LRRK2 gene mutation associated with autosomal dominant Parkinson's disease. *Lancet*, 365, 412-415, doi:10.1016/S0140-6736(05)17829-5 (2005).

[44] Gilks, WP; et al. A common LRRK2 mutation in idiopathic Parkinson's disease. *Lancet*, 365, 415-416, doi:10.1016/S0140-6736(05)17830-1 (2005).

[45] Lesage, S; Brice, A. Parkinson's disease: from monogenic forms to genetic susceptibility factors. *Human molecular genetics*, 18, R48-59, doi:10.1093/hmg/ddp012 (2009).

[46] Nichols, WC; et al. Genetic screening for a single common LRRK2 mutation in familial Parkinson's disease. *Lancet*, 365, 410-412, doi:10.1016/S0140-6736(05)17828-3 (2005).

[47] Kachergus, J; et al. Identification of a novel LRRK2 mutation linked to autosomal dominant parkinsonism: evidence of a common founder across European populations. *American journal of human genetics*, 76, 672-680, doi:10.1086/429256 (2005).

[48] Healy, DG; et al. Phenotype, genotype, and worldwide genetic penetrance of LRRK2-associated Parkinson's disease: a case-control study. *The Lancet. Neurology*, 7, 583-590, doi:10.1016/S1474-4422(08)70117-0 (2008).

[49] Carballo-Carbajal, I; et al. Leucine-rich repeat kinase 2 induces alpha-synuclein expression via the extracellular signal-regulated kinase pathway. *Cellular signalling*, 22, 821-827, doi:10.1016/ j.cellsig.2010.01.006 (2010).

[50] Nguyen, HN; et al. LRRK2 mutant iPSC-derived DA neurons demonstrate increased susceptibility to oxidative stress. *Cell stem cell*, 8, 267-280, doi:10.1016/ j.stem.2011.01.013 (2011).

[51] Leroy, E; et al. The ubiquitin pathway in Parkinson's disease. *Nature*,395, 451-452, doi:10.1038/26652 (1998).

[52] Setsuie, R; et al. Dopaminergic neuronal loss in transgenic mice expressing the Parkinson's disease-associated UCH-L1 I93M mutant. *Neurochemistry international*, 50, 119-129, doi:10.1016/j.neuint. 2006.07.015 (2007).

[53] Shimura, H; et al. Ubiquitination of a new form of alpha-synuclein by parkin from human brain: implications for Parkinson's disease. *Science*, 293, 263-269, doi:10.1126/science.1060627 (2001).

[54] Lucking, CB; et al. Association between early-onset Parkinson's disease and mutations in the parkin gene. *The New England journal of medicine*, 342, 1560-1567, doi:10.1056/NEJM200005253422103 (2000).

[55] Narendra, D; Tanaka, A; Suen, DF; Youle, RJ. Parkin is recruited selectively to impaired mitochondria and promotes their autophagy. *The Journal of cell biology*, 183, 795-803, doi:10.1083/jcb.200809125 (2008).

[56] Rakovic, A; et al. Phosphatase and tensin homolog (PTEN)-induced putative kinase 1 (PINK1)-dependent ubiquitination of endogenous Parkin attenuates mitophagy: study in human primary fibroblasts and induced pluripotent stem cell-derived neurons. *The Journal of biological chemistry*, 288, 2223-2237, doi:10.1074/jbc.M112.391680 (2013).

[57] Vives-Bauza, C; et al. PINK1-dependent recruitment of Parkin to mitochondria in mitophagy. *Proceedings of the National Academy of Sciences of the United States of America*, 107, 378-383, doi:10.1073/pnas.0911187107 (2010).

[58] Rosner, S; Giladi, N; Orr-Urtreger, A. Advances in the genetics of Parkinson's disease. *Acta pharmacologica Sinica*, 29, 21-34, doi:10.1111/j.1745-7254.2008.00731.x (2008).

[59] Samaranch, L; et al. PINK1-linked parkinsonism is associated with Lewy body pathology. *Brain : a journal of neurology*, 133, 1128-1142, doi:10.1093/brain/awq051 (2010).

[60] Ramirez, A; et al. Hereditary parkinsonism with dementia is caused by mutations in ATP13A2, encoding a lysosomal type 5 P-type ATPase. *Nature genetics*, 38, 1184-1191, doi:10.1038/ng1884 (2006).

[61] Najim al-Din, AS; Wriekat, A; Mubaidin, A; Dasouki, M; Hiari, M. Pallido-pyramidal degeneration, supranuclear upgaze paresis and dementia: Kufor-Rakeb syndrome. *Acta neurologica Scandinavica*, 89, 347-352 (1994).

[62] Drozdzik, M; et al. Polymorphism in the P-glycoprotein drug transporter MDR1 gene: a possible link between environmental and genetic factors in Parkinson's disease. *Pharmacogenetics*, 13, 259-263, doi:10.1097/01.fpc.0000054087.48725.d9 (2003).

[63] Furuno, T; et al. Expression polymorphism of the blood-brain barrier component P-glycoprotein (MDR1) in relation to Parkinson's disease. *Pharmacogenetics*, 12, 529-534 (2002).

[64] Lee, CG; et al. MDR1, the blood-brain barrier transporter, is associated with Parkinson's disease in ethnic Chinese. *Journal of medical genetics*, 41, e60 (2004).

[65] Tan, EK; et al. Effect of MDR1 haplotype on risk of Parkinson disease. *Archives of neurology*, 62, 460-464, doi:10.1001/archneur.62.3.460 (2005).

[66] Westerlund, M; et al. Association of a polymorphism in the ABCB1 gene with Parkinson's disease. *Parkinsonism & related disorders*, 15, 422-424, doi:10.1016/j.parkreldis.2008.11.010 (2009).

[67] Dutheil, F; Beaune, P; Tzourio, C; Loriot, MA; Elbaz, A. Interaction between ABCB1 and professional exposure to organochlorine insecticides in Parkinson disease. *Archives of neurology*, 67, 739-745, doi:10.1001/archneurol.2010.101 (2010).

[68] Kiyohara, C; et al. MDR1 C3435T polymorphism and interaction with environmental factors in risk of Parkinson's disease: a case-control study in Japan. *Drug metabolism and pharmacokinetics*, 28, 138-143 (2013).

[69] Zschiedrich, K; et al. MDR1 variants and risk of Parkinson disease. Association with pesticide exposure? *Journal of neurology*, 256, 115-120, doi:10.1007/s00415-009-0089-x (2009).

[70] Schinkel, AH; et al. Disruption of the mouse mdr1a P-glycoprotein gene leads to a deficiency in the blood-brain barrier and to increased sensitivity to drugs. *Cell*, 77, 491-502 (1994).

[71] Cannon, JR; Greenamyre, JT. The role of environmental exposures in neurodegeneration and neurodegenerative diseases. *Toxicological sciences : an official journal of the Society of Toxicology*, 124, 225-250, doi:10.1093/toxsci/kfr239 (2011).

[72] Tanner, CM; et al. Rotenone, paraquat, and Parkinson's disease. *Environmental health perspectives*, 119, 866-872, doi:10.1289/ehp.1002839 (2011).

[73] Cannon, JR; Greenamyre, JT. Neurotoxic in vivo models of Parkinson's disease recent advances. *Progress in brain research*, 184, 17-33, doi:10.1016/S0079-6123(10)84002-6 (2010).

[74] Allen, MT; Levy, LS. Parkinson's disease and pesticide exposure--a new assessment. *Critical reviews in toxicology*, 43, 515-534, doi:10.3109/10408444.2013.798719 (2013).

[75] Bhatt, MH; Elias, MA; Mankodi, AK. Acute and reversible parkinsonism due to organophosphate pesticide intoxication: five cases. *Neurology*, 52, 1467-1471 (1999).

[76] Costello, S; Cockburn, M; Bronstein, J; Zhang, X; Ritz, B. Parkinson's disease and residential exposure to maneb and paraquat from agricultural applications in the central valley of California. *American journal of epidemiology*, 169, 919-926, doi:10.1093/aje/kwp006 (2009).

[77] Hancock, DB; et al. Pesticide exposure and risk of Parkinson's disease: a family-based case-control study. *BMC neurology*, 8, 6, doi:10.1186/1471-2377-8-6 (2008).

[78] Meco, G; Bonifati, V; Vanacore, N; Fabrizio, E. Parkinsonism after chronic exposure to the fungicide maneb (manganese ethylene-bis-dithiocarbamate). *Scandinavian journal of work, environment & health*, 20, 301-305 (1994).

[79] Tanner, CM; et al. Occupation and risk of parkinsonism: a multicenter case-control study. *Archives of neurology*, 66, 1106-1113, doi:10.1001/archneurol.2009.195 (2009).

[80] Brown, TP; Rumsby, PC; Capleton, AC; Rushton, L; Levy, LS. Pesticides and Parkinson's disease--is there a link? *Environmental health perspectives*, 114, 156-164 (2006).

[81] van der Mark, M; et al. Is pesticide use related to Parkinson disease? Some clues to heterogeneity in study results. *Environmental health perspectives*, 120, 340-347, doi:10.1289/ehp.1103881 (2012).

[82] Gao, HM; Hong, JS. Gene-environment interactions: key to unraveling the mystery of Parkinson's disease. *Progress in neurobiology*, 94, 1-19, doi:10.1016/j.pneurobio.2011.03.005 (2011).

[83] Dexter, DT; et al. Alterations in the levels of iron, ferritin and other trace metals in Parkinson's disease and other neurodegenerative diseases affecting the basal ganglia. *Brain : a journal of neurology*, 114 (Pt, 4), 1953-1975 (1991).

[84] Martin, WR. Quantitative estimation of regional brain iron with magnetic resonance imaging. *Parkinsonism & related disorders*, 15 Suppl, 3, S215-218, doi:10.1016/S1353-8020(09)70818-1 (2009).

[85] Santamaria, AB; Sulsky, SI. Risk assessment of an essential element: manganese. *Journal of toxicology and environmental health. Part A* 73, 128-155, doi:10.1080/15287390903337118 (2010).

[86] Gash, DM; et al. Trichloroethylene: Parkinsonism and complex 1 mitochondrial neurotoxicity. *Annals of neurology*, 63, 184-192, doi:10.1002/ana.21288 (2008).

[87] Guehl, D; et al. Trichloroethylene and parkinsonism: a human and experimental observation. *European journal of neurology : the official journal of the European Federation of Neurological Societies*, 6, 609-611 (1999).

[88] Kochen, W; Kohlmuller, D; De Biasi, P; Ramsay, R. The endogeneous formation of highly chlorinated tetrahydro-beta-carbolines as a possible causative mechanism in idiopathic Parkinson's disease. *Advances in experimental medicine and biology*, 527, 253-263 (2003).

[89] Liu, M; et al. Trichloroethylene induces dopaminergic neurodegeneration in Fisher 344 rats. *Journal of neurochemistry*, 112, 773-783, doi:10.1111/j.1471-4159.2009.06497.x (2010).

[90] Akundi, RS; et al. 1-trichloromethyl-1,2,3,4-tetrahydro-beta-carboline-induced apoptosis in the human neuroblastoma cell line SK-N-SH. *Journal of neurochemistry*, 91, 263-273, doi:10.1111/j.1471-4159.2004.02710.x (2004).

[91] Bringmann, G; et al. Bromal-derived tetrahydro-beta-carbolines as neurotoxic agents: chemistry, impairment of the dopamine metabolism, and inhibitory effects on mitochondrial respiration. *Bioorganic & medicinal chemistry*, 8, 1467-1478 (2000).

[92] Riederer, P; et al. Biochemical and pharmacological characterization of, 1-trichloromethyl-1,2,3,4-tetrahydro-beta-carboline: a biologically relevant neurotoxin? *European journal of pharmacology*, 442, 1-16 (2002).

[93] Lock, EA; Zhang, J; Checkoway, H. Solvents and Parkinson disease: a systematic review of toxicological and epidemiological evidence. *Toxicology and applied pharmacology*, 266, 345-355, doi:10.1016/ j.taap.2012.11.016 (2013).

[94] Gralewicz, S; Dyzma, M. Organic solvents and the dopaminergic system. *International journal of occupational medicine and environmental health*, 18, 103-113 (2005).

[95] Langston, JW; Irwin, I; Langston, EB; Forno, LS. 1-Methyl-4-phenylpyridinium ion (MPP+): identification of a metabolite of MPTP, a toxin selective to the substantia nigra. *Neuroscience letters*, 48, 87-92 (1984).

[96] Markey, SP; Johannessen, JN; Chiueh, CC; Burns, RS; Herkenham, MA. Intraneuronal generation of a pyridinium metabolite may cause drug-induced parkinsonism. *Nature*, 311, 464-467 (1984).

[97] Ramsay, RR; Salach, JI; Dadgar, J; Singer, TP. Inhibition of mitochondrial NADH dehydrogenase by pyridine derivatives and its possible relation to experimental and idiopathic parkinsonism. *Biochemical and biophysical research communications*, 135, 269-275 (1986).

[98] Hwang, O. Role of oxidative stress in Parkinson's disease. *Experimental neurobiology*, 22, 11-17, doi:10.5607/en.2013.22.1.11 (2013).

[99] Jenner, P. Oxidative stress in Parkinson's disease. *Annals of neurology*, 53 Suppl, 3, S26-36; discussion S36-28, doi:10.1002/ana.10483 (2003).

[100] Zhou, C; Huang, Y; Przedborski, S. Oxidative stress in Parkinson's disease: a mechanism of pathogenic and therapeutic significance. *Annals of the New York Academy of Sciences*, 1147, 93-104, doi:10.1196/ annals.1427.023 (2008).

[101] Betarbet, R; et al. Chronic systemic pesticide exposure reproduces features of Parkinson's disease. *Nature neuroscience*, 3, 1301-1306, doi:10.1038/81834 (2000).

[102] Sherer, TB; et al. Mechanism of toxicity in rotenone models of Parkinson's disease. *The Journal of neuroscience : the official journal of the Society for Neuroscience*, 23, 10756-10764 (2003).

[103] Brooks, AI; Chadwick, CA; Gelbard, HA; Cory-Slechta, DA; Federoff, HJ. Paraquat elicited neurobehavioral syndrome caused by dopaminergic neuron loss. *Brain research*, 823, 1-10 (1999).

[104] Chun, HS; et al. Dopaminergic cell death induced by MPP(+), oxidant and specific neurotoxicants shares the common molecular mechanism. *Journal of neurochemistry*, 76, 1010-1021 (2001).

[105] Fisher, HK; Clements, JA; Wright, RR. Enhancement of oxygen toxicity by the herbicide paraquat. *The American review of respiratory disease*, 107, 246-252 (1973).

[106] Ilett, KF; Stripp, B; Menard, RH; Reid, WD; Gillette, JR. Studies on the mechanism of the lung toxicity of paraquat: comparison of tissue distribution and some biochemical parameters in rats and rabbits. *Toxicology and applied pharmacology*, 28, 216-226 (1974).

[107] McCormack, AL; Di Monte, DA. Effects of L-dopa and other amino acids against paraquat-induced nigrostriatal degeneration. *Journal of neurochemistry*, 85, 82-86 (2003).

[108] McCormack, AL; et al. Environmental risk factors and Parkinson's disease: selective degeneration of nigral dopaminergic neurons caused by the herbicide paraquat. *Neurobiology of disease*, 10, 119-127 (2002).

[109] Zecca, L; et al. Iron and other metals in neuromelanin, substantia nigra, and putamen of human brain. *Journal of neurochemistry*, 62, 1097-1101 (1994).

[110] Griffiths, PD; Dobson, BR; Jones, GR; Clarke, DT. Iron in the basal ganglia in Parkinson's disease. An in vitro study using extended X-ray absorption fine structure and cryo-electron microscopy. *Brain : a journal of neurology*, 122 (Pt, 4), 667-673 (1999).

[111] Jellinger, K; et al. Iron-melanin complex in substantia nigra of parkinsonian brains: an x-ray microanalysis. *Journal of neurochemistry*, 59, 1168-1171 (1992).

[112] Riederer, P; et al. Distribution of iron in different brain regions and subcellular compartments in Parkinson's disease. *Annals of neurology*, 32 Suppl, S101-104 (1992).

[113] Sofic, E; Paulus, W; Jellinger, K; Riederer, P; Youdim, MB. Selective increase of iron in substantia nigra zona compacta of parkinsonian brains. *Journal of neurochemistry*, 56, 978-982 (1991).

[114] Winterbourn, CC. Toxicity of iron and hydrogen peroxide: the Fenton reaction. *Toxicology letters*, 82-83, 969-974 (1995).

[115] Cadet, JL; Brannock, C. Free radicals and the pathobiology of brain dopamine systems. *Neurochemistry international*, 32, 117-131 (1998).

[116] Esposito, G; Ana Clara, F; Verstreken, P. Synaptic vesicle trafficking and Parkinson's disease. *Developmental neurobiology*, 72, 134-144, doi:10.1002/dneu.20916 (2012).

[117] Miyazaki, I; Asanuma, M. Dopaminergic neuron-specific oxidative stress caused by dopamine itself. *Acta medica Okayama*, 62, 141-150 (2008).

[118] Hald, A; Lotharius, J. Oxidative stress and inflammation in Parkinson's disease: is there a causal link? *Experimental neurology*, 193, 279-290, doi:10.1016/j.expneurol.2005.01.013 (2005).

[119] Mosley, RL; et al. Neuroinflammation, Oxidative Stress and the Pathogenesis of Parkinson's Disease. *Clinical neuroscience research*, 6, 261-281, doi:10.1016/j.cnr.2006.09.006 (2006).

[120] Janetzky, B; et al. Unaltered aconitase activity, but decreased complex I activity in substantia nigra pars compacta of patients with Parkinson's disease. *Neuroscience letters*, 169, 126-128 (1994).

[121] Mann, VM; et al. Complex I, iron, and ferritin in Parkinson's disease substantia nigra. *Annals of neurology*, 36, 876-881, doi:10.1002/ ana.410360612 (1994).

[122] Schapira, AH; et al. Mitochondrial complex I deficiency in Parkinson's disease. *Journal of neurochemistry*, 54, 823-827 (1990).

[123] Schapira, AH; et al. Mitochondrial complex I deficiency in Parkinson's disease. *Lancet*, 1, 1269 (1989).

[124] Bindoff, LA; Birch-Machin, M; Cartlidge, NE; Parker, WD; Jr; Turnbull, DM. Mitochondrial function in Parkinson's disease. *Lancet*, 2, 49 (1989).

[125] Mizuno, Y; et al. Deficiencies in complex I subunits of the respiratory chain in Parkinson's disease. *Biochemical and biophysical research communications*, 163, 1450-1455 (1989).

[126] Parker, WD; Jr., Boyson, SJ; Parks, JK. Abnormalities of the electron transport chain in idiopathic Parkinson's disease. *Annals of neurology*, 26, 719-723, doi:10.1002/ ana.410260606 (1989).

[127] Schapira, AH. Mitochondria in the aetiology and pathogenesis of Parkinson's disease. *The Lancet. Neurology*, 7, 97-109, doi:10.1016/ S1474-4422(07)70327-7 (2008).

[128] Shamoto-Nagai, M; et al. An inhibitor of mitochondrial complex I, rotenone, inactivates proteasome by oxidative modification and induces aggregation of oxidized proteins in SH-SY5Y cells. *Journal of neuroscience research*, 74, 589-597, doi:10.1002/jnr.10777 (2003).

[129] Barrientos, A; Moraes, CT. Titrating the effects of mitochondrial complex I impairment in the cell physiology. *The Journal of biological chemistry*, 274, 16188-16197 (1999).

[130] Kweon, GR; et al. Distinct mechanisms of neurodegeneration induced by chronic complex I inhibition in dopaminergic and non-dopaminergic cells. *The Journal of biological chemistry*, 279, 51783-51792, doi:10.1074/jbc.M407336200 (2004).

[131] Menke, T; et al. Coenzyme Q10 reduces the toxicity of rotenone in neuronal cultures by preserving the mitochondrial membrane potential. *BioFactors*, 18, 65-72 (2003).

[132] Moon, SO; et al. Angiopoietin-1 reduces iopromide-induced endothelial cell apoptosis through activation of phosphatidylinositol 3'-kinase/p70 S6 kinase. *International journal of tissue reactions*, 27, 115-124 (2005).

[133] Chinta, SJ; Mallajosyula, JK; Rane, A; Andersen, JK. Mitochondrial alpha-synuclein accumulation impairs complex I function in dopaminergic neurons and results in increased mitophagy in vivo. *Neuroscience letters*, 486, 235-239, doi:10.1016/j.neulet. 2010.09.061 (2010).

[134] Devi, L; Raghavendran, V; Prabhu, BM; Avadhani, NG; Anandatheerthavarada, HK. Mitochondrial import and accumulation of alpha-synuclein impair complex I in human dopaminergic neuronal cultures and Parkinson disease brain. *The Journal of biological chemistry*, 283, 9089-9100, doi:10.1074/jbc.M710012200 (2008).

[135] Parihar, MS; Parihar, A; Fujita, M; Hashimoto, M; Ghafourifar, P. Mitochondrial association of alpha-synuclein causes oxidative stress. *Cellular and molecular life sciences : CMLS*, 65, 1272-1284, doi:10.1007/s00018-008-7589-1 (2008).

[136] Shavali, S; Brown-Borg, HM; Ebadi, M; Porter, J. Mitochondrial localization of alpha-synuclein protein in alpha-synuclein overexpressing cells. *Neuroscience letters*, 439, 125-128, doi:10.1016/j.neulet.2008.05.005 (2008).

[137] Song, DD; Shults, CW; Sisk, A; Rockenstein, E; Masliah, E. Enhanced substantia nigra mitochondrial pathology in human alpha-synuclein transgenic mice after treatment with MPTP. *Experimental neurology*, 186, 158-172, doi:10.1016/S0014-4886(03)00342-X (2004).

[138] Biskup, S; et al. Localization of LRRK2 to membranous and vesicular structures in mammalian brain. *Annals of neurology*, 60, 557-569, doi:10.1002/ana.21019 (2006).

[139] Sanders, LH; et al. LRRK2 mutations cause mitochondrial DNA damage in iPSC-derived neural cells from Parkinson's disease patients: reversal by gene correction. *Neurobiology of disease*, 62, 381-386, doi:10.1016/ j.nbd.2013.10.013 (2014).

[140] Winklhofer, KF; Haass, C. Mitochondrial dysfunction in Parkinson's disease. *Biochimica et biophysica acta*, 1802, 29-44, doi:10.1016/ j.bbadis.2009.08.013 (2010).

[141] Dexter, DT; Jenner, P. Parkinson disease: from pathology to molecular disease mechanisms. *Free radical biology & medicine*, 62, 132-144, doi:10.1016/j.free radbiomed.2013.01.018 (2013).

[142] Ariga, H; et al. Neuroprotective function of DJ-1 in Parkinson's disease. *Oxidative medicine and cellular longevity*, 2013, 683920, doi:10.1155/ 2013/683920 (2013).

[143] Glick, D; Barth, S; Macleod, KF. Autophagy: cellular and molecular mechanisms. *The Journal of pathology*, 221, 3-12, doi:10.1002/ path.2697 (2010).

[144] Mizushima, N; Levine, B; Cuervo, AM; Klionsky, DJ. Autophagy fights disease through cellular self-digestion. *Nature*, 451, 1069-1075, doi:10.1038/nature06639 (2008).

[145] McNaught, KS; Olanow, CW; Halliwell, B; Isacson, O; Jenner, P. Failure of the ubiquitin-proteasome system in Parkinson's disease. *Nature reviews. Neuroscience*, 2, 589-594, doi:10.1038/35086067 (2001).

[146] Park, C; Cuervo, AM. Selective autophagy: talking with the UPS. *Cell biochemistry and biophysics*, 67, 3-13, doi:10.1007/s12013-013-9623-7 (2013).

[147] Shen, YF; Tang, Y; Zhang, XJ; Huang, KX; Le, WD. Adaptive changes in autophagy after UPS impairment in Parkinson's disease. *Acta pharmacologica Sinica*, 34, 667-673, doi:10.1038/aps.2012.203 (2013).

[148] Ebrahimi-Fakhari, D; et al. Distinct roles in vivo for the ubiquitin-proteasome system and the autophagy-lysosomal pathway in the degradation of alpha-synuclein. *The Journal of neuroscience : the official journal of the Society for Neuroscience*, 31, 14508-14520, doi:10.1523/JNEUROSCI.1560-11.2011 (2011).

[149] Dehay, B; et al. Pathogenic lysosomal depletion in Parkinson's disease. *The Journal of neuroscience : the official journal of the Society for Neuroscience*, 30, 12535-12544, doi:10.1523/JNEUROSCI.1920-10.2010 (2010).

[150] Jesenberger, V; Jentsch, S. Deadly encounter: ubiquitin meets apoptosis. *Nature reviews. Molecular cell biology*, 3, 112-121, doi:10.1038/nrm731 (2002).

[151] Qiu, JH; et al. Proteasome inhibitors induce cytochrome c-caspase-3-like protease-mediated apoptosis in cultured cortical neurons. *The Journal of neuroscience : the official journal of the Society for Neuroscience*, 20, 259-265 (2000).
[152] Dagda, RK; Das Banerjee, T; Janda, E. How Parkinsonian toxins dysregulate the autophagy machinery. *International journal of molecular sciences*, 14, 22163-22189, doi:10.3390/ijms141122163 (2013).
[153] Ghee, M; Fournier, A; Mallet, J. Rat alpha-synuclein interacts with Tat binding protein, 1, a component of the 26S proteasomal complex. *Journal of neurochemistry*, 75, 2221-2224 (2000).
[154] Tanaka, Y; et al. Inducible expression of mutant alpha-synuclein decreases proteasome activity and increases sensitivity to mitochondria-dependent apoptosis. *Human molecular genetics*, 10, 919-926 (2001).
[155] Tanik, SA; Schultheiss, CE; Volpicelli-Daley, LA; Brunden, KR; Lee, VM. Lewy body-like alpha-synuclein aggregates resist degradation and impair macroautophagy. *The Journal of biological chemistry*, 288, 15194-15210, doi:10.1074/jbc.M113.457408 (2013).
[156] Winslow, AR; et al. alpha-Synuclein impairs macroautophagy: implications for Parkinson's disease. *The Journal of cell biology*, 190, 1023-1037, doi:10.1083/jcb.201003122 (2010).
[157] Plowey, ED; Cherra, SJ; 3rd, Liu, YJ; Chu, CT. Role of autophagy in G2019S-LRRK2-associated neurite shortening in differentiated SH-SY5Y cells. *Journal of neurochemistry*, 105, 1048-1056, doi:10.1111/ j.1471-4159.2008.05217.x (2008).
[158] Manzoni, C; et al. Inhibition of LRRK2 kinase activity stimulates macroautophagy. *Biochimica et biophysica acta*, 1833, 2900-2910, doi:10.1016/j.bbamcr.2013.07.020 (2013).
[159] Alegre-Abarrategui, J; et al. LRRK2 regulates autophagic activity and localizes to specific membrane microdomains in a novel human genomic reporter cellular model. *Human molecular genetics*, 18, 4022-4034, doi:10.1093/hmg/ddp346 (2009).
[160] Tong, Y; et al. Loss of leucine-rich repeat kinase 2 causes impairment of protein degradation pathways, accumulation of alpha-synuclein, and apoptotic cell death in aged mice. *Proceedings of the National Academy of Sciences of the United States of America*, 107, 9879-9884, doi:10.1073/ pnas.1004676107 (2010).
[161] Falkenburger, BH; Schulz, JB. Limitations of cellular models in Parkinson's disease research. *Journal of neural transmission. Supplementum*, 261-268 (2006).
[162] Beevers, JE; Caffrey, TM; Wade-Martins, R. Induced pluripotent stem cell (iPSC)-derived dopaminergic models of Parkinson's disease. *Biochemical Society transactions*, 41, 1503-1508, doi:10.1042/ BST20130194 (2013).
[163] Byers, B; Lee, HL; Reijo Pera, R. Modeling Parkinson's disease using induced pluripotent stem cells. *Current neurology and neuroscience reports*, 12, 237-242, doi:10.1007/s11910-012-0270-y (2012).
[164] Kriks, S; et al. Dopamine neurons derived from human ES cells efficiently engraft in animal models of Parkinson's disease. *Nature*, 480, 547-551, doi:10.1038/nature10648 (2011).
[165] Takahashi, K; et al. Induction of pluripotent stem cells from adult human fibroblasts by defined factors. *Cell*, 131, 861-872, doi:10.1016/ j.cell.2007.11.019 (2007).

[166] Devine, MJ; et al. Parkinson's disease induced pluripotent stem cells with triplication of the alpha-synuclein locus. *Nature communications*, 2, 440, doi:10.1038/ncomms1453 (2011).

[167] Ryan, SD; et al. Isogenic human iPSC Parkinson's model shows nitrosative stress-induced dysfunction in MEF2-PGC1alpha transcription. *Cell*, 155, 1351-1364, doi:10.1016/j.cell.2013.11.009 (2013).

[168] Soldner, F; et al. Generation of isogenic pluripotent stem cells differing exclusively at two early onset Parkinson point mutations. *Cell*, 146, 318-331, doi:10.1016/j.cell.2011.06.019 (2011).

[169] Cho, EG; et al. MEF2C enhances dopaminergic neuron differentiation of human embryonic stem cells in a parkinsonian rat model. *PloS one*, 6, e24027, doi:10.1371/journal.pone.0024027 (2011).

[170] Clark, J; Simon, DK. Transcribe to survive: transcriptional control of antioxidant defense programs for neuroprotection in Parkinson's disease. *Antioxidants & redox signaling*, 11, 509-528, doi:10.1089/ ARS.2008.2241 (2009).

[171] Reinhardt, P; et al. Genetic correction of a LRRK2 mutation in human iPSCs links parkinsonian neurodegeneration to ERK-dependent changes in gene expression. *Cell stem cell*, 12, 354-367, doi:10.1016/ j.stem.2013.01.008 (2013).

[172] Sanchez-Danes, A; et al. Disease-specific phenotypes in dopamine neurons from human iPS-based models of genetic and sporadic Parkinson's disease. *EMBO molecular medicine*, 4, 380-395, doi:10.1002/emmm.201200215 (2012).

[173] Cai, Q; Zakaria, HM; Simone, A; Sheng, ZH. Spatial parkin translocation and degradation of damaged mitochondria via mitophagy in live cortical neurons. *Current biology : CB*, 22, 545-552, doi:10.1016/j.cub.2012.02.005 (2012).

[174] Narendra, DP; et al. PINK1 is selectively stabilized on impaired mitochondria to activate Parkin. *PLoS biology*, 8, e1000298, doi:10.1371/journal.pbio.1000298 (2010).

[175] Van Laar, VS; et al. Bioenergetics of neurons inhibit the translocation response of Parkin following rapid mitochondrial depolarization. *Human molecular genetics*, 20, 927-940, doi:10.1093/hmg/ddq531 (2011).

[176] Imaizumi, Y; et al. Mitochondrial dysfunction associated with increased oxidative stress and alpha-synuclein accumulation in PARK2 iPSC-derived neurons and postmortem brain tissue. *Molecular brain*, 5, 35, doi:10.1186/1756-6606-5-35 (2012).

[177] Jiang, H; et al. Parkin controls dopamine utilization in human midbrain dopaminergic neurons derived from induced pluripotent stem cells. *Nature communications*, 3, 668, doi:10.1038/ncomms1669 (2012).

[178] Seibler, P; et al. Mitochondrial Parkin recruitment is impaired in neurons derived from mutant PINK1 induced pluripotent stem cells. *The Journal of neuroscience : the official journal of the Society for Neuroscience*, 31, 5970-5976, doi:10.1523/JNEUROSCI. 4441-10.2011 (2011).

[179] Cuervo, AM; Stefanis, L; Fredenburg, R; Lansbury, PT; Sulzer, D. Impaired degradation of mutant alpha-synuclein by chaperone-mediated autophagy. *Science*, 305, 1292-1295, doi:10.1126/science.1101738 (2004).

[180] Vogiatzi, T; Xilouri, M; Vekrellis, K; Stefanis, L. Wild type alpha-synuclein is degraded by chaperone-mediated autophagy and macroautophagy in neuronal cells. *The Journal of biological chemistry*, 283, 23542-23556, doi:10.1074/jbc.M801992200 (2008).

[181] Murphy, KE; et al. Reduced glucocerebrosidase is associated with increased alpha-synuclein in sporadic Parkinson's disease. *Brain : a journal of neurology*, 137, 834-848, doi:10.1093/brain/awt367 (2014).

[182] Neumann, J; et al. Glucocerebrosidase mutations in clinical and pathologically proven Parkinson's disease. *Brain : a journal of neurology*, 132, 1783-1794, doi:10.1093/brain/awp044 (2009).

[183] Sidransky, E; et al. Multicenter analysis of glucocerebrosidase mutations in Parkinson's disease. *The New England journal of medicine*, 361, 1651-1661, doi:10.1056/NEJMoa 0901281 (2009).

[184] Mazzulli, JR; et al. Gaucher disease glucocerebrosidase and alpha-synuclein form a bidirectional pathogenic loop in synucleinopathies. *Cell*, 146, 37-52, doi:10.1016/j.cell.2011.06.001 (2011).

[185] Panicker, LM; et al. Induced pluripotent stem cell model recapitulates pathologic hallmarks of Gaucher disease. *Proceedings of the National Academy of Sciences of the United States of America*, 109, 18054-18059, doi:10.1073/pnas.1207889109 (2012).

[186] Tiscornia, G; et al. Neuronopathic Gaucher's disease: induced pluripotent stem cells for disease modelling and testing chaperone activity of small compounds. *Human molecular genetics*, 22, 633-645, doi:10.1093/hmg/dds471 (2013).

[187] Kruger, R; et al. Ala30Pro mutation in the gene encoding alpha-synuclein in Parkinson's disease. *Nature genetics*, 18, 106-108, doi:10.1038/ng0298-106 (1998).

[188] Funayama, M; et al. A new locus for Parkinson's disease (PARK8) maps to chromosome 12p11.2-q13.1. *Annals of neurology*, 51, 296-301 (2002).

[189] Kitada, T; et al. Mutations in the parkin gene cause autosomal recessive juvenile parkinsonism. *Nature*, 392, 605-608, doi:10.1038/33416 (1998).

[190] West, AB; Maidment, NT. Genetics of parkin-linked disease. *Human genetics*, 114, 327-336, doi:10.1007/s00439-003-1074-6 (2004).

[191] Valente, EM; et al. Localization of a novel locus for autosomal recessive early-onset parkinsonism, PARK6, on human chromosome1p35-p36. *American journal of human genetics*, 68, 895-900 (2001).

[192] Bonifati, V; et al. DJ-1(PARK7), a novel gene for autosomal recessive, early onset parkinsonism. *Neurological sciences : official journal of the Italian Neurological Society and of the Italian Society of Clinical Neurophysiology*, 24, 159-160, doi:10.1007/s10072-003-0108-0 (2003).

[193] Bonifati, V; et al. Mutations in the DJ-1 gene associated with autosomal recessive early-onset parkinsonism. *Science*, 299, 256-259, doi:10.1126/science.1077209 (2003).

[194] van Duijn, CM; et al. Park7, a novel locus for autosomal recessive early-onset parkinsonism, on chromosome 1p36. *American journal of human genetics*, 69, 629-634, doi:10.1086/322996 (2001).

[195] Vilarino-Guell, C; et al. VPS35 mutations in Parkinson disease. *American journal of human genetics*, 89, 162-167, doi:10.1016/ j.ajhg.2011.06.001 (2011).

[196] Wider, C; et al. Autosomal dominant dopa-responsive parkinsonism in a multigenerational Swiss family. *Parkinsonism & related disorders*, 14, 465-470, doi:10.1016/j.parkreldis.2007.11.013 (2008).

[197] Chartier-Harlin, MC; et al. Translation initiator EIF4G1 mutations in familial Parkinson disease. *American journal of human genetics*, 89, 398-406, doi:10.1016/j.ajhg. 2011.08.009 (2011).

[198] Paisan-Ruiz, C; et al. Widespread Lewy body and tau accumulation in childhood and adult onset dystonia-parkinsonism cases with PLA2G6 mutations. *Neurobiology of aging*, 33, 814-823, doi:10.1016/ j.neurobiolaging.2010.05.009 (2012).

[199] Spatola, M; Wider, C. Genetics of Parkinson's disease: the yield. *Parkinsonism & related disorders*, 20 Suppl, 1, S35-38, doi:10.1016/S1353-8020(13)70011-7 (2014).

[200] Gomez-Garre, P; et al. Systematic mutational analysis of FBXO7 in a Parkinson's disease population from southern Spain. *Neurobiology of aging*, 35, 727 e725-727, doi:10.1016/j.neurobiolaging.2013.09.011 (2014).

[201] Ross, OA; et al. Genetic variation of Omi/HtrA2 and Parkinson's disease. *Parkinsonism & related disorders*, 14, 539-543, doi:10.1016/j.parkreldis.2008.08.003 (2008).

[202] Simon-Sanchez, J; Singleton, AB. Sequencing analysis of OMI/HTRA2 shows previously reported pathogenic mutations in neurologically normal controls. *Human molecular genetics*, 17, 1988-1993, doi:10.1093/hmg/ddn096 (2008).

[203] Strauss, KM; et al. Loss of function mutations in the gene encoding Omi/HtrA2 in Parkinson's disease. *Human molecular genetics*, 14, 2099-2111, doi:10.1093/hmg/ ddi215 (2005).

[204] Fong, CS; Cheng, CW; Wu, RM. Pesticides exposure and genetic polymorphism of paraoxonase in the susceptibility of Parkinson's disease. *Acta neurologica Taiwanica*, 14, 55-60 (2005).

[205] Taylor, MC; Le Couteur, DG; Mellick, GD; Board, PG. Paraoxonase polymorphisms, pesticide exposure and Parkinson's disease in a Caucasian population. *Journal of neural transmission*, 107, 979-983 (2000).

[206] Manthripragada, AD; Costello, S; Cockburn, MG; Bronstein, JM; Ritz, B. Paraoxonase, 1, agricultural organophosphate exposure, and Parkinson disease. *Epidemiology*, 21, 87-94, doi:10.1097/ EDE.0b013e3181c15ec6 (2010).

[207] Fong, CS; et al. Pesticide exposure on southwestern Taiwanese with MnSOD and NQO1 polymorphisms is associated with increased risk of Parkinson's disease. *Clinica chimica acta; international journal of clinical chemistry*, 378, 136-141, doi:10.1016/j.cca.2006.11.006 (2007).

[208] Kelada, SN; et al. 5' and 3' region variability in the dopamine transporter gene (SLC6A3), pesticide exposure and Parkinson's disease risk: a hypothesis-generating study. *Human molecular genetics*, 15, 3055-3062, doi:10.1093/hmg/ddl247 (2006).

[209] Ritz, BR; et al. Dopamine transporter genetic variants and pesticides in Parkinson's disease. *Environmental health perspectives*, 117, 964-969, doi:10.1289/ehp.0800277 (2009).

[210] Elbaz, A; et al. CYP2D6 polymorphism, pesticide exposure, and Parkinson's disease. *Annals of neurology*, 55, 430-434, doi:10.1002/ana.20051 (2004).

[211] Deng, Y; Newman, B; Dunne, MP; Silburn, PA; Mellick, GD. Further evidence that interactions between CYP2D6 and pesticide exposure increase risk for Parkinson's disease. *Annals of neurology*, 55, 897, doi:10.1002/ana.20143 (2004).

[212] De Palma, G; Mozzoni, P; Mutti, A; Calzetti, S; Negrotti, A. Case-control study of interactions between genetic and environmental factors in Parkinson's disease. *Lancet*, 352, 1986-1987 (1998).

[213] Fitzmaurice, AG; Rhodes, SL; Cockburn, M; Ritz, B; Bronstein, JM. Aldehyde dehydrogenase variation enhances effect of pesticides associated with Parkinson disease. *Neurology*, 82, 419-426, doi:10.1212/ WNL.0000000000000083 (2014).

[214] Checkoway, H; et al. A genetic polymorphism of MAO-B modifies the association of cigarette smoking and Parkinson's disease. *Neurology*, 50, 1458-1461 (1998).

[215] Hancock, DB; Martin, ER; Vance, JM; Scott, WK. Nitric oxide synthase genes and their interactions with environmental factors in Parkinson's disease. *Neurogenetics*, 9, 249-262, doi:10.1007/s10048-008-0137-1 (2008).

[216] Levecque, C; et al. Association between Parkinson's disease and polymorphisms in the nNOS and iNOS genes in a community-based case-control study. *Human molecular genetics*, 12, 79-86 (2003).

[217] McCulloch, CC; et al. Exploring gene-environment interactions in Parkinson's disease. *Human genetics*, 123, 257-265, doi:10.1007/ s00439-008-0466-z (2008).

[218] Rhodes, SL; et al. Pesticides that inhibit the ubiquitin-proteasome system: effect measure modification by genetic variation in SKP1 in Parkinsons disease. *Environmental research*, 126, 1-8, doi:10.1016/ j.envres.2013.08.001 (2013).

[219] Chen, ML; Lin, CH; Lee, MJ; Wu, RM. BST1 rs11724635 interacts with environmental factors to increase the risk of Parkinson's disease in a Taiwanese population. *Parkinsonism & related disorders*, 20, 280-283, doi:10.1016/j.parkreldis.2013.11.009 (2014).

[220] Dauer, W; Przedborski, S. Parkinson's disease: mechanisms and models. *Neuron*, 39, 889-909 (2003).

[221] Luthman, J; Fredriksson, A; Sundstrom, E; Jonsson, G; Archer, T. Selective lesion of central dopamine or noradrenaline neuron systems in the neonatal rat: motor behavior and monoamine alterations at adult stage. *Behavioural brain research*, 33, 267-277 (1989).

[222] Ungerstedt, U. 6-Hydroxy-dopamine induced degeneration of central monoamine neurons. *European journal of pharmacology*, 5, 107-110 (1968).

[223] Ungerstedt, U. Stereotaxic mapping of the monoamine pathways in the rat brain. *Acta physiologica Scandinavica. Supplementum*, 367, 1-48 (1971).

[224] Ungerstedt, U; Arbuthnott, GW. Quantitative recording of rotational behavior in rats after, 6-hydroxy-dopamine lesions of the nigrostriatal dopamine system. *Brain research*, 24, 485-493 (1970).

[225] Betarbet, R; et al. Intersecting pathways to neurodegeneration in Parkinson's disease: effects of the pesticide rotenone on DJ-1, alpha-synuclein, and the ubiquitin-proteasome system. *Neurobiology of disease*, 22, 404-420, doi:10.1016/j.nbd.2005.12.003 (2006).

[226] Liu, GH; et al. Progressive degeneration of human neural stem cells caused by pathogenic LRRK2. *Nature*, 491, 603-607, doi:10.1038/nature11557 (2012).

[227] Waseem, M; Ali, M; Dogra, S; Dutta, KK; Kaw, JL. Toxicity of trichloroethylene following inhalation and drinking contaminated water. *Journal of applied toxicology : JAT*, 21, 441-444 (2001).

[228] Borie, C; et al. Association study between iron-related genes polymorphisms and Parkinson's disease. *Journal of neurology*, 249, 801-804, doi:10.1007/s00415-002-0704-6 (2002).

[229] Dekker, MC; et al. Mutations in the hemochromatosis gene (HFE), Parkinson's disease and parkinsonism. *Neuroscience letters*, 348, 117-119 (2003).

[230] Gerlach, M; Double, KL; Youdim, MB; Riederer, P. Potential sources of increased iron in the substantia nigra of parkinsonian patients. *Journal of neural transmission. Supplementum*, 133-142 (2006).

[231] Mastroberardino, PG; et al. A novel transferrin/TfR2-mediated mitochondrial iron transport system is disrupted in Parkinson's disease. *Neurobiology of disease*, 34, 417-431, doi:10.1016/j.nbd.2009.02.009 (2009).

[232] Oakley, AE; et al. Individual dopaminergic neurons show raised iron levels in Parkinson disease. *Neurology*, 68, 1820-1825, doi:10.1212/01.wnl.0000262033.01945.9a (2007).

[233] Corrigan, FM; Murray, L; Wyatt, CL; Shore, RF. Diorthosubstituted polychlorinated biphenyls in caudate nucleus in Parkinson's disease. *Experimental neurology*, 150, 339-342, doi:10.1006/exnr.1998.6776 (1998).

[234] Corrigan, FM; Wienburg, CL; Shore, RF; Daniel, SE; Mann, D. Organochlorine insecticides in substantia nigra in Parkinson's disease. *Journal of toxicology and environmental health. Part A* 59, 229-234 (2000).

[235] Fleming, L; Mann, JB; Bean, J; Briggle, T; Sanchez-Ramos, JR. Parkinson's disease and brain levels of organochlorine pesticides. *Annals of neurology*, 36, 100-103, doi:10.1002/ana.410360119 (1994).

In: Gene-Environment Interactions and Human Diseases
Editor: Lu Qi
ISBN: 978-1-63483-680-7

Chapter 10

GENE-ENVIRONMENT INTERACTIONS IN MULTIPLE SCLEROSIS

Roberto Alvarez-Lafuente*

Servicio de Neurología. Hospital Clínico San Carlos, Instituto de Investigación Sanitaria del Hospital Clínico San Carlos (IdISSC), Madrid. Spain

ABSTRACT

Multiple sclerosis (MS) is an inflammatory demyelinating disease of the central nervous system. Although the etiology of MS is unknown, it is generally believed that genetic and environmental factors are involved. Multiple genetic factors have been found to play a role in the development of MS; certain alleles of the major histocompatibility complex have been linked to a higher risk of developing MS, particularly the HLA-DRB1 locus. Many different environmental factors have also been linked to MS, including Vitamin D, sunlight and UV exposure, smoking, and viruses. Viruses have long been suspected to play a potential role in the development of MS, and there seems to be increasing evidence of a link between MS and several different viruses. Numerous viruses have been implicated in the development of MS, including measles, mumps, rubella, Epstein-Barr virus (EBV), human herpesvirus 6 (HHV-6), Varicella-zoster virus (VZV) and human endogenous retroviruses (HERVs) (retroviruses which entered into the human genome millions of years ago). Some of the most convincing data for an association between viruses and MS is found in EBV and HHV-6, two different herpes viruses. The observation of an association between MS and latitude has led to studies on the nature of Vitamin D and UV exposure in relation to MS. Vitamin D levels have been found to be low in MS patients at baseline, and there also seems to an association between low Vitamin D levels and an increased risk of relapse. The objective of this review is to summarize studies performed in MS patients investigating: interactions between the host genes and the levels of vitamin D, interactions between host genes and EBV and HHV-6, and finally HERVs sequences and their interactions with other viruses.

* Corresponding author: Dr. Roberto Álvarez Lafuente. Hospital Clínico San Carlos. Unidad de Investigación Experimental (Planta -1). Laboratorio de Investigación de Esclerosis Múltiple (Laboratorios 3 y 6). C./ Profesor Martín Lagos s/n. 28040 Madrid. España. Tel. +34-91-3303000 Ext. 7437. Fax. +34-91-3303457. E-mail. ralvarezlafuente@yahoo.es.

INTRODUCTION

Multiple sclerosis (MS) is a common inflammatory disease of the central nervous system characterized by myelin loss, axonal pathology, and progressive neurological dysfunction, and it is also a leading cause of disability in young adults [1]. Clinically, MS patients show a variety of neurological signs and symptoms attributed to white matter lesions which are disseminated in time and space, that may occur in sudden attacks or be insidious and progressive. MS is twice as common in females. It typically presents in young adults between 20 and 50 years of age, with a peak occurring at 30 years of age, although childhood or older age cases occasionally occur [1]. The etiology of MS is unknown, however it is clear that genetic and environmental components are important [1, 2]. It has its greatest prevalence amongst individuals of Northern European ancestry [3] and is moderately heritable [4], with a sibling relative recurrence risk (λs) of ~ 6.3 [5]. Genetic risk factors for MS includes primarily immune genes, [6] with human leukocyte antigen (HLA)-DRB1*15 as the strongest risk allele [7, 8] and HLA-A*02 as a protective allele [7, 9]. Aside from the early success in demonstrating the important effects exerted by variants in the HLA genes from the Major Histocompatibility Complex (MHC) [10], there was little progress in unraveling the genetic architecture underlying MS susceptibility prior to the advent of genome-wide association studies (GWAS). Over the last decade, the International Multiple Sclerosis Genetics Consortium (IMSGC) has performed several GWAS and meta-analyses in large cohorts [11-13], cumulatively identifying a high number of non-MHC susceptibility alleles [14].

Furthermore, many different environmental factors have also been linked to MS, including Vitamin D [15], sunlight and UV exposure [16], smoking [17], and viruses (Table 1). Viruses have long been suspected to play a potential role in the development of MS, and there seems to be increasing evidence of a link between MS and several different viruses. In the last decades, numerous viruses have been implicated in the development of MS, including measles, [18] mumps, [19] rubella, [20] Epstein-Barr virus (EBV), [21] human herpesvirus 6 (HHV-6), [22] Varicella-zoster virus (VZV) [23] and human endogenous retroviruses (HERVs) [24-27] (retroviruses which entered into the human genome millions of years ago). Some of the most convincing data for an association between viruses and MS is found in EBV and HHV-6, two different herpes viruses [28]. The observation of an association between MS and latitude has led to studies on the nature of Vitamin D and UV exposure in relation to MS [29]. Vitamin D levels have been found to be low in MS patients at baseline, and there also seems to be an association between low Vitamin D levels and an increased risk of relapse [30].

Table 1. Viruses that have been associated with MS pathogenesis

Family	Virus
Herpesviridae	Herpes simples virus (HSV)
	Varicella zoster virus (VZV)
	Epstein-Barr virus (EBV)
	Human herpesvirus 6 (HHV-6)
	Marek disease virus (MDV)
Togaviridae	Rubella virus

Family	Virus
Retroviridae	Human T-cell leucemia virus type 1 (HTLV-1)
	Human endogenous retroviruses (HERV-Fc1, HERV-K18, HERV-H, HERV-W)
Paramyxoviridae	Measles virus
	Mumps virus
	Parainfluenza virus type 1
	Canine distemper virus
Anelloviridae	Torque teno virus
Coronaviridae	Coronavirus
Papovaviridae	JC virus
Bornaviridae	Borna disease virus

But, what are the links between these environmental and genetic risk factors in MS? In the last years, many authors have been focused on this topic; here, we are going to summarize studies performed in MS patients investigating interactions between host genes and levels of vitamin D, interactions between host genes and EBV and HHV-6, and HERVs sequences and their interactions with other viruses, and how can they affect to the development and prognosis of the disease.

1. Sun Exposure, Vitamin D and Multiple Sclerosis

1.1. Introduction

Vitamin D is a secosteroid hormone synthesized in the skin or ingested in the diet. Intake from dietary sources accounts for a much smaller proportion of total vitamin D, mainly owing to its rarity in foods [32, 33]. During exposure to sunlight, ultraviolet B (UVB) radiation (290–315 nm) is responsible for photolyzing 7-dehydrocholesterol, the precursor of vitamin D3, to previtamin D3 which, in turn, rapidly spontaneously isomerizes to vitamin D3 [32, 33]. Vitamin D3 is biologically inert and requires hydroxylation in the liver to 25-hydroxyvitamin D3 (25(OH)D). Once formed, this major circulating form of vitamin D3 is further hydroxylated in the kidney to its active form, 1,25-dihydroxyvitamin D3 (1,25(OH)2D), by 25-hydroxyvitamin D-1a-hydroxylase (1-OHase).

In the Northern Hemisphere, MS prevalence shows a north-south gradient, mirrored by a south-north gradient in the southern hemisphere [33]. In accordance with the disease geography, sunlight, specifically through its role in generating active vitamin D, has been proposed as a key environmental factor for the disease [34]. Circumstantial evidence to support this comes from studies showing that MS patients are deficient in vitamin D [35] and that dietary vitamin intake reduces disease risk [15]. Additionally, a pooled analysis of over 40,000 patients from Canada, Great Britain, Denmark, and Sweden showed that fewer people with MS were born in November and more in May [36], highlighting a risk factor that varies seasonally. Vitamin D is primarily known for its critical role in calcium homeostasis; however, recent evidence has highlighted many actions on immune and central nervous system development and function [37].

1.2. Vitamin D and HLA-DR in MS Patients

Early studies had provided evidence for an effect of vitamin D on HLA gene expression. In 1990, Rigby et al. [38] show that 1,25(OH),-D treatment of human monocytes consistently and specifically reduces their expression of HLA-DR and CD4 antigens; in contrast, pretreatment of monocytes with 1,25(OH),-D markedly impaired their ability to induce antigen-dependent proliferation of purified T lymphocytes. These results were confirmed by other groups [39].

Most biological effects of 1,25-dihydroxyvitamin D3 or calcitriol, are mediated by the vitamin D receptor (VDR). This receptor is a member of the steroid receptor super-family and influences the rate of transcription of vitamin D responsive genes by acting as a ligand activated transcription factor that binds to vitamin D response elements (VDREs) in gene promoters [37]. Recently, Ramagopalan et al. [40] examined the hypothesis of a direct interaction between vitamin D and MS associated MHC class II genes; the authors identified and functionally characterized a vitamin D response element (VDRE) in the HLA-DRB1 promoter region. The intriguing possibility that vitamin D responsiveness rather than any antigen specificity determines the increased MS risk of the HLA-DRB1*15 haplotype warrants consideration and can be tested in the infrequent haplotypes bearing the VDRE on other non-HLADRB1*15 haplotypes. Thus, it can be speculated that a general reduction in the expression of disease-associated class II alleles including HLA-DRB1*15 in the thymus during early life might result in loss of central tolerance, perhaps increasing the risk of autoimmunity in later life [41].

In the last years, more studies have attempted to confirm the association between vitamin D levels and HLA in MS patients. Lonergan et al.42 analyzed the MS prevalence, and 25(OH)D and HLA status in different counties of Ireland; the prevalence of MS was significantly higher in Donegal (northwest) than in Wexford (southeast) and South Dublin. Furthermore, levels of 25(OH)D (mean 50.74 nmol/l) were significantly higher in South Dublin (area with lowest prevalence) ($p < 0.0001$) than in Donegal or Wexford, and HLA DRB1*15 occurred most frequently in Donegal (greatest MS prevalence) and least frequently in South Dublin. Agliardi et al. [43] analyzed SNPs within the VDR gene in association with the HLA-DRB1 locus in 641 MS patients and 558 age- and sex-matched healthy controls, to verify possible correlations between the vitamin D/VDR complex, HLA-DRB1, and susceptibility to MS; analyses performed on HLA-DRB1*15-positive MS patients and healthy controls alone confirmed the protective role of rs731236 TT VDR genotype (p(y)=0.004; OR: 0.53; 95% CI: 0.33-0.83). Authors conclude that rs731236 TT VDR genotype modulates VDR expression and confers protection against MS in HLA-DRB1*15-positive individuals; therefore, results offer a model justifying the interaction between the major genetic (HLA-DRB*15) and environmental (vitamin D) factors associated with MS onset.

1.3. Vitamin D and non-HLA Susceptibility Genes

Genes associated to MS in the last GWAS have been discovered to be regulated by vitamin D:

- IRF8, that was associated with MS susceptibility, [44] encodes a transcription factor known to function in type I interferon signaling, and is associated with higher mRNA expression of interferon-response pathway genes in subjects with MS. In a study of Ramagopalan et al. [45] described novel intronic VDR binding sites involving IRF8, highlighting a role for vitamin D in gene regulation that had not previously been proposed.
- CD226 has also been associated with MS recently [46]. CD226 (also known as DNAX accessory molecule 1, DNAM-1) is a 67 kDa type I membrane protein involved in the adhesion and co-stimulation of T cells; it belongs to the immunoglobulin supergene family of receptors, containing two Ig-like domains in the extracellular region and is constitutively expressed on the majority of natural killer (NK) cells, CD4+ and CD8+ T cells, monocytes, platelets and a subset of B cells [47]. Similarly to IR8, two VDR binding sites has been described for CD226 [45].

Finally, other genes associated with MS susceptibility, like DHCR7, [48] have been involved in the regulation of vitamin D levels. DHCR7 codes for the enzyme delta-7-sterol reductase (EC 1.3.1.21), which converts 7-dehydrocholesterol (7-DHC) to cholesterol. Vitamin D is made in the skin when 7-DHC reacts with ultraviolet light. Thus, the conversion of 7-DHC by the gene product of DHCR7 removes the substrate 7-DHC from the synthetic pathway for the production of vitamin D. This study was the first to demonstrate a genetic link between an important regulator of vitamin D levels, DHCR7, and risk for MS.

Table 2. EBV serology in MS patients and controls

Year	Authors	MS patients				Healthy controls			
		Pos.		Neg	%	Pos.		Neg	%
1997	Munch et al.	137	/	1	99,3%	124	/	14	89,9%
1998	Myhr et al.	144	/	0	100,0%	162	/	8	95,3%
2000	Wagner et al.	107	/	0	100,0%	153	/	10	93,9%
2001	Ascherio et al.	143	/	1	99,3%	269	/	18	93,7%
2004	Haahr et al.	153	/	0	100,0%	50	/	3	94,3%
2004	Sundström et al.	234	/	0	100,0%	693	/	9	98,7%
2005	Ponsonby et al.	136	/	0	100,0%	252	/	9	96,6%
2006	DeLorenze et al.	42	/	4	91,3%	79	/	8	90,8%
2007	Banwell et al.	78	/	18	81.3%	61	/	35	63,5%
2008	Lünemann et al.	21	/	2	91,3%	11	/	6	64,7%
2010	Ingram et al.	70	/	5	93,3%	18	/	7	72,0%
2010	Jafari et al.	108	/	6	94,7%	51	/	9	85,0%
2010	Lindsey et al.	78	/	2	97,5%	74	/	6	92,5%
2011	Waubant et al.	167	/	22	88,4%	36	/	30	54,5%
2011	Lucas et al.	199	/	7	96,6%	198	/	19	91,2%
2011	Villegas et al.	66	/	10	86,8%	62	/	13	82,7%
2012	Sundquist et al.	580	/	5	99,1%	616	/	48	92,8%

2. Epstein-Barr Virus and Human Herpesvirus 6 Infections and Multiple Sclerosis

2.1. Introduction

EBV (also known as human herpes virus 4) is the causative agent of infectious mononucleosis; EBV has been of interest as a potential cause or trigger of MS since the 1970's, but the association with MS has been strengthened over more recent years in light of growing research [49]. EBV seropositivity has been found to be more common in patients with MS compared to controls (Table 2), with a seroprevalence of 99% in MS patients and 89% in controls [28]. A meta-analysis studying infectious mononucleosis and MS risk found that individuals who are not infected with EBV have a risk of MS close to zero, and MS risk increases after infection with EBV; this risk was further increased if individuals developed the infection in adolescence or adulthood [50].

HHV-6 is a beta herpesvirus which was discovered in 1986 when it was isolated from immunocompromised patients with HIV and lymphoproliferative disorders [51]. HHV-6 was first implicated in MS in the early 1990's. In 1993, a significantly higher HHV-6 serum antibody titers by immunofluorescence analysis in MS patients was found compared to controls [52]. Shortly after, immunocytochemistry directed against HHV-6 proteins revealed protein expression in MS brains, but not in control cases; moreover, the expression was more precisely localized to the oligodendrocytes [53], Since then, more studies have additionally supported a link between HHV-6 and MS: HHV-6 DNA has been found more frequently in CNS tissue, CSF and in fluids outside of the CNS (Table 3) of MS patients than in controls, [54-56] and there are also serological evidences for an association between HHV-6 and MS [57].

Table 3. Prevalence of HHV-6 DNA by PCR in serum samples of MS patients and controls

Year	Authors	MS patients				Healthy controls			
		Pos.		N	%	Pos.		N	%
1994	Wilborn et al.	0	/	21	0,0%	0	/	16	0,0%
1997	Martin et al.	0	/	20	0,0%	0	/	20	0,0%
1997	Soldan et al.	15	/	50	30,0%	0	/	47	0,0%
1998	Fillet et al.	2	/	32	6,3%	1	/	34	2,9%
1999	Goldberg et al.	1	/	24	4,2%	0	/	30	0,0%
1999	Mirandola et al.	0	/	32	0,0%	0	/	12	0,0%
2000	Akhyani et al.	8	/	34	23,5%	0	/	19	0,0%
2001	Tomsone et al.	14	/	38	36,8%	0	/	43	0,0%
2002	Tejada-Simon et al.	22	/	33	66,7%	7	/	21	33,3%
2003	Al-Shammari et al.	0	/	24	0,0%	1	/	33	3,0%
2004	Alvarez-Lafuente et al.	17	/	105	16,2%	0	/	49	0,0%
2005	Fögdell-Hahn et al.	4	/	42	9,5%	0	/	123	0,0%
2006	Alvarez-Lafuente et al.	16	/	63	25,4%	0	/	63	0,0%
2007	Martinez et al.	22	/	99	22,2%	-	/	-	-

Year	Authors	MS patients			Healthy controls		
		Pos.	N	%	Pos.	N	%
2008	Kuusisto et al.	0	/ 34	0,0%	-	/ -	-
2009	Ahram M, et al.	8	/ 30	26,7%	8	/ 33	24,2%
2009	Franciotta D et al.	0	/ 54	0,0%	0	/ 25	0,0%
2011	Nora-Krukle et al.	8	/ 17	32,0%	-	/ -	-
2011	Garcia-Montojo et al.	18	/ 35	33,9%	-	/ -	-
2013	Ben Fredj et al.	4	/ 56	6,7%	2	/ 61	3,2%

2.2. EBV / HHV-6 and HLA-DR in MS Patients

2.2.1. Epstein-Barr Virus

Although the first studies found that the association between anti-EBNA-1 antibody titers and MS risk was not affected by adjustment for DR15 and was similar in DR15-positive and DR15-negative women, and then the authors considered both as independent risk factors of MS, [58] later studies have shown EBV-HLA interactions. Zivadinov et al. [59] performed HLA typing and testing for antibodies against EBV to determine the role of gene-environmental interactions between the Class I and Class II HLA alleles and the humoral anti-EBV responses in the development of brain injury and clinical disability in MS patients; the authors found that the presence of HLA B7 was associated with increased anti-EBV VCA IgG levels, higher disability (EDSS) and more destructive MRI parameters, while the presence of HLA A2 was associated with lower EDSS and a trend toward decreased anti-EBV VCA IgG levels, suggesting that gene-environment interactions between specific HLA Class I loci and EBV exposure are associated with MRI markers of lesion injury and brain atrophy in MS patients. Sundström et al. [60] analyzed epitope reactivity of MS plasma antibodies against the EBNA-1 and its association with HLA DRB1*1501 status; they found that MS cases had increased antibody reactivity against several EBNA-1 domains, of which antibodies against EBNA-1 (amino acid 385-420) in HLA DRB1*1501 positive individuals were associated with a 24-fold risk increase for MS, suggesting a role for this epitope in the autoimmune pathogenesis of MS. Furthermore, Sundqvist et al. [61] analyzed possible interactions between antibodies against EBNA1 or EBNA1 fragments, presence of DRB1*15 and absence of A*02; all EBNA1 fragment IgGs were associated with MS risk, however, EBNA1 fragment 385-420 IgG levels were more strongly associated to MS than total EBNA1 IgG, OR=3.60 (2.75-4.72 95%CI), and also interacted with both DRB1*15 and absence of A*02, suggesting that the mechanism through which HLA genes influence the risk of MS may, at least in part, involve the immune control of EBV infection. Another interesting study performed by Rubicz R et al. [62] use an integrative genomics approach in order to localize genetic factors influencing levels of EBNA-1 IgG antibodies, as a measure of history of infection with this pathogen; genome-wide evidence of both significant linkage and association was obtained on chromosome 6 in the HLA region and replicated in an independent cohort (minimum p-value in combined analysis of both datasets is 1.4×10(-15)) for SNPs rs477515 and rs2516049, revealing the underlying mechanisms by which the immune system regulates infection with this pathogen. Finally, a recent study on pediatric MS has been performed. Waubant et al. [63] analyzed antibody titers of EBV, CMV and HSV-1 in pediatric MS patients and also assessed whether HLA-DRB1*1501 genotype influenced viral antibody levels; although similar antibody concentrations were found for all the viruses between pediatric EBNA-1 humoral responses were higher in HLA-DRB1 positive individuals (p=0.005) whereas other viral humoral responses were similar in HLA-DRB1 positive and negative individuals MS patients and controls, suggesting that genotype might influence the humoral response to EBV.

2.2.2. *Human Herpesvirus 6*

Regarding HHV-6, Martinez et al. [64] analyzed CIITA, a transcription factor controlling the expression of MHC class II genes (the main genetic determinants of MS susceptibility), and also a target of the immunoevasive strategies; two polymorphisms in MHC2TA gene (rs4,774G/C and rs3,087,456A/G) were studied in two groups: one in MS patients with active HHV-6A replication, and the other of MS patients with no detectable HHV-6 active infection after two-years follow-up, and an association of the rs4774C allele with the HHV-6-positive group was found when compared with the HHV-6A-negative (47.7% vs 18.8%, p = 0.0001; odds ratio = 3.94) and also with the control group (47.7% vs 25.5%, p = 0.001, odds ratio = 2.67), suggesting a strong gene-environment interaction. These results were replicated some years later for the same group, and also a different clinical behavior for MS patients with HHV-6 active infection and minor allele C was described [65]. It has also been recently published that a non-productive infection of HHV-6 in dendritic cells leads to the up-regulation of HLA-ABC, via autocrine IFN-α signaling, as well as the up-regulation of HLA-DR, although HHV-6 exposure reduces IL-8 secretion by dendritic cells and their capacity to stimulate allogenic T cell proliferation; the ability to suppress dendritic cells functions important for activation of innate and adaptive immune responses might be one successful strategy by which HHV-6 avoids the induction of appropriate host defense mechanisms, and thus facilitating persistent infection [66]. Finally, Engdahl et al. [67] characterized the plasma IgG antibody response against HHV-6 in MS patients and healthy controls, and the relationship between MS susceptibility factors and the anti-HHV-6 response; interestingly, carriership of HLA-A(*)02 was associated with lower anti-HHV-6 IgG levels (p=0.0017), possibly reflecting a relation between autoimmunity and how the immune system handles viral infections.

2.3. EBV/HHV-6 and Non-HLA Susceptibility Genes

As it has been previously mentioned, the etiology of MS involves a complex genetic trait and environmental triggers that include viral infections. Among the risk alleles that have repeatedly been identified by genome-wide association studies, three are located near the Casitas B-lineage lymphoma proto-oncogene b gene (CBLB). The CBLB protein (CBL-B) is a key regulator of peripheral immune tolerance by limiting T cell activation and expansion and hence T cell-mediated autoimmunity through its ubiquitin E3-ligase activity. A recent study of Stürner et al. [67] shows that CBL-B expression is reduced in CD4+ T cells from MS patients during relapse. The MS risk-related SNP of CBLB rs12487066 is associated with diminished CBL-B expression levels and alters the effects of type I IFNs on human CD4+ T cell proliferation. Mechanistically, the CBLB rs12487066 risk allele mediates increased binding of the transcription factor C/EBPβ and reduced CBL-B expression in human CD4+ T cells. These data suggest a role of the CBLB rs12487066 variant in the interactions of a genetic risk factor and IFN function during viral infections in MS.

While non-HLA susceptibility genes have not been analyzed for possible interactions with EBV in MS patients, some studies have been published for HHV-6:

- IRF5. Members of the interferon regulatory factor (IRF) family have critical roles in cytokine signaling, hematopoietic differentiation, cell cycle arrest and apoptosis. IRF pathways also participate in the regulation of the antiviral innate response. IRF5-induced innate antiviral responses appear to result in a broad alteration of the transcriptional profile of cellular genes. In MS, two SNPs (rs4728142 and rs3807306) have reached significant association in the three independent cohort tested [68]. Vandenbroeck et al. [68] addressed whether these polymorphisms were associated with the active replication of HHV-6, and association was observed between rs3807306T and infection with HHV-6 [p=0.05, OR (95% CI)=1.56 (1.00-2.44)]. As we have previously mentioned, IRF8, another member of the IRF family, has been associated with MS susceptibility [44].
- TNFRSF14 and TNFRSF6B. Two main superfamilies of co-signaling molecules based on their structure modulate the T cell response: the immunoglobulin superfamily and the tumour necrosis factor receptor superfamily (TNFRSF). One of these TNFRSF members is HVEM (herpes virus entry mediator or TNFRSF14), a molecular switch that regulates the host immune response depending on which ligand it binds. The costimulatory ligand LIGHT is shared by another member of the TNFRSF, DcR3 (decoy receptor 3 or TNFRSF6B) [69]. In a study published by Blanco-Kelly et al. [69], the authors aimed to ascertain the role of these genes in MS susceptibility and to investigate the relationship of the gene encoding the widely expressed HVEM receptor with the active replication of HHV6 found in some MS patients; both polymorphisms were associated with MS predisposition, with stronger effect in patients with HHV6 active replication-TNFRSF6B-rs4809330(*)A: P=0.028, OR=1.13, and TNFRSF14-rs6684865(*)A: overall P=0.0008, OR=1.2. Another member of this superfamily (TNFRSF1A) has also been recently associated with MS susceptibility in the last GWAS [70].

3. Human Endogenous Retroviruses and Multiple Sclerosis

3.1. Introduction

Retroviruses can exist either as exogenous or endogenous entities, or both [71]. HERVs are assumed to be remnants of ancient retroviral infections of our ancestors' germ-line cells, resulting in vertical transmission in a Mendelian fashion. The HERVs were first described in 1981, [72] and can be classified in at least 31 families, and numerous solitary LTRs are currently known. Around 8% of the human genome consists of HERV sequences; [73] the majority of them are common to humans, Old World monkeys, and apes, indicating that the exogenous virus that gave rise to them infected our ancestors at least 25 million years ago. As a result of their longevity within the host genome, many of these HERVs contain mutations such as deletions, stop codons, or frameshifts, ensuring that no protein or viable viral particles are produced [74]. However, it has been described that mRNA from several HERV families are ubiquitously expressed in white blood cells, [75-76] and the extent of expression varies from tissue to tissue and also between individuals [77]. Virions containing HERV sequences are very rarely found, but have been reported for certain HERV families in placental tissues, [78] and in Sjögren's syndrome, [79] seminomas, [80] melanomas, [81] schizophrenia, [82] and MS [83].

Although, as we have previously mentioned, several families of HERVs have been described, only for some of them, both mRNA transcripts and proteins, can be produced from part or all of the provirus; these include ERV-3, HERV-W, and a number of members of the HERV-K family, suggesting their potential for involvement in autoimmunity [74]. However, in the last years, the most compelling evidences of possible a relationship between MS and HERVs have been found in studies of MS-associated retrovirus (MSRV) / HERV-W endogenous retrovirus. At this point, it should also be made clear that for the HERVs, disease-associations have been established, but causal effects have yet to be clearly demonstrated.

3.2. MSRV/HERV-W and Multiple Sclerosis

Multiple Sclerosis Associated Retrovirus (MSRV) is a member of type-W Human Endogenous Retroviruses (HERV-W), an HERV family suggested to be involved in Multiple Sclerosis (MS) pathogenesis. The study of the relationship between HERVs and MS began in 1989, when retroviral particles were discovered in cultures of leptomeningeal cells and monocytes from MS patients [83-84]. We have focused on HERV-W because (i) MSRV is able to cause a T cell-mediated neuropathology in vivo, [85] (ii) MSRV/HERV-W envelope protein is able to induce a potent activation of innate immunity and subsequent release of pro-inflammatory cytokines, through Toll-like receptor (TLR-4), [86] and (iii) acts as superantigen, producing polyclonal T-lymphocyte activation [87]. Moreover, it has been described that HERV-W env protein can induce oligodendrocyte toxicity inflammation-mediated [88]. Finally, it has also been described that: depending on the cohort studied, the percentage of MS serum or plasma samples positive to MSRV ranges from 50% to 100% [89-90]; enhanced expression of MSRV was also detected in MS plaques in independent post-mortem brain studies [91-93]; and increased MSRV copy number in MS blood DNA was associated with a poorer MS prognosis [94-95].

3.3. MSRV/HERV-W and Human Herpesviruses

Since HERVs can be considered as an environmental factor, but also as a susceptibility genetic factor in MS, their possible interactions with other environmental factors like the human herpesviruses have been studied.

In 2007, Brudek et al. [96] analyzed the possible role of herpesviruses as contributing factors in HERV activation. They demonstrate the ability of HSV-1, HHV-6, and VZV antigens to induce higher reverse transcriptase (RT) activity in peripheral lymphocytes from MS patients vs. controls during the first 6 days post-antigen stimulation; for HHV-6 they found significantly higher RT activity in MS patient PBMCs compared with healthy controls on day 2 (P = 0.018) and day 4 (P = 0.025). The authors conclude that these results contribute to the theory that herpesviruses and HERVs, by their interactions, may play a role in the pathogenesis of MS.

Studies analyzing the possible interactions between HERV-W and EBV have also been performed. Mameli et al. [97] searched the possible links between the expression of HERV-W/MSRV/syncytin-1, with/without exposure to EBV or to EBV glycoprotein350 (EBVgp350), on peripheral blood mononuclear cells from healthy volunteers and MS patients, and on astrocytes, by discriminatory env-specific RT-PCR assays, and by flow cytometry; the authors found that, in vitro, EBV activates the potentially immunopathogenic and neuropathogenic HERV-W/MSRV/syncytin-1, in cells deriving from blood and brain, although in vivo, pathogenic outcomes would depend on abnormal situations, as in late EBV primary infection, that is often symptomatic, or/and in the presence of particular host genetic backgrounds. They suggest a possible model for MS pathogenesis that include EBV as initial trigger of future MS, years later, and HERV-W/MSRV/syncytin-1 as actual contributor to MS pathogenicity, in striking parallelism with disease behavior. The same group published a new study deepening in this interaction one year later [98]. Since the ascertained links between EBV and MS are history of late primary infection, possibly leading to infectious mononucleosis, and high titers of pre-onset IgG against EBV nuclear antigens, the authors analyzed whether HERV-W might be activated in vivo, in hospitalized young adults with IM symptoms; the results showed that activation of HERV-W/MSRV occurs in blood mononuclear cells of infectious mononucleosis patients, indicating that the two main links between EBV and MS (infectious mononucleosis and high anti-EBNA-1-IgG titers) are paralleled by activation of the potentially neuropathogenic HERV-W/MSRV.

Summary and Perspectives

Although there are no definitive evidences linking the previously mentioned environmental factors with the pathogenesis of MS, the results accumulated in the last years seem to support the possible involvement of one or more of these agents in the disease. As it is generally accepted that MS is a polygenic disease, with different genes contributing to disease susceptibility, perhaps we should accept that more than one of these environmental factors could contribute to the disease or interact each other to trigger the autoimmune process leading to MS.

Finally, seems to exist a close relationship between the possible environmental factors related to MS and the genes that confer susceptibility to the disease, modulating the autoimmune response and contributing to the heterogeneity that characterizes MS. Therefore, future studies will produce more meaningful and convincing results only if we take into account these two factors, environmental and genetic, and their possible interactions, to better understand the pathogenesis of the disease.

References

[1] Noseworthy, JH; et al. Multiple sclerosis. *N. Engl. J. Med.*, 343, 938–952 (2000).

[2] Ebers, GC. Environmental factors and multiple sclerosis. *Lancet Neurol.*, 7, 268–277 (2008)

[3] Compston, A; et al. McAlpine's Multiple Sclerosis. Churchill Livingstone; London (2006).

[4] Dyment, DA; et al. Multiple sclerosis in stepsiblings: recurrence risk and ascertainment. *J. Neurol. Neurosurg. Psychiatry.*, 77, 258–259 (2006).

[5] Hemminki, K; et al. Risk for multiple sclerosis in relatives and spouses of patients diagnosed with autoimmune and related conditions. *Neurogenetics.*, 10, 5–11 (2009).

[6] Sawcer, S. et al. Genetic risk and a primary role for cell-mediated immune mechanisms in multiple sclerosis. *Nature.*, 476, 214–219 (2011).

[7] Brynedal, B; et al. HLA-A confers an HLA-DRB1 independent influence on the risk of multiple sclerosis. *PLoS One.*, 2, e664 (2007).

[8] Hillert, J; et al. Multiple sclerosis is associated with genes within or close to the HLA-DR-DQ subregion on a normal DR15, DQ6, Dw2 haplotype. *Neurology.*, 43, 163–168 (1993).

[9] Fogdell-Hahn, A; et al. Multiple sclerosis: a modifying influence of HLA class I genes in an HLA class II associated autoimmune disease. *Tissue Antigens.*, 55, 140–148 (2000).

[10] Jersild, C; et al. HL-A antigens and multiple sclerosis. *Lancet.*, 1, 1240–1241 (1972).

[11] IMSGC. Risk Alleles for Multiple Sclerosis Identified by a Genomewide Study. *N. Engl. J. Med.*, 357, 851–862 (2007).

[12] De Jager, PL; et al. Meta-analysis of genome scans and replication identify CD6, IRF8 and TNFRSF1A as new multiple sclerosis susceptibility loci. *Nat. Genet.*, 41, 776–782 (2009).

[13] Patsopoulos, NA; et al. Genome-wide meta-analysis identifies novel multiple sclerosis susceptibility loci. *Ann. Neurol.*, 70, 897–912 (2011).

[14] IMSGC. Analysis of immune-related loci identifies 48 new susceptibility variants for multiple sclerosis. *Nat. Genet.*, 45, 1353-1360 (2013).

[15] Munger, KL; et al. Serum 25-hydroxyvitamin D levels and risk of multiple sclerosis. *JAMA.*, 296, 2832–2838 (2006).

[16] Vukusic, S; et al. Regional variations in the prevalence of multiple sclerosis in French farmers. *J. Neurol. Neurosurg. Psychiatry.*, 78, 707-709 (2007).

[17] Hedstrom, AK; et al. Tobacco smoking, but not Swedish snuff use, increases the risk of multiple sclerosis. *Neurology.*, 73, 696–701 (2009).

[18] Vandvik, B; et al. Oligoclonal measles virus-specific IgG antibodies isolated from cerebrospinal fluids, brain extracts, and sera from patients with subacute sclerosing panencephalitis and multiple sclerosis. *Scand. J. Immunol.*, 5, 979–992 (1976).

[19] Norrby, E; et al. Comparison of antibodies against different viruses in cerebrospinal fluid and serum samples from patients with multiple sclerosis. *Infect. Immun.*, 10, 688-694 (1974).

[20] Brettschneider, J; et al. IgG antibodies against measles, rubella, and varicella zoster virus predict conversion to multiple sclerosis in clinically isolated syndrome. *PLoS One.*, 4, e7638 (2009).
[21] Munch, M; et al. The implications of Epstein-Barr virus in multiple sclerosis--a review. *Acta Neurol. Scand. Suppl.*, 169, 59-64 (1997).
[22] Challoner, PB; et al. Plaque-associated expression of human herpesvirus 6 in multiple sclerosis. *Proc. Natl. Acad. Sci. USA.*, 92, 7440-7444 (1995).
[23] Mancuso, R; et al. Increased prevalence of varicella zoster virus DNA in cerebrospinal fluid from patients with multiple sclerosis. *J. Med. Virol.*, 79, 192-199 (2007).
[24] Perron, H; et al. Multiple sclerosis retrovirus particles and recombinant envelope trigger an abnormal immune response in vitro, by inducing polyclonal Vbeta16 T-lymphocyte activation. *Virology.*, 287, 321–332 (2001).
[25] Christensen, T; et al. Molecular characterization of HERV-H variants associated with multiple sclerosis. *Acta Neurol. Scand.*, 101, 229-238 (2000).
[26] Tai, AK; et al. Human endogenous retrovirus-K18 Env as a risk factor in multiple sclerosis. *Mult. Scler.*, 14, 1175-1180 (2008).
[27] Nexø, BA; et al. The etiology of multiple sclerosis: genetic evidence for the involvement of the human endogenous retrovirus HERV-Fc1. *PLoS One.*, 6, e16652 (2011).
[28] Tselis, A. Evidence for viral etiology of multiple sclerosis. *Semin. Neurol.*, 31, 307-316 (2011).
[29] Van der Mei, IA; et al. Past exposure to sun, skin phenotype, and risk of multiple sclerosis: case-control study. *BMJ.*, 327, 316 (2003).
[30] Weinstock-Guttman, B; et al. Vitamin D and multiple sclerosis. *Neurologist.*, 18, 179-813 (2012).
[31] Feldman, D; et al. Vitamin D. San Diego; London: Academic Press (2005).
[32] Holick, MF. Vitamin D: A millenium perspective. *J. Cell. Biochem.*, 88, 296–307 (2003).
[33] Pugliatti, M; et al. The worldwide prevalence of multiple sclerosis. *Clin. Neurol. Neurosurg.*, 104, 182–191 (2002).
[34] Acheson, ED; et al. Some comments on the relationship of the distribution of multiple sclerosis to latitude, solar radiation, and other variables. Acta Psychiatr. *Scand. Suppl*, 35, 132–147 (1960).
[35] Nieves, J; et al. High prevalence of vitamin D deficiency and reduced bone mass in multiple sclerosis. *Neurology.*, 44, 1687–1692 (1994).
[36] Willer, CJ; et al. Timing of birth and risk of multiple sclerosis: population based study. *BMJ.*, 330, 120 (2005).
[37] Smolders, J; et al. Vitamin D as an immune modulator in multiple sclerosis, a review. *J. Neuroimmunol.*, 194, 7-17(2008).
[38] Rigby, WF; et al. Regulation of human monocyte HLA-DR and CD4 antigen expression, and antigen presentation by 1,25-dihydroxyvitamin D3. *Blood.*, 76, 189-197 (1990).
[39] Tokuda, N; et al. 1,25-Dihydroxyvitamin D3 down-regulation of HLA-DR on human peripheral blood monocytes. *Immunology.*, 75, 349-354 (1992).
[40] Ramagopalan, SV; et al. Expression of the multiple sclerosis-associated MHC class II Allele HLA-DRB1*1501 is regulated by vitamin D. *PLoS Genet.*, 5, e1000369 (2009).

[41] Handunnetthi, L; et al. Multiple sclerosis, vitamin D, and HLA-DRB1*15. *Neurology.*, 74, 1905-1910 (2010).

[42] Lonergan, R; et al. Multiple sclerosis prevalence in Ireland: relationship to vitamin D status and HLA genotype. *J. Neurol. Neurosurg. Psychiatry.*, 82, 317-322 (2011).

[43] Agliardi, C; et al. Vitamin D receptor (VDR) gene SNPs influence VDR expression and modulate protection from multiple sclerosis in HLA-DRB1*15-positive individuals. *Brain Behav. Immun.*, 25, 1460-1467 (2011).

[44] De Jager, PL; et al. Meta-analysis of genome scans and replication identify CD6, IRF8 and TNFRSF1A as new multiple sclerosis susceptibility loci. *Nat. Genet.*, 41, 776-782 (2009).

[45] Ramagopalan, SV; et al. A ChIP-seq defined genome-wide map of vitamin D receptor binding: Associations with disease and evolution. *Genome Res.*, 20, 1352-1360 (2010).

[46] Hafler, JP; et al. CD226 Gly307Ser association with multiple autoimmune diseases. *Genes Immun.*, 10, 5-10 (2009).

[47] Shibuya, A; et al. DNAM-1, a novel adhesion molecule involved in the cytolytic function of T lymphocytes. *Immunity.*, 4, 573–581 (1996).

[48] Alloza, I; et al. ANKRD55 and DHCR7 are novel multiple sclerosis risk loci. *Genes Immun.*, 13, 253-257 (2012).

[49] Ascherio, A. Environmental factors in multiple sclerosis. *Expert Rev Neurother.*, 13 (12 Suppl), 3-9 (2013).

[50] Thacker, EL. Infectious mononucleosis and risk for multiple sclerosis: a meta-analysis. *Ann Neurol.*, 59, 499-503 (2006).

[51] Salahuddin, SZ; et al. Isolation of a new virus, HBLV, in patients with lymphoproliferative disorders. *Science.*, 234, 596-601 (1986).

[52] Sola, P; et al. Human herpesvirus 6 and multiple sclerosis: survey of anti-HHV-6 antibodies by immunofluorescence analysis and of viral sequences by polymerase chain reaction. *J Neurol Neurosurg Psychiatry.*, 56, 917-919 (1993).

[53] Challoner, PB; et al. Plaque-associated expression of human herpesvirus 6 in multiple sclerosis. *Proc Natl Acad Sci U S A.*, 92, 7440-7444 (1995).

[54] Cermelli, C; et al. High frequency of human herpesvirus 6 DNA in multiple sclerosis plaques isolated by laser microdissection. *J Infect Dis.*, 187, 1377-1387 (2003).

[55] Tejada-Simon, MV; et al. Detection of viral DNA and immune responses to the human herpesvirus 6 101-kilodalton virion protein in patients with multiple sclerosis and in controls. *J Virol.*, 76, 6147-6154 (2002).

[56] Alvarez-Lafuente, R; et al. Active human herpesvirus 6 infection in patients with multiple sclerosis. *Arch Neurol.*, 59, 929-933 (2002).

[57] Ortega-Madueño, I; et al. Anti-Human Herpesvirus 6A/B IgG Correlates with Relapses and Progression in Multiple Sclerosis. *PLoS One.*, 9, e104836 (2014).

[58] De Jager, PL; et al. Integrating risk factors: HLA-DRB1*1501 and Epstein-Barr virus in multiple sclerosis. *Neurology.*, 70, 1113-1118 (2008).

[59] Zivadinov, R; et al. Gene-environment interactions between HLA B7/A2, EBV antibodies are associated with MRI injury in multiple sclerosis. *J Neuroimmunol.*, 209, 123-130 (2009).

[60] Sundström, P; et al. Antibodies to specific EBNA-1 domains and HLA DRB1*1501 interact as risk factors for multiple sclerosis. *J Neuroimmunol.*, 215, 102-107 (2009).

[61] Sundqvist, E; et al. Epstein-Barr virus and multiple sclerosis: interaction with HLA. *Genes Immun.*, 13, 14-20 (2012).

[62] Waubant, E; et al. Antibody response to common viruses and human leukocyte antigen-DRB1 in pediatric multiple sclerosis. *Mult Scler.*, 19, 891-895 (2013).

[63] Martínez, A; et al. Environment-gene interaction in multiple sclerosis: human herpesvirus 6 and MHC2TA. *Hum Immunol.*, 68, 685-689 (2007).

[64] Alvarez-Lafuente, R; et al. MHC2TA rs4774C and HHV-6A active replication in multiple sclerosis patients. *Eur J Neurol.*, 17, 129-135 (2010).

[65] Gustafsson, RK; et al. Human herpesvirus 6A partially suppresses functional properties of DC without viral replication. *PLoS One.*, 8, e58122 (2013).

[66] Engdahl, E; et al. HLA-A(*)02, gender and tobacco smoking, but not multiple sclerosis, affects the IgG antibody response against human herpesvirus 6. *Hum Immunol.*, 75, 524-530 (2014).

[67] Stürner, KH; et al. A Multiple Sclerosis-Associated Variant of CBLB Links Genetic Risk with Type I IFN Function. *J Immunol.*, pii: 1303077 [Epub ahead of print]. (2014).

[68] Vandenbroeck, K; et al. Validation of IRF5 as multiple sclerosis risk gene: putative role in interferon beta therapy and human herpes virus-6 infection. *Genes Immun.*, 12, 40-45 (2011).

[69] Blanco-Kelly, F; et al. Members 6B and 14 of the TNF receptor superfamily in multiple sclerosis predisposition. *Genes Immun.* 12, 145-148 (2011).

[70] International Multiple Sclerosis Genetics Consortium (IMSGC). et al. Analysis of immune-related loci identifies 48 new susceptibility variants for multiple sclerosis. *Nat Genet.*, 45, 1353-1360 (2013).

[71] Christensen, T. Association of human endogenous retroviruses with multiple sclerosis and possible interactions with herpes viruses. *Rev Med Virol.*, 15, 179-211 (2005).

[72] Martin, MA; et al. Identification and cloning of endogenous retroviral sequences present in human DNA. *Proc Natl Acad Sci USA.*, 78, 4892-4896 (1981).

[73] Prak, ET; et al. Mobile elements and the human genome. *Nature Rev Genet.*, 1, 134-144 (2000).

[74] Moyes, DL; et al. The distribution of the endogenous retroviruses HERV-K113 and HERV-K115 in health and disease. *Genomics.*, 86, 337-341 (2005).

[75] Medstrand, P; et al. Expression of human endogenous retroviral sequences in peripheral blood mononuclear cells of healthy individuals. *J Gen Virol.*, 73, 2463-2466 (1992).

[76] Rasmussen, HB; et al. Expression of endogenous retroviruses in blood mononuclear cells and brain tissue from multiple sclerosis patients. *Acta Neurol Scand.*, 95(Suppl. 169), 38-44 (1997).

[77] Andersson, ML; et al. Differential expression of human endogenous retroviral sequences similar to mouse mammary tumor virus in normal peripheral blood mononuclear cells. *AIDS Res Hum Retroviruses.*, 12, 833-840 (1996).

[78] Muir, A; et al. Expression and functions of human endogenous retroviruses in the placenta: an update. *Placenta.*, 25(Suppl. A), S16-S25 (2004).

[79] Garry, RF; et al. Detection of a human intracisternal A-type retroviral particle antigenically related to HIV. *Science.*, 250, 1127-1129 (1990).

[80] Herbst, H. et al. Expression of human endogenous retrovirus K elements in germ cell and trophoblastic tumors. *Am J Pathol.*, 149, 1727-1735 (1996).

[81] Muster, T; et al. An endogenous retrovirus derived from human melanoma cells. *Cancer Res.*, 63, 8735-8741 (2003).
[82] Karlsson, H; et al. Retroviral RNA identified in the cerebrospinal fluids and brains of individuals with schizophrenia. *Proc Natl Acad Sci USA.*, 98, 4634-4639 (2001).
[83] Perron, H; et al. Leptomeningeal cell line from multiple sclerosis with reverse transcriptase activity and viral particles. *Res Virol.*, 140, 551-561 (1989).
[84] Perron, H; et al. Isolation of retrovirus from patients with multiple sclerosis. *Lancet.*, 337, 8623863 (1991).
[85] Firouzi, R; et al. Multiple sclerosis-associated retrovirus particles cause T lymphocyte-dependent death with brain hemorrhage in humanized SCID mice model. *J Neurovirol.*, 9, 79-93 (2003).
[86] Rolland, A; et al. The envelope protein of a human endogenous retrovirus-W family activates innate immunity through CD14/TLR4 and promotes Th1-like responses. *J Immunol.*, 176, 7636-7644 (2006).
[87] Perron, H; et al. Multiple sclerosis retrovirus particles and recombinant envelope trigger an abnormal immune response in vitro, by inducing polyclonal Vbeta16 T-lymphocyte activation. *Virology.*, 287, 321-332 (2001).
[88] Antony, JM; et al. Human endogenous retrovirus glycoprotein-mediated induction of redox reactants causes oligodendrocyte death and demyelination. *Nat Neurosci.*, 7, 1088-1095 (2004).
[89] Garson, JA; et al. Detection of virion-associated MSRV-RNA in serum of patients with multiple sclerosis. *Lancet.*, 351, 33 (1998).
[90] Serra, C; et al. Multiple sclerosis and multiple sclerosis-associated retrovirus in Sardinia. *Neurol Sci.*, 22, 171-173 (2001).
[91] Perron, H; et al. Human endogenous retrovirus (HERV)-W ENV and GAG proteins: physiological expression in human brain and pathophysiological modulation in multiple sclerosis lesions. *J Neurovirol.*, 11, 23-33 (2005).
[92] Mameli, G; et al. Brains and peripheral blood mononuclear cells of multiple sclerosis (MS) patients hyperexpress MS-associated retrovirus/HERV-W endogenous retrovirus, but not Human herpesvirus 6. *J Gen Virol.*, 88, 264-274 (2007).
[93] Antony, JM; et al. Quantitative analysis of human endogenous retrovirus-W env in neuroinflammatory diseases. *AIDS Res Hum Retroviruses.*, 22, 1253-1259 (2006).
[94] Garcia-Montojo, M; et al. The DNA copy number of human endogenous retrovirus-W (MSRV-type) is increased in multiple sclerosis patients and is influenced by gender and disease severity. *PLoS One.*, 8, e53623 (2013).
[95] Zawada, M; et al. MSRV pol sequence copy number as a potential marker of multiple sclerosis. *Pol J Pharmacol.*, 55, 869-875 (2003).
[96] Brudek, T; et al. Activation of endogenous retrovirus reverse transcriptase in multiple sclerosis patient lymphocytes by inactivated HSV-1, HHV-6 and VZV. *J Neuroimmunol.*, 187, 147-55 (2007).
[97] Mameli, G; et al. Expression and activation by Epstein Barr virus of human endogenous retroviruses-W in blood cells and astrocytes: inference for multiple sclerosis. *PLoS One.*, 7, e44991 (2012).
[98] Mameli, G; et al. Activation of MSRV-type endogenous retroviruses during infectious mononucleosis and Epstein-Barr virus latency: the missing link with multiple sclerosis? *PLoS One.*, 8, e78474 (2013).

In: Gene-Environment Interactions and Human Diseases
Editor: Lu Qi
ISBN: 978-1-63483-680-7

Chapter 11

THE ROLE OF GENE-ENVIRONMENT INTERACTIONS IN PANCREATITIS

***Li-Zsa Tan*[1] *and Chee Yee Ooi*[2,*]**

[1]Department of Gastroenterology, Sydney Children's Hospital, Randwick, NSW Australia

[2]Discipline of Paediatrics, School of Women's and Children's Health, Faculty of Medicine, University of New South Wales, and Department of Gastroenterology, Sydney Children's Hospital Randwick, Sydney, Australia

ABSTRACT

Complex interactions between the ductular and acinar components of the pancreas are present in the development of pancreatitis. Each component is further influenced by various genetic and environmental modifier factors. Genetic factors have only recently been discovered to play an important role in the pathophysiology of pancreatitis. These include the cationic trypsinogen (PRSS1) gene, serine protease inhibitor Karzal type 1 (SPINK1) gene, cystic fibrosis transmembrane regulator (CFTR) gene mutations, chymotrypsinogen C (CTRC) gene, calcium sensing receptor (CASR) gene, and more recently the Claudin-2 (CLDN2) gene. The extent of which these defects cause, contribute or confer susceptibility to disease is variable. The disease penetrance of the various mutations varies according to the affected gene as well as the functional consequences of the mutation. SPINK1, CTRC and CASR, for instance, have roles in the control of trypsin activity within the pancreas yet the presence of isolated variants alone is unlikely to cause significant disease. In contrast, hereditary pancreatitis has been linked to the presence of PRSS1 gain of function mutations. Environmental factors such as alcohol and cigarette smoking have long been recognized as risk factors for the development of pancreatitis. This review covers the literature on how environmental factors may interact with known genetic mutations to cause pancreatitis.

* Corresponding author: Dr C.Y. Ooi, Discipline of Paediatrics, School of Women's and Children's Health, Faculty of Medicine, University of New South Wales, Australia.

1. Introduction

Pancreatitis poses a significant burden of disease and is a source of substantial morbidity, mortality and cost. An increase in the incidence of acute pancreatitis (AP) has been reported in adults and children worldwide [1]. In the Netherlands, the incidence of AP increased from 13.2 to 14.7 per 100,000 between 1995 and 2005 [2]. An even greater increase was reported in Wales, with the incidence of AP rising from 27.6 to 36.4 from 1999 to 2010 [3]. AP was cited as the most common gastrointestinal reason for hospitalization in 2012, costing the US an estimated 2.6 billion dollars per year [4]. Similiarly, the incidence of chronic pancreatitis (CP) is also increasing, particularly in light of rising alcohol consumption [5].

The definition of AP in adults was revised in 2013 [6]. The diagnosis requires 2 of 3 features: (i) abdominal pain consistent with acute pancreatitis (acute onset of a severe epigastric pain radiating to the back; (ii) serum lipase/amylase activity at least 3 times greater than the upper limit of normal, (iii) characteristic findings on imaging (Computer Tomography (CT), Magnetic Resonance Imaging (MRI) or ultrasound).

The definition of CP has also been revised since the original 1963 Marseilles classification [7] and is characterized by irreversible morphological changes that may be associated with pain and/or pancreatic insufficiency [8]. The diagnosis is made through a combination of histologic, clinical or functional findings.

2. Pathophysiology of Acute and Chronic Pancreatitis: Two Ends of the Same Spectrum

2.1. Physiology

The exocrine pancreas is comprised of the acinar and ductal cells. Together, these cells produce a pancreatic secretion which is 97% water and electrolytes and 3% protein in the form of digestive enzymes i.e., protease, amylase, lipase and nuclease [9]. The acinar cell synthesizes, stores and excretes pancreatic digestive enzymes, usually in their inactive forms, termed zymogens. The ductal cell produces a bicarbonate rich secretion which helps to flush the digestive enzymes through to the duodenum and neutralizes gastric acid in the digestive tract [11].

2.2. Theories of Pathophysiology

There are complex interactions between the ductular and acinar components of the pancreas in the development of pancreatitis. It has been theorized that the central cause of AP is the inappropriate activation of proteolytic pancreatic enzymes within the pancreas itself rather than the normal physiological site in the small intestine. The premature cleavage of zymogens [12] (particularly trypsinogen to trypsin) within the pancreas leads to the activation of other zymogens, triggering a cascade which results in autodigestion of the pancreas and an acute inflammatory response [12]. This theory seems to be supported by: [a] the auto-digested appearance of the pancreas first made by Chiari in 1896 at necropsy [11] and [b] observations that the majority of genetic susceptibility factors to acute recurrent pancreatitis (ARP) and CP centre on the control of trypsin activity within the pancreas [13]. These genetic factors include Cationic Trypsinogen gene (PRSS1), the Serine Protein Inhibitor Kazal type 1 (SPINK 1), the Chymotrypsinogen gene (CTRC) and the Calcium Sensing Receptor gene (CASR).

Following the acute insult, healing occurs. Anti inflammatory cytokines such as IL-10 and Tumour Necrosis Factor alpha feature prominently in the healing process [14]. In the presence of chronic insult and repeated exposure to metabolic stressors (e.g., in RAP or repeated heavy alcohol intake), active deposition of collagen and matrix proteins continues. This results in pancreatic fibrosis which interferes with normal pancreatic homeostasis and causes additional metabolic stress. This feedback loop ultimately results in a perpetuating cycle of injury, fibrosis, metabolic stress and more injury, causing chronic pancreatitis [13, 14].

Pancreatic stellate cells play an important role in pancreatic fibrogenesis (a prominent histological feature of both CP and pancreatic cancer) and progression from AP to CP [15,16]. Pancreatic stellate cells are located adjacent to pancreatic acinar cells, around pancreatic ducts, and small blood vessels. When injured by exogenous factors (e.g., cytokines, oxidant stress and ethanol), these cells are activated resulting in excessive synthesis and deposition of extracellular matrix proteins. New findings suggest that pancreatic stellate cells may also play an intermediary role in the regulation of cholecystokinin induced exocrine pancreatic secretion [16].

2.3. Other Theories

Activation of trypsin may not be the only mechanism for triggering acute pancreatitis. Studies have explored other potential mechanisms; among them is the discovery that genetic activation of the nuclear factor kB (NF-kB) transcription factor initiates a widespread inflammatory response in the pancreas and may be the end result of an alternate common pathway causing pancreatitis [17]. Other theories include an increase in Ras signaling and endoplasmic reticulum stress, and impaired autophagy due to lysosomal dysfunction [18].

2.4. A Multifactorial Disease

Not unlike other complex diseases in humans, pancreatitis is thought to be a complex multifactorial disease involving interplay between genetic and environmental factors. It is likely that a combination of genetic and environmental factors are responsible not only for the initial event but the eventual propagation of the entire disease spectrum spanning from AP to ARP to CP [13, 19]. Environmental factors such as alcohol, smoking and hypertriglyceridemia are the most commonly studied environmental risk factors and will be discussed in this chapter.

3. Environmental Factors

3.1. Alcohol

The primary environmental etiology associated with CP is heavy alcohol intake [20]. The evidence for this is derived from both animal studies and human genetic and epidemiological studies.

In animal models, chronic alcohol exposure results in metabolic and oxidative stress in acinar cells. In murine studies, ethanol in high concentrations increases intra-acinar calcium levels. Following non-oxidative ethanol degradation, fatty acid ethyl esters are produced, leading to a sustained increase in intracellular calcium concentrations as well as ATP depletion [21]. A murine study by Deng et al. in 2005 [22] compared pancreatic histology between control rats and rats fed with 5% ethanol. Acute recurrent pancreatitis was then induced in both groups. The ethanol-fed rats with recurrent acute pancreatitis had exaggerated immune response changes compared to controls; anti-inflammatory cytokines and inflammatory cell infiltrates were markedly elevated. Histologically, only the ethanol-fed group with recurrent pancreatitis exhibited microabscesses, severe necrosis and early calcification [22]. Other animal studies have shown that ethanol and acetylaldehyde triggers mitogen activated protein kinases on pancreatic stellate cells which led to pancreatic fibrosis and CP [22, 23, 24]. Ethanol also appears to work synergistically with cholecystokinin resulting in an up to six fold increase in activation of trypsinogen and chymotrypsinogen, thus lowering the threshold for pancreatitis [25, 26].

The NAPS2 study, a multicenter consortium which studied > 1000 patients with recurrent acute pancreatitis and CP, confirmed the relationship between alcohol consumption and chronic pancreatitis. A significant increase in chronic pancreatitis at or above the threshold of 5 alcoholic drinks per day was reported in this study [20]. Previous epidemiological studies had attributed up to 56% of chronic pancreatitis cases to alcohol consumption [27, 28, 29]. However the NAPS2 study found only 38.4% of men and 11% of women with chronic pancreatitis were heavy drinkers [20].

Despite this, CP is actually rare among alcoholics. The risk of developing CP amongst heavy drinkers has been estimated at less than 5% [20]. There is also heterogeneity between races and gender, with higher rates found in blacks and men [20, 30, 31]. This provides the basis for a genetic predisposition for developing alcoholic pancreatitis.

3.2. Smoking

Epidemiological studies have linked cigarette smoking to CP; [32, 33] however, as alcohol ingestion is closely associated with habitual cigarette smoking this may confound the relationship.

Animal studies have been performed to simulate cigarette smoking in rats. Wittel et al. studied rats which were exposed to two filter cigarettes a day for twelve weeks. In 58% of these rats there was destruction of acinar cells with reorganization of acinar cells to connective tissue [34, 35]. Infiltration by TGF beta positive immune cells and other inflammatory mediators such as protein 1 alpha and interleukin 1 beta further suggested this process was the result of ongoing inflammation [36, 37]. Wittel et al. [34] further observed an upregulation in trypsinogen and chymotrypsinogen expression, whereas expression of Serine Protein Inhibitor Kazal type 1 (SPINK1), which prevents premature activation of trypsin, remained the same. This imbalance theoretically induces a shift towards pancreatic autodigestion and pancreatitis.

In terms of human epidemiological studies, Law et al. [38] performed a single centre, cross sectional study of 79 patients diagnosed with CP. The authors found an association with active cigarette smoking with CP even after adjusting for alcohol consumption and other risk factors. The number of pack years appears to be associated with an increased risk of CP. Analysis of a small subset confirmed findings from previous studies [28, 39] that ex-smokers have a decreased risk of CP compared to current smokers. Yadav et al. analysed the NAPS2 cohort and also confirmed that smoking was an independent yet underestimated risk factor of CP (OR 2.43-3.08) [40].

In a 2011 retrospective analysis, van Geenan et al. [41] collected human pancreatic tissue during autopsies of patients. Patients with documented pancreatic and liver disease, excessive alcohol intake and a history of major abdominal or gastrointestinal surgery were excluded. Of 111 samples, active smokers at the time of death had significantly more moderate – severe total pancreatic fibrosis compared to subjects who never smoked ($p=0.027$). Active smokers at time of death also had more intralobular pancreatic fibrosis than never smokers ($p=0.013$). Basic information detailing daily consumption and number of pack years was inadequate, and so the authors were unable to subanalyse the relationship between chronicity of smoking and pancreatic fibrosis.

To date however, there is little data describing genetic associations between smoking and pancreatitis.

3.3. Hypertriglyceridemia

Hypertryglyceridemia may be responsible for up to 1-7% of pancreatitis [42, 43]. However, the true incidence of hypertriglyceridemia induced pancreatitis is difficult to determine as alcoholism and hypertriglyceriemia often exist co-morbidly. Furthermore, there are large variations in the incidence of pancreatitis in patients with hypertriglyceridemia. Some patients rarely develop pancreatitis even with marked hypertriglyceridemia [44], supporting the role of genetic and/or environmental modifiers.

The pathophysiology of hyperglyceridemia induced pancreatitis is uncertain. Triglyceride-rich lipoprotein particles form chylomicrons, which appear in the circulation when serum triglyceride exceeds 10mmol/L [45]. One hypothesis is that these chylomicrons may restrict blood flow while traversing the pancreatic capillary bed, resulting in ischemia and disruption of pancreatic acini [46]. Another theory suggests that when these chylomicrons are degraded by pancreatic lipase, the resulting free fatty acids release inflammatory mediators and free radicals which cause further inflammation, edema and haemorrhage [47, 48]. Yet another alternate hypothesis is that phospholipids on the surfaces of the chylomicrons activate pancreatic proteases. The resulting high concentration of activated proteases may overwhelm trypsin regulatory mechanisms, triggering the zymogen activation cascade [49].

4. Genetic Factors

Alcohol and gallstone disease have long been recognized as the most common causes of acute pancreatitis in adults, accounting for two-thirds of diagnoses [50]. Other recognized etiologies include metabolic, autoimmune, anatomical abnormalities, infection, drug induced, trauma and genetics. Yet there is wide variability in severity and chronicity of disease even between patients who evidently share similar disease etiologies. This observation led to the hypothesis of an underlying predisposition or "modifier" factor, and resulted in further investigations into the influence of genetics on the phenotype of pancreatic disease.

One of the earliest genetic studies in 1952 elucidated the familial pedigree of a family with chronic recurrent pancreatitis [51]. Advances in genetic research gradually enabled linkage analysis of multi-generational kindreds [52, 53] confirming the association between genetic mutations and hereditary pancreatitis.

The development of high density single nucleotide polymorphism (SNP) chips for genome wide association studies (GWAS) has further revolutionized the field of human genetics. This technique allows for investigation of the entire genome and identification of gene-disease associations in contrast to the more limited but hypothesis-driven gene specific candidate studies. In 2012, the first GWAS for adult pancreatitis [54] was published by the North American Pancreatitis Study 2 (NAPS2) research group, identifying two single-nucleotide polymorphisms (SNPs) significantly associated with the risk of recurrent acute pancreatitis and chronic pancreatitis. These findings will be discussed later in the chapter.

In the future, the demand for next generation sequencing (NGS) analysis may allow for quicker and massive parallelization of the sequencing and resequencing process at a lower cost.

Genetic mutations which are central in the pathophysiology of pancreatitis can be considered in two groups, i.e., those which contribute toward premature trypsinogen activation in the pancreatic acinar cell (PRSS1, SPINK1, CTRC, CASR) and those which cause imbalance in normal pancreatic ductal homeostasis (CFTR) [10,18].

4.1. Cationic Trypsinogen Gene or Serine Protease 1 (PRSS 1)

The discovery of hereditary pancreatitis stemmed from observations that some cases of chronic pancreatitis tended to recur within different generations of the same families. In 1996, a mutation in the gene encoding cationic trypsinogen (also known as Serine Protease 1 or PRSS 1) was identified by Bodic et al. as the underlying defect in five families with CP in an autosomal dominant fashion [52]. Other PRSS 1 mutations have since been described [55]. The majority of mutations identified are 'gain of function' mutations for trypsin, ie increasing susceptibility for autoactivation. Some have even hypothesized that trypsinogen may be sensitive to gene dosage effect. Le et al. [56] reported a triplication of all trypsinogen genes on chromosome 7 in affected individuals in 34 French families with hereditary pancreatitis. This triplication was absent in healthy controls of the same population.

R122H, the most common PRSS 1 mutation [57], is a gain of function mutation which significantly enhances autoactivation of cationic trypsinogen to trypsin. This activates other digestive enzymes in an enzymatic cascade. R122H also appears to inhibit trypsin autolysis

[58] which perpetuates the resulting acute inflammatory response and activates pancreatic stellate cells [59]. Recurrent episodes then lead to further fibrosis. Conversely, loss of function mutations have also been discovered. One mutation at p.A121T is located near p.R122H, and thus was initially assumed to share a similar mechanism of action. This p.A121T mutation however proved to significantly increase trypsin cleavage rate, thus conferring loss of function of trypsin [60].

Due to its close resemblance to cationic trypsinogen, anionic trypsinogen (PRSS 2) was initially an obvious choice for gene candidate study. Instead, a mutation on the PRSS 2 gene – the G191R variant – was discovered to cause a loss of function by generating a new cleavage site on anionic trypsinogen, predisposing to autolysis, hence is a protective factor rather than a risk for pancreatitis. Witt et al. [61] investigated three missense variants of PRSS2 in 200 German patients with chronic pancreatitis and 240 control subjects by single strand conformation polymoprhism (SSCP) analysis. Two of the variants were equally represented in the both CP and control population. The third, G191R, was to their surprise more common in control subjects, and was found in none in the CP group. The study was expanded to include 918 CP patients and 4440 control subjects from nine other European countries and confirmed that the frequency in controls was significantly higher compared to CP patients (OR 0.41; P = 0.00097) [61].

4.1.1. Alcohol

The association between PRSS1 and alcohol was studied in the NAPS2 study. Based on more than 1000 patients enrolled in the North American Pancreatitis Study 2 (NAPS2) [54, 62] GWAS, the C allele at rs10273639 of the PRSS1-PRSS 2 locus does alter expression of the primary trypsinogen gene. The risk for alcohol-related pancreatitis was greater for rs10273639 (PRSS1-PRSS2 locus) than those for non-alcohol related pancreatitis, thus establishing the association between this locus and an increased risk of developing alcoholic pancreatitis.

4.2. Serine Protease Inhibitor Kazal-type1 or SPINK1

The discovery of PRSS1 was a catalyst for genetic research into other genetic causes of CP. Serine Protease Inhibitor Kazal-type 1 (SPINK1) also known as Pancreatic Secretory Trypsin Inhibitor, is a protease inhibitor which functions to inactivate intrapancreatic trypsin activity [57]. SPINK1 codes for a polypeptide which serves as a target substrate for trypsin, preventing premature activation and autodigestion. SPINK1 is estimated to inhibit up to 20% of trypsinogen in the pancreas but inhibition is not permanent [63]. Termed “temporary inhibition,” trypsin activity is restored with prolonged incubation of SPINK1 and trypsin together. In the event of premature activation within the pancreas, SPINK1 covalently binds to the active trypsin molecule and the complex passes harmlessly into the duodenum to be degraded later [64].

The N34S variant appears to be the most common SPINK1 mutation while the M1T variant, although extremely rare, causes severe disease [57, 65, 66]. In 2000, Witt et al. [65] performed DNA sequencing analysis on 96 unrelated children with chronic pancreatitis and found SPINK1 mutations in 23%. 18 of these patients had the N34S mutation. One patient with hereditary pancreatitis had a heterozygous M1T mutation which destroyed the start codon of SPINK1. The mutation was also found in the index patient's unaffected father and affected grandfather. The patient's (deceased) great grandfather suffered from pancreatitis. Futhermore, the M1T destroys the only initiation codon and knocks out one allele, causing a dominant pattern of disease. Based on these findings, Witt et al. hypothesized that M1T had an autosomal dominant inheritance [57].

SPINK1 mutations however, are common in the general population. Studies have shown that up to 1% of the healthy population carry SPINK1 mutations [67] and [68]. Pfutzer et al. [68] performed genetic linkage studies in 5 families with familial pancreatitis. Although the authors confirmed SPINK1 mutations were clearly associated with pancreatitis and noted similar frequencies of the mutation in the healthy control population, the authors disagreed that SPINK1 mutations followed any Mendelian pattern of inheritance. There was no difference in age of onset or severity of disease between homozygous or heterozygous patients. The authors concluded that SPINK1 mutations alone are insufficient to cause disease but rather behaved as a disease modifier; either lowering the threshold of pancreatitis or increasing the severity of disease initiated by other genetic or environmental factors]. This was confirmed in a later study by Pfutzer [69] looking at 192 patients with known PRSS1 mutations and found 26 patients also carrying the SPINK1 mutation. They found that the patients with both PRSS1 and SPINK1 mutations had significantly earlier age of onset of disease ($p<0.001$).

4.2.1. Alcohol

A 2008 meta-analysis by Aoun et al. [70] studied 24 reports of SPINK1 N34S in CP (2421 cases, 4857 controls) stratified by etiological factors and found that there was only a small (but still significant) association between the SPINK1 mutation N34S and alcoholic CP (OR 4.98, 95% CI:3.16-7.85). This suggests that alcohol may cause pancreatitis through mechanisms other than through trypsin dependent pathways, perhaps by directly affecting the immune system and stellate cells.

4.3. Cystic Fibrosis Transmembrane Conductor Regulator (CFTR)

The cystic fibrosis transmembrane conductor regulator (CFTR) protein is found on the apical membranes of epithelial cells in the pancreas, intestine, liver, airway, vas deferens and sweat glands. It functions as a cyclic AMP dependant chloride channel, a bicarbonate channel and a modulator of other ion channels [71]. The exocrine pancreas produces a bicarbonate rich alkaline fluid which functions to flush enzymes into the gastrointestinal tract where digestion can take place. It also provides the appropriate pH for digestive enzyme activity. With CFTR mutations affecting the pancreas, less fluid is produced. This predisposes to the precipitation of protein rich secretions and obstruction of the small pancreatic ducts, which in turn leads to inflammation and fibrosis [71]. Reduced luminal pH may also predispose to pancreatitis by means of trypsin autoactivation and subsequent loss of tight junction integrity allowing leakage of enzymes into the duct lumen and interstitial spaces [71, 72, 73, 74].

The relationship between severity of CFTR genotypes and risk of AP was recently elucidated [71]. Patients with cystic fibrosis (CF) who carry severe mutations on both CFTR alleles are pancreatic insufficient (termed PI), and do not have adequate residual pancreatic acinar tissue to develop symptomatic AP. Patients who have a mild mutation on at least one allele however, usually maintain sufficient exocrine function and are at risk of developing AP. There is also a significantly increased risk of developing CP among those identified with CF-causing CFTR variants [75]. Of the pancreatic sufficient CF patients, approximately 20% develop pancreatitis [71,73]. As not all PS CF patients develop pancreatitis, non CFTR genetic and/or environmental modifiers likely play a contributing role in the development of disease. For instance, patients with both CFTR and SPINK1 mutations have been reported to be at higher risk of developing pancreatitis [59, 71].

4.3.1. Alcohol

While the effect of alcohol on pancreatic acinar and stellate cells have previously been elucidated, the effects of alcohol on pancreatic ducts are less defined. Animal studies in ductal physiology [76, 77] observed that different *n*-alcohols have either inhibitory or stimulatory effects on pancreatic ductal fluid secretion. Judak et al. [78] then hypothesized that as the CFTR Cl^- channel plays a key role in pancreatic duct epithelial cell (PDEC) physiology, ethanol might specifically affect CFTR to impair PDEC function. The authors then went on to perform a 2014 animal study which showed that ethanol and its non-oxidative metabolites did indeed inhibit CFTR chloride channel activity by reducing intracellular ATP levels, thus affecting pancreatic fluid secretion. Mutations in CFTR may therefore play a major role in predisposing to alcohol induced pancreatitis.

4.3.2. Smoking

Smoking has been shown to decrease expression of the CFTR gene, protein, and function. A 2006 study by Cantin et al. [79] observed via in vitro studies that cigarette smoking decreased CFTR expression and function, and theorized that these observed effects were not cell or tissue specific, ie not confined to the respiratory tract. The authors further performed in vivo transepithelial potential difference measurements of the nasal respiratory epithelium of 26 men (9 smokers with 17 non-smokers as controls) with no detectable CFTR gene mutations. The nasal potential difference of the cigarette smoking cohort showed a pattern typical to CFTR deficiency, further underlining that cigarette smoking results in an acquired CFTR deficiency.

Raju et al. [80] then performed ex vivo studies analyzing CFTR activity by intestinal current measurement from the rectal biopsy samples of healthy smokers and never smokers. A 60% reduction in cAMP-dependant chloride transport was observed in the smokers cohort reflecting reduced CFTR activity in the gastrointestinal tract of patients without CFTR mutations. A natural conclusion from this would be that cigarette smoking has the potential to cause CFTR dysfunction in other sites of the gastrointestinal tract, including pancreatic duct epithelial cells. Smokers with CFTR mutations may then be at higher risk of developing pancreatitis.

4.3.3. Hypertriglyceridemia

In a 2008 gene candidate study of 146 adults, Chang et al. [81] reported that the CFTR c.1408A > G (M470V) variant was found significantly more frequently (OR 3.70 (95% CI, 2.9-15.5) in hypertriglyceridemic patients who developed hyperlipidemic pancreatitis. Similar results were noted with individuals carrying the TNF promoter 863A variant (OR 4.71 (95%

CI, 2.14-10.38)). In patients identified with both these mutations, the OR was higher (5.18: 95% CI 2.21-12.16).

4.4. Calcium Sensing Receptor (CASR)

The Calcium Sensing Receptor (CASR) is a plasma membrane bound G protein coupled receptor which plays an important role in calcium homeostasis [82]. It is expressed in the parathyroid glands, renal tubules as well as the pancreas. CASR senses high calcium levels in pancreatic juice and interacts with CFTR to activate bicarbonate secretions. This flushes out the ducts and lowers the risk of pancreatitis because trypsinogen is activated when high concentrations of calcium are present in pancreatic juices [83].

Several CASR variants have been identified. One of these mutations, CASR R990G, is associated with an increased risk of chronic pancreatitis. The exact mechanism is unknown [82]. In addition to this, loss of function CASR mutations are often found in the pancreas with CFTR pathogenic variants, but the association between the two variants and their contribution towards the risk of pancreatitis is again unclear [59].

4.4.1. Alcohol

A 2008 study by Muddana et al. [82] screened 219 subjects with chronic pancreatitis for the three most common SNPs in the CASR gene exon 7. They found that the CASR R990G mutation is not only significantly associated with CP, but also strongly associated with CP in moderate and heavy alcohol drinkers. The presence of the R990G mutation on its own doubles the risk of developing CP, but in individuals who are moderate to heavy drinkers, this risk increases by three-fold. While previous studies closely associated CASR gene mutations to SPINK1 N34S mutations in cases of CP [84, 85] it is still unclear if the concurrent presence of both mutations behave synergistically to predispose to alcoholic pancreatitis.

4.5. Claudin-2 (CLDN2)

Regulation of water and ions between endothelial cells occur at the tight junction, which are formed by proteins called claudins. Claudin 2 is normally expressed in the proximal pancreatic duct and forms ion and water channels between cells. This facilitates sodium and water movement into the pancreatic duct to match the bicarbonate and chloride that are secreted through CFTR [59, 83]. Claudin 2 can also be inserted into and removed from tight junctions rapidly. Gene expression is enhanced in cells during injury and stress [83].

Whitcomb et al. identified a CLDN 2 variant which was strongly associated with alcohol related CP, (but not acute recurrent pancreatitis (RAP)). As the high risk variant is common, this may indicate that CLDN 2 is another disease modifier which increases the rate of progression of acute to chronic pancreatitis [59, 83, 54].

4.5.1. Alcohol

CLDN2, as previously described, encodes for the claudin-2 protein. The same GWAS study by Whitcomb et al. in 2012 [54], which described and confirmed the association between the PRSS1-PRSS2 loci and alcohol related pancreatitis, found a stronger association between the CLDN2 locus found at Xq23.3 and chronic pancreatitis. Of the two loci studied, CLDN2 also had the strongest association with alcohol related pancreatitis (case only analysis, p = 4x 10^7). The risk allele is located on the X chromosome and could partially account for the higher incidence of alcoholic pancreatitis in men, as it would manifest as a dominant trait in heterozygous males. Because the CLDN 2 variant is strongly associated with alcohol related CP but not AP, The authors concluded that it may act as a disease modifier by accelerating transition from ARP to CP [54, 59].

4.6. Chymotrypsin C (CTRC)

Chymotrypsin C (CTRC) degrades trypsin and trypsinogen, thus loss of function mutations in the CTRC gene increases the risk of pancreatitis [86, 87]. CTRC variants cause loss of CTRC function by one or more of three mechanisms: (i) reduced secretion of CTRC, (ii) a catalytic defect, or (iii) increased degradation of CTRC by trypsin [86]. CTRC loss of function variants have been associated with CP, as well as alcoholic pancreatitis, idiopathic pancreatitis and tropical pancreatitis [87, 88, 89]. CTRC mutations alone are not thought to cause CP, rather it appears more likely that CTRC interacts with other genetic variants to cause disease [75].

4.6.1. Hypertriglyceridemia

To study the possible modifying effect of CTRC and SPINK1 mutations in these patients, Tremblay et al. [90] controlled for hypertriglyceridemia by examining a homogenous cohort (n=38) of patients who carried the same Lipoprotein Lipase (LPL) gene defect. They concluded that patients with concurrent inheritance of CTRC-SPINK1 variant and LPL causing hypertriglyceridemia were significantly more likely to present with recurrent pancreatitis. This suggests that protease regulating genes such as CTRC-SPINK1 have a modifying role in hypertriglyceridemia induced pancreatitis.

4.7. Lipoprotein Lipase Gene (LPL) and Hypertriglyceridemia

Lipoprotein Lipase (LPL) is the rate limiting enzyme involved in the lipolysis of chylomicron and very low density lipoproteins (VLDL). Familial Lipoprotein Lipase (LPL) deficiency is inherited as an autosomal recessive disease. While homozygous mutations of LPL result in severe hypertriglyceridemia, Murase et al. [91] reported 15 subjects with heterozygous LPL deficiency that, LPL activity also appears to be reduced by 25% in heterozygote mutations, specifically in females. Familial LPL deficiency manifests clinically with hyperchylomicronemia and ARP [92]. Clinical disease may be aggravated by fatty meals, diabetes, alcohol, pregnancy and steroids [49, 93]. By 2009, almost 100 mutations in the LPL gene causing LPL deficiency had been identified [94].

4.8. Gamma Glutamyl Transferase 1 (GGT1)

Gamma Glutamyl Transferase 1 (GGT1) plays a key role in maintaining levels of intracellular glutathione, which protects cells against free radical and oxidative stress [95]. Oxidative stress can damage pancreatic cells directly and activate proinflammatory pathways leading to pancreatitis. Brand et al. (2013) [96] compared associations between four common GGT1 SNP polymorphisms and CP in 496 patients with CP and 465 controls. The authors reported that two SNPs located in the GGT gene (rs8135987 and rs48205999) were significantly associated with risk of CP. Subjects with at least one minor allele were at increased risk (OR, 1.36; 95% CI, 1.03-1.80 and OR, 1.39; 95% CI, 1.04-1.84 respectively). Results were adjusted for age, gender, race, cigarette smoking, and alcohol use. There were no significant trends associated with alcohol and cigarette smoking.

Table 1. Summary of genetics-environment studies in pancreatitis

Genetics	Environmental modifier	Studies
PRSS1-PRSS2	Alcohol	Whitcomb et al. 2012 [54] - GWAS on NAPS2 cohort > 1000 patients reports association between PRSS1-PRSS2 loci and alcoholic pancreatitis
SPINK1	Alcohol	Aoun et al. 2008 [70] - Meta-analysis showing small but significant association between SPINK1 mutation (N34S) and alcoholic CP (OR 4.48, 95% CI:3.16-7.85)
CFTR	Hypertriglyceridemia	Cheng et al. 2008 [81] - Gene candidate study of 146 adults reporting increased frequency of CFTR c.1408A>G (M470V)variant in patients with hyperlipidemic pancreatitis
CASR	Alcohol	Muddana et al. 2008 [82] - Gene candidate study showing 3 fold association of CASR R990G mutation with alcoholic CP (moderate – heavy drinkers)
CLDN2	Alcohol	Whitcomb et al. 2012 [54] - GWAS of NAPS2 cohort finds strong association between CLDN2 locus at X923.3 and alcohol related pancreatitis
CTRC	Hypertriglyceridemia	Tremblay et al. 2014 [90] - Gene sequencing and genotype stratified analyses of 38 subjects with hypertriglyceridemia secondary to LPL deficiency shows positive association between CTRC-SPINK1 and hypertriglyceridemic pancreatitis

Conclusion

Acute and chronic pancreatitis likely results from interactions between genetic and environmental influences. Genetic variants associated with pancreatitis have been shown to disrupt exocrine pancreatic homeostasis. While there are some mutations that are highly penetrant, many of the currently identified variants appear to have modifier roles, behave cumulatively with other genes and/or environmental factors to cause pancreatitis. The study of this complex interplay between genes and the environment is challenging due to the multifactorial nature, including the numerous potential candidate genes. In addition, the study of environmental factors is confounded by the fact that the main three influences upon disease (i.e., smoking, alcohol ingestion and hypertriglyceridemia) tend to coexist in many subjects. The study of recurrent and chronic pancreatitis in childhood, which is less likely associated with the aforementioned environmental factors, may provide additional unique insights. A better understanding of mechanisms of disease in pancreatitis will open up new therapeutic options for treating this disease.

References

[1] Yadav, D; Lowenfels, AB. Trends in the epidemiology of the first attack of acute pancreatitis: a systemic review. *Pancreas*, 2006 Nov, 33(4), 323-30.

[2] Marcel Spanier, BW; Bruno, MJ; Dijkgraaf, MGW. Incidence and mortality of acute and chronic pancreatitis in the Netherlands: A nationwide record linked cohort study for the years 1995-2005. *World J Gastroenterol*, 2013, 19(20), 3018-3026.

[3] Roberts, SE; Akbari, A; Thorne, K; Atkinson, M; Evans, PA. The incidence of acute pancreatitis:impact of social deprivation, alcohol consumption, seasonal and demographic factors. *Aliment Pharmacol Ther*, 2013, 38(5), 539-48.

[4] Peery, AF; Dellon, ES; Lund, J; Crockett, SD; McGowan, CE; et al. Burden of Gastrointestinal Disease in the United States: 2012 Update. *Gastroenterology*, 2012, 143(5), 1179-1187.

[5] Jupp, J; Fine, D; Johnson, CD. The epidemiology and socioeconomic impact of chronic pancreatitis. *Best Pract Res Clin Gastroenterol*, 2010 Jun, 24(3), 219-31.

[6] Banks, PA; Bollen, TL; Dervenis, C; Gooszen, HG; Johnson, CD; Sarr, MG; Tdotos, GG; Vege, SS. Classification of acute pancreatitis – 2012: revision of the Atlanta classification and definitions by international consensus. *Gut*, 2013, 62, 102-111.

[7] Sarles, H. *Pancreatitis: Symposium of Marseille*, 1963. Basel, Switzerland: Karger, 1965.

[8] Etemad, B; Whitcomb, DC. Chronic Pancreatitis: Diagnosis, Classification, and New Genetic Developments. *Gastroenterology*, 2001, 120, 682-707.

[9] Gullo, L; Pezzilli, R; Priori, P; Baldoni, F; Paparo, F; Mattioli, G. Pure pancreatic juice collection over 24 consecutive hours. *Pancreas*, 1987, 2, 620-623.

[10] Hegyi, P; Pandol, S; Venglovecz, V; Rakonczay, Jr. Z. The acinar-ductal tango in the pathogenensis of acute pancreatitis. *Gut*, 2011, 60, 544-552.

[11] Chiari, H. Über die selbstverdauung des menschlichen pankreas. *Zeitschrift für Heilkunde*, 1896, 69-96.

[12] Mooren, F; Hlouschek, V; Finkes, T; et al. Early changes in pancreatic acinar cell calcium signaling after pancreatic duct obstruction. *J Biol Chem*, 2003, 278, 9361-69.
[13] Ooi, CY; Gonska, T; Durie, PR; Freedman, SD. Genetic testing in pancreatitis. *Gastroenterology*, 2010, 138, 2202-2206.
[14] Whitcomb, DC. Genetic Predisposition to Alcoholic Chronic Pancreatitis. *Pancreas*, 2003, 27, 321-326.
[15] Apte, MV; Pirola, RC; Wilson, JS. The fibrosis of chronic pancreatitis: new insights into the role of pancreatic stellate cells. *Antioxid. Redox Signal*, 2011, 15, 2711-2722.
[16] Apte, MV; Pirola, RC; Wilson, JS. Pancreatic stellate cells: a starring role in normal and diseased pancreas. *Frontiers in Physiology*, August 2012, 3 (344), 1-13.
[17] Chen, X; Ji, B; Han, B; Ernst, SA; Simeone, D; Logsdon, CD. Nf-kappab activiation in pancreas induces pancreatic and systemic inflammatory response. *Gastroenterology*, 2002, 122, 448-57.
[18] Meyer, A; Coffey, MJ; Oliver, MR; Ooi, CY. Contrasts and comparisons between childhood and adult onset acute pancreatitis. *Pancreatology*, 2013, 13, 429-435.
[19] Whitcomb, DC. Genetic aspects of pancreatitis. *Annual review of medicine*, 2010, 61, 413-424.
[20] Yadav, D; Hawes, RH; Brand, RD; Anderson, MA; Money, ME; Banks, PA; Bishop, MD; Bailie, J; Sherman, S; DiSario, J; Burton, FR; Gardner, TB; Amann, ST; Gelrud, A; Lawrence, C; Elinoff, B; Greer, JB; O'Connell, M; Barmada, MM; Slivka, A; Whitcomb, DC. for the North American Pancreatic Study Group. Alcohol Consumption, Cigarette Smoking and the Risk of Recurrent Acute and Chronic Pancreatitis. *Arch Intern Med*, 2009, 169(11), 1035-1045
[21] Criddle, DN; Rarty, MG; Neoptolemos, JP; Tepikin, AV; Peterson, OH; Sutton, R. Ethanol toxicitiy in pancreatic acinar cells: mediation by nonoxidative fatty acid metabolites. *Proc Natl Acad Sci USA*, 2004, 101, 10738-43.
[22] Deng, X; Wang, L; Elm, MS; Gabazadeh, D; Diorio, GJ; Eagon, PK; Whitcomb, DC. Chronic Alcohol Consumption Accelerates Fibrosis in Response to Cerulein-Induced Pancreatitis in Rats. *American Journal of Pathology*, 2005, 166(1), 93-106.
[23] Apte, MV; Pirola, RC; Wilson, JS. Molecular mechanisms of alcoholic pancreatitis. *Dig Dis*, 2005, 23, 232-40.
[24] Apte, MV; Wilson, JS. Stellate cell activation in alcoholic pancreatitis. *Pancreas*, 2003, 27, 316-20.
[25] Ponnappa, BC; Marciniak, R; Schneider, T; Hoek, JB; Rubin, E. Ethanol consumption and susceptibility of the pancreas to cerulein-induced pancreatitis. *Pancreas*, 1997, 14, 150-157.
[26] Katz, M; Carangelo, R; Miller, LJ; Gorelick, F. Effect of ethanol on cholecystokinin-stimulated zymogen conversion in pancreatic acinar cells. *Am J Physiol*, 1996, 270, G171—G175.
[27] Layer, P; Yamamoto, H; Kalthoff, L; Clain, JE; Bakken, LJ; DiMagno, EP. The different courses of early- and late-onset idiopathic and alcoholic chronic pancreatitis. *Gastroenterology*, 1994, 107, 1481-1487.
[28] Cote, GA; Yadav, D; Slivka, A; Hawes, RH; Anderson, MA; Burton, FR; Brand, RE; Banks, PA; Lewis, MD; Disario, JA; Gardner, TB; Gelrud, A; Amann, ST; Bailie, J; Money, ME; O'Connell, M; Whitcomb, DC; Sherman, S. Alcohol and smoking as risk

factors in an epidemiology study of patients with chronic pancreatitis. *Clin Gastro Hepatol*, 2011, 9, 266-273.

[29] Yadav, D; Timmons, L; Benson, JT; Dierkhising, RA; Chiari, ST. Incidence, prevalence, and survival of chronic pancreatitis: a population based study. *Am J Gastroenterol*, 2011,106, 2192-2199.

[30] Lowenfels, AB; Maisonneuve, P; Cavallini, G; Ammann, RW; Lankisch, PG; Andersen, JR; DiMagno, EP; Andren-Sandberg, A; Domellof, L; Di Francesco, V. Prognosis of chronic pancreatitis: an international multicenter study. *Am J Gastroenterol*, 1994 Sep, 89(9), 1467-71.

[31] Herreros-Villaneuva, M; Hijona, E; Banales, JM; Cosme, A; Bujanda, L. Alcohol consumption on pancreatic diseases. *World J Gastroenterol*, 2013, 19(5), 638-647.

[32] Bourliere, M; Barthet, M; Berthezene, P; Durbec, JP; Sarles, H. Is tobacco a risk factor for chronic pancreatitis and alcoholic cirrhosis? *Gut*, 1991, 32, 1392-1395.

[33] Lin, Y; Tamakoshi, A; Hayakawa, T; Ogawa, M; Ohno, Y. Cigarette smoking is a risk factor for chronic pancreatitis: a case control study in Japan. Research Committee on Intractable Pancreatic Diseases. *Pancreas*, 2000, 21, 109-114.

[34] Wittel, UA; Singh, AP; Henley, BJ; Andrianifahanana, M; Akheter, MP; Cullen, DM; Batra, SK. Cigarette smoke-induced differential expression of the genes involved in exocrine function of the rat pancreas. *Pancreas*, 2006, 33, 364-370.

[35] Wittel, UA; Pandey, KK; Andrianifahanana, M; Johansson, SL; Cullen, DM; Akheter, MP; Brand, RE; Prokopczyk, B; Batra, SK. Chronic pancreatic inflammation induced by environmental tobacco smoke inhalation in rats. *Am J Gastroenterol*, 2006, 101, 148-159.

[36] Shi, MM; Chong, IW; Long, NC; Love, JA; Godleski, JJ; Paulaskis, JD. Functional characterization of recombinant rat macrophage inflammatory protein-1 alpha and mRNA expression in pulmonary inflammation 1998. *Inflammation*, 22, 29-43

[37] Rau, B; Paszkowski, A; Lillich, S; Baumgart, K; Moller, P; Beger, HG. Differential effects of caspase-1/interleukin-1 beta-converting enzyme on acinar cell necrosis and apoptosis in severe acute experimental pancreatitis. *Lab Invest*, 2001, 81, 1001-1013.

[38] Law, R; Parsi, M; Lopez, R; Zuccaro, G; Stevens, T. Cigarette smoking is Independently Associated with Chronic Pancreatitis. *Pancreatology*, 2010, 10, 54-59.

[39] Talamini, G; Bassi, C; Falconi, M; Sartori, N; Vaona, B; Bovo, P; Benini, L; Cavallini, G; Pederzoli, P; Vantini, I. Smoking cessation at the clinical onset of chronic pancreatitis and risk of pancreatic calcifications. *Pancreas*, 2007, 35, 320-326.

[40] Yadav, D; Slivka, A; Sherman, S; Hawes, RH; Anderson, MA; Burton, FR; Brand, RE; Lewis, MD; Gardner, TB; Gelrud, A; DiSario, J; Amann, ST; Baillie, J; Lawrence, C; O'Connell, MO; Lowenfels, AB; Banks, PA; Whitcomb, DC. Smoking is Underrecognized as a Risk Factor for Chronic Pancreatitis. *Pancreatology*, 2010, 10, 713-719.

[41] van Geenan, EJM; Smits, MM; Schreuder, TCMA; van der Peet, DL; Bloemena, E; Mulder, CJJ. Smoking Is Related to Pancreatic Fibrosis in Humans. *Am J Gastroenterol*, 2011, 106, 1161-1166.

[42] Toskes, PP. Hyperlipidemic pancreatitis. *Gastroenterol Clin North Am*, 1990, 19, 783-791.

[43] Dominguez-Munoz, JE; Malfertheiner, P; Ditschuneit, HH; Blanco-Chavez, J; Uhl, W; Buchler, M; Ditschuneit, H. Hyperlipidemia in acute pancreatitis. Relationship with etiology, onset, and severity of the disease. *Int J Pancreatol*, 1991, 10, 261-267.
[44] Yadav, D; Pichumoni, CS. Issues in hyperlipidemic pancreatitis. *J Clin Gastroenterol*, 2003, 36, 54-62.
[45] Johansen, CT; Kathiresan, S; Hegele, RA. Genetic determinants of plasma triglycerides. *J. Lipid Res*, 2011, 52, 189-206.
[46] Gan, SI; Edwards, AL; Symonds, CJ; Beck, PL. Hypertriglyceridemia-induced pancreatitis: A case-based review. *World J Gastroenterol*, 2006, 12(44), 7197-7202.
[47] Havel, RJ. Pathogenesis, differentiation and management of hypertriglyceridemia. *Adv Intern Med*, 1969, 15, 117-154.
[48] Saharia, P; Margolis, S; Zuidema, GD; Cameron, JL. Acute pancreatitis with hyperlipidemia: studies with an isolated perfused canine pancreas. *Surgery*, 1977, 82, 60-67
[49] Frossard, JL; Steer, ML; Pastor, CM. Acute pancreatitis. *Lancet*, 2008, 371, 143-152.
[50] Steinberg, W; Tenner, S. Acute pancreatitis. *N Engl J Med*, 1994, 330, 1198-1210.
[51] Comfort, MW; Steinberg, AG. Pedigree of a family with hereditary chronic relapsing pancreatitis. *Gastroenterology*, 1952 May, 21(1), 54-63.
[52] Bodic, LL; Bignon, J; Raguenes, O; Mercier, B; Georgelin, T; Schnee, M; Soulard, F; Gagne, K; Bonneville, F; Muller, J; Bachner, L; Ferec, C. The hereditary pancreatitis gene maps to long arm of chromosome 7. *Human Molecular Genetics*, 1996, 5(4), 549-554.
[53] Gorry, MC; Gabbaizedeh, D; Furey, W; Gates, LK; Jr. Preston, RA; Aston, CE; Zhang, Y; Ulrich, C; Ehrlich, GD; Whitcomb, DC. Mutations in the cationic trypsinogen gene are associated with recurrent acute and chronic pancreatitis. *Gastroenterology*, 1997 Oct, 113(4), 1063-8.
[54] Whitcomb, DC; LaRusch, J; Krasinskas, AM; Klei, L; Smith, JP; Brand, RE; Neoptolemos, JP; Lerch, MM; Tector, M; Sandhu, BS; Guda, NM; Orlichenko, L. Alzheimer's Disease Genetics Consortium, Alkaade, S; Amann, ST; Anderson, MA; Baillie, J; Banks, PA; Conwell, D; Coté, GA; Cotton, PB; DiSario, J; Farrer, LA; Forsmark, CE; Johnstone, M; Gardner, TB; Gelrud, A; Greenhalf, W; Haines, JL; Hartman, DJ; Hawes, RA; Lawrence, C; Lewis, M; Mayerle, J; Mayeux, R; Melhem, NM; Money, ME; Muniraj, T; Papachristou, GI; Pericak-Vance, MA; Romagnuolo, J; Schellenberg, GD; Sherman, S; Simon, P; Singh, VP; Slivka, A; Stolz, D; Sutton, R; Weiss, FU; Wilcox, CM; Zarnescu, NO; Wisniewski, SR; O'Connell, MR; Kienholz, ML; Roeder, K; Barmada, MM; Yadav, D; Devlin, B. Common genetic variants in the CLDN2 and PRSS1-PRSS2 loci alter risk for alcohol-related and sporadic pancreatitis. *Nat Genet.*, 2012, 44, 1349–1354.
[55] Derikx, MHM; Drenth, JPH. Genetics in chronic pancreatitis; implications for diagnosis, management and progress. *Best Practice and Research Clinical Gastroenterology*, 2010, 24, 251-270
[56] Le Marechal, C; Masson, Emmanuelle; Chen, JM; Morel, F; Ruszniewski, P; Levy, P; Ferec, C. Hereditary pancreatitis caused by triplication of the trypsinogen locus. *Nature Genetics*, 2006, 38(12), 1372-4.
[57] Witt, H. Gene Mutations in Children with Pancreatitis. *Pancreatology*, 2001, 1, 432-438.

[58] Sahin-Toth, M; Toth, M. Gain-of-function mutations associated with hereditary pancreatitis enhance autoactivation of human cationic trypsinogen. *Biochem Biophys Res Commun.*, 2000, 278, 286-289.

[59] LaRusch, J; Solomon, S; Whitcomb, DC. *Pancreatitis Overview.*, 2014 Mar 13. In: Pagon RA, Adam MP, Ardinger HH, et al., editors. GeneReviews® [Internet]. Seattle (WA): University of Washington, Seattle; 1993-2014. Available from: http:// www.ncbi.nlm.nih.gov/ books/NBK190101/

[60] Felderbauer, P; Schnekenburger, J; Lebert, R; Bulut, K; Parry, M; Meister, T; et al. A novel A121 T mutation in human cationic trypsinogen associated with hereditary pancreatitis: functional data indicating a loss-of-function mutation influencing the R122 trypsin cleavage site. *J Med Genet*, 2008 Aug, 45(8), 507-12.

[61] Witt, H; Sahin-Toth, M; Landt, O; Chen, JM; Kahne, T; Drenth, JPH; et al. A degradation-sensitive anionic trypsinogen (PRSS2) variant protects against chronic pancreatitis. *Nat Genet*, 2006 June, 38(6), 668-673.

[62] Whitcomb, DC; et al. Multicentre approach to recurrent acute and chronic pancreatitis in the United States: the North American Pancreatitis Study 2 (NAPS2). *Pancreatology*, 2008, 8, 520-31.

[63] Laskowski, M; Wu, FC. Temporary inhibition of trypsin. *J Biol Chem*, 1953, 204, 797-805.

[64] Schneider, A. Serine Protease Inhibitor Karzal Type 1 Mutations and Pancreatitis. *Clin Lab Med*, 2005, 25, 61-78.

[65] Witt, H; Luck, W; Hennies, HC; et al. Mutations in the gene encoding the serine protease inhibitor, Kazal type 1 are associated with chronic pancreatitis. *Nat Genet*, 2000, 25, 213-216.

[66] Drenth, JPH; te Morsche, R; Jansen, JBM. Mutations in serine protease inhibitor Kazal type 1 are strongly associated with chronic pancreatitis. *Gut*, 2002, 50, 687-692.

[67] Chen, J-M; Mercier, B; Audrezet, M-P; Ferec, C. Mutational analysis of the human pancreatic secretory trypsin inhibitor (PSTI) gene in hereditary and sporadic chronic pancreatitis. *J Med Genet*, 2000, 37, 67-69.

[68] Pfutzer, RH; Barmada, MM; Brunskil, AP; Finch, R; Hart, PS; Neoptolemos, J; et al. SPINK1/PSTI polymorphisms act as disease modifiers in familial and idiopathic chronic pancreatitis. *Gastroenterology*, 2000, 119, 615-23.

[69] Pfutzer, RH; Myers, E; Finch, R; et al. SPINK1 mutations modify phenotypic expression in hereditary pancreatitis caused by cationic trypsinogen (PRSS1) mutations R122H and N21. *Gastroenterology*, 2001, 120, A33.

[70] Aoun, E; Chang, CH; Greer, JB; Papachristou, GI; Barmada, MM; Whitcomb, DC' (2008) Pathways to Injury in Chronic Pancreatitis: Decoding the Role of the High-Risk SPINK1 N34S Haplotype Using Meta Analysis. PLoS ONE 3(4): e2003. doi:10.1371/ journal. pone.0002003.

[71] Ooi, CY; Durie, PR. Cystic fibrosis transmembrane conductance regulator (CFTR) gene mutations in pancreatitis. *Journal of Cystic Fibrosis*, 2012, 11, 355-362.

[72] Pallagi, P; Venglovecz, V; Rakonczay, Jr. Z; et al. Trypsin reduces pancreatic ductal bicarbonate secretion by inhibiting CFTR Cl- channels and luminal anion exchangers. *Gastroenterology*, 2011, 141(6), 2228-39.

[73] Ooi, CY; Dorfman, R; Cipolli, M; Gonska, T; Castellani, C; Keenan, K; Freedman, SD; et al. Type of CFTR mutation determines risk of pancreatitis in patients with cystic fibrosis. *Gastroenterology*, 2011, 140,153-161.

[74] Ooi, CY; Dorfman, R; Keenan, K; et al. SPINK1 mutation N34S increases risk of pancreatitis in patient with CF & CFTR related disorders. *Paediatric Pulmonology*, 2011, S34, 390.

[75] Rosendahl, J; Landt, O; Bernadova, J; Kovacs, P; Teich, N; Bödeker, H; Keim, V; Ruffert, C; Mössner, J; Kage, A; Stumvoll, M; Groneberg, D; Krüger, R; Luck, W; Treiber, M; Becker, M; Witt, H. CFTR, SPINK1, CTRC and PRSS1 variants in chronic pancreatitis: is the role of mutated CFTR overestimated? *Gut.*, 2013, 62, 582–92.

[76] Yamamoto, A; Ishiguro, H; Ko, SB; et al. Ethanol induces fluid hypersecretion from guinea-pig pancreatic duct cells. *J Physiol*, 2003, 551, 917-26.

[77] Hamada, H; Ishiguro, H; Yamamoto, A; et al. Dual effects of n-alcohols on fluid secretion from guinea pig pancreatic ducts. *Am J Phsiol Cell Physiol*, 2005, 288, C1431-C1439.

[78] Judak, L; Hegyi, P; Rakonczay, Z; Jr. Maleth, J; Grady, MA; Venglovecz, V. Ethanol and its non-oxidative metabolites profoundly inhibit CFTR function in pancreatic epithelial cells which is prevented by ATP supplementation. *Pflugers Arch – Eur J Physiol*, 2014, 466, 549-562.

[79] Cantin, AM; Hanrahan, JW; Bilodeau, G; Ellis, L; Dupuis, A; Liao, J; Zielenski, J; Durie, P. Cystic fbrosis transmembrane conductance regulator function is suppressed in cigarette smokers. *Am J Respir Crit Care Med*, 2006, 176, 1139-1144.

[80] Raju, SV; Jackson, PL; Courville, CA; McNicholas, CM; Sloane, PA; Sabbatini, G; Tidwell, S; Tang, LP; Liu, B; Fortenberry, JA; Jones, CW; Boydston, JA; Clancy, JP; Bowen, LE; Accurso, FJ; Blalock, JE; Dransfield, MT; Rowe, SM. Cigarette Smoke Induces Systemic Defects in Cystic Fibrosis Transmembrane Conductance Regulator Function. *Am J Respir Crit Care Med*, 2013, 188(11), 1321-1330.

[81] Chang, YT; Chang, MC; Su, TC; Liang, PC; Su, YN; Kuo, CH; Wei, SC; Wong, JM. Association of Cystic Fibrosis Transmembrane Conductance Regulator (CFTR) Mutation/Variant/Haplotype and Tumor Necrosis Factor (TNF) *Clin Chem*, 2008 Jan, 54(1), 131-8.

[82] Muddana, V; Lamb, J; Greer, JB; Elinoff, B; Hawes, RH; Cotton, PB; Anderson, MA; Brand, RE; Slivka, A; Whitcomb, DC. Association between calcium sensing receptor gene polymorphisms and chronic pancreatitis in the US population: Role of serine protease inhibitor Kazal 1 type and alcohol. *World J Gastroenterol*, 2008, 14(28), 4486-4491.

[83] LaRusch, J; Whitcomb, DC. Genetics of pancreatitis with a focus on the pancreatic ducts. *Minerva Gastroentol Dietol.*, 2012 Dec, 58(4), 299-308.

[84] Felderbauer, P; Klein, W; Bulut, K; Ansorge, N; Dekomien, G; Werner, I; Epplen, JT; Schmitz, F; Schmidt, WE. Mutations in the calcium-sensing receptor: a new genetic risk factor for chronic pancreatitis? *Scand J Gastroenterol*, 2006, 41, 343-348.

[85] Murugaian, EE; Premkumar, RM; Radhakrishnan, L; Vallath, B. Novel mutations in the calcium sensing receptor gene in tropical chronic pancreatitis in India. *Scand J Gastroenterol*, 2008, 43, 117-121.

[86] Beer, S; Zhou, J; Szabo, A; Keiles, S; Chandak, GR; Witt, H; Sahin-Toth, M. Comprehensive functional analysis of chymotrypsin C (CTRC) variants reveals distinct

loss-of-function mechanisms associated with pancreatitis risk. *Gut*, 2013, 62, 1616-1624.

[87] Rosendahl, J; Witt, H; Szmola, R; Bhatia, E; Oszvari, B; et al. Chymotrypsin C (CTRC) alterations that diminish activity or secretion are associated with chronic pancreatitis. *Nat Genetics*, 2008 January, 40(1), 78-82.

[88] Masson, E; Chen, JM; Scotet, V; Le Marechal, C; Ferec, C. Association of rare chymotrypsinogen C (CTRC) gene variations in patients with idiopathic chronic pancreatitis. *Hum Genet.*, 2008a, 123, 83–91.

[89] Chang, MC; Chang, YT; Wei, SC; Liang, PC; Jan, IS; Su, YN; Kuo, CH; Wong, JM. Association of novel chymotrypsin C gene variations and haplotypes in patients with chronic pancreatitis in Chinese in Taiwan. *Pancreatology.*, 2009, 9, 287–92.

[90] Tremblay, K; Dubois-Bouchard, C; Brisson, D; Gaudet, D. Association of CTRC and SPINK1 gene variants with recurrent hospitalizations for pancreatitis or acute abdominal pain in lipoprotein lipase deficiency. *Frontiers in genetics*, 2014, 5, 1-9.

[91] Murase, T; Ebara, T; Okubo, M. Lipoprotein lipase activity in heterozygotes for lipoprotein lipase gene mutations reveals a gender bias. *Ann Clin Biochem*, 2014 Mar, 51(Pt2), 294-7 (abstract).

[92] Santamarina-Foho, S. The familial chylomicronemia syndrome. *Endocrinol Metab Clin North Am.*, 1998, 27, 551-567.

[93] Kielson, LM; Vary, CP; Sprecher, DL; et al. Hyperlipidemia and pancreatitis during pregnancy in two sisters with a mutation in the lipoprotein lipase gene. *Ann Intern Med*, 1996, 124, 425-428.

[94] Rahalkar, AR; Giffen, F; Har, B; Ho, J; Morrison, KM; Hill, J; Wang, J; Hegele, RA; Joy, T. Novel LPL mutations associated with lipoprotein lipase deficiency: two case reports and a literature review. *Can J Physiol Pharmacol.*, 2009 Mar, 87(3), 151-60.

[95] Whitfield, JB. Gamma Glutamyl transferase. *Crit Rev Clin Lab Sci*, 2001, 38(4), 263-355.

[96] Brand, H; Diergaarde, B; O'Connell, MR; Whitcomb, DC; Brand, RE. Variation in the Gamma Glutamyltransferase 1 Gene and Risk of Chronic Pancreatitis. *Pancreas*, 2013, 42, 836-840.

In: Gene-Environment Interactions and Human Diseases
Editor: Lu Qi
ISBN: 978-1-63483-680-7

Chapter 12

GENETICS AND GENE-ENVIRONMENT INTERACTIONS ON LONGEVITY AND LIFESPAN

***Chao-Qiang Lai*[1], *Yiyi Ma*[2] *and Laurence D. Parnell*[1]**
[1]Jean Mayer USDA Human Nutrition Research Center
on Aging at Tufts University, Boston, MA, US
[2]Department of Medicine, Boston University, Boston, MA, US

ABSTRACT

Longevity is a complex trait and highly associated with healthspan – lifespan without major diseases. In human populations there is a large amount of variation in longevity, which can be attributed to genetics, environment, and interactions between them. The genetic contribution to longevity is about 25%. Candidate gene approaches and genome-wide association studies (GWAS) have identified many putative genetic factors that may have a role in lifespan. However, only two variants in the *APOE* and *FOXO3* genes have shown consistent association with longevity in several studies. Many known genetic variants interact with environmental factors, particularly with diet and lifestyle, influencing risk of age-related diseases, such cardiovascular diseases (CVD), type 2 diabetes, and cancer. Little is known, however, about genotype by environment (GxE) interactions on longevity directly. A paradox has arisen with a few studies showing that long-lived individuals carry similar numbers of risk alleles as the individuals who have a normal lifespan. However, evidence suggests that long-lived individuals may carry protective genetic factors that delay the detrimental effects of risk alleles. Such genetic factors remain to be identified. Environmental factors can modify epigenetic changes, which in turn may affect the rate of aging and longevity. Understanding the molecular mechanisms of epigenetic changes on aging and longevity is important and remains to be explored.

1. INTRODUCTION

Age-related diseases are one of the major burdens of public health. The objective of studying aging and longevity in humans is to reduce the rate of age-related diseases and achieve long and healthy lifespan. The increase in mean lifespan in developed countries over the past several decades has been attributed to reduction in early life mortality and in later life mortality of the elderly who are over 70 years [1]. The reduction in mortality is owed to the environmental changes in the control of infectious diseases and health care, and medical advances in disease diagnostics and treatment. Yet, there is a large amount of variation in lifespan within and among human populations. Genetic contribution to the total phenotypic variation, estimated as about 25% [2, 3], is believed to be one of the important factors in longevity – age at death. Thus far, candidate gene approaches and GWAS have identified only a very few genetic variants that contribute to longevity [4, 5]. On the other hand, the environment can exert significant influence on lifespan and longevity [6-8]. For example, declines in functional capacity, of an organ, body process or the individual as a whole, with age can be offset by a low-risk lifestyle, including sufficient physical activity and abstinence from tobacco smoking [9, 10]. Environmental factors can interact with genotypes to affect lifespan, although the direct evidence of impact by the genotype by environment interaction on longevity is scant. In addition, environmental factors can induce epigenetic changes that in turn affect gene expression, aging and age-related diseases, hence act on healthspan (healthy life without major diseases) and longevity [11, 12]. For this chapter, we will examine genetic variants with effects on longevity and explore their interactions with environmental factors on age-related diseases in humans.

2. Genetics of Longevity in Humans

The complexity of longevity limits its study design and genetic analysis in human populations. Despite the fact that genetic factors of longevity and lifespan are of great interest, identification of longevity genetic factors in humans is becoming possible only in the recent history of human population studies [4, 13]. While many approaches have been tested, the most fruitful studies are candidate gene and genome-wide association studies [14].

Candidate Gene Approach

Candidate gene association studies are hypothesis-driven approaches to link genetic variants in genes that are known contributors of longevity either in humans or model organisms to major human diseases and premature aging, such as CVD. For CVD, genetic variants at *APOE, CETP, LDLR, APOC3, APOB, MTTP* [15], *ACE* [16], have been tested for their associations with longevity [5, 13, 17]. While most of these variants show some, but inconsistent correlation with lifespan or healthspan, only *APOE* was illustrated to have the most consistent association with lifespan across many populations, where the *APOE E4* allele was associated with shorter lifespan and increased risk of CVD and Alzheimer's disease [18]. For premature aging diseases, genetic variants at *WRN* [19, 20] and *KLOTHO* (*KL*)[21] are reported to be associated with longevity. Certain variants in *KL* are associated with chronic renal failure [22] and coronary artery disease (CAD) [23] and often at ages of early onset, phenomena that can be described justifiably as age-related.

Lifespan genes known in model organisms include genes in the insulin/IGF signaling pathway, oxidative stress of aging, and telomere length hypothesis [24]. *FOXO3A*, a constituent of the insulin/IGF1 signaling pathway, carriers of the major allele of rs2802292 are exceptionally long-lived across populations and have lower prevalence of cancer, CVD, and T2D [25-27]. Among genes that are involved in processes of oxidative stress, *SOD1* and *SOD2* have been tested for their contributions to lifespan in humans [28], but inconsistent associations have been observed across populations [28]. In addition, mitochondrial variants were also tested for their associations with longevity. Variants C150T was associated with longevity in Japanese and replicated in Italian populations [29, 30]. For genes that regulate telomere length, *TERT* variants show association with longevity in Ashkenazi Jewish and Danish populations [31, 32].

Overall, candidate gene approaches have identified only a few genetic variants that show consistent association with longevity. Most potential variants show inconsistent correlation. This suggests that longevity is a complex phenotype, and environmental factors and gene by environment interactions could exert effects on longevity.

GWAS of Longevity

In contrast to candidate gene approaches, GWAS is a hypothesis-free method that identifies genetic factors that contribute to a particular phenotype. Instead of testing a few variants in individual candidate genes, GWAS interrogates the whole genome for genetic variants that associate with longevity. So far, there are about 10 GWAS on longevity mostly in populations of European descent. Among them, only three GWAS [33-35] consistently demonstrated two variants, rs4420638 at *APOE* and rs2075650 at *TOMM40*, are associated with longevity at the genome-wide level ($5x10^{-8}$). In particular, Sebastiani et al. [35] conducted a discovery GWAS in 801 New England centenarians and 914 genetically matched controls for 243,980 SNPs, and replicated findings in two additional and independent case/control samples. Only one SNP, rs2075650, near *APOE* reached genome-wide significance. That a high proportion of cases are centenarians is a common characteristic shared by these three GWAS. On the other hand, for other GWAS most contain mainly nonagenarians and none of the associations reached genome-wide significance for longevity ($P > 5x10^{-8}$). These GWAS without genome-wide significant association are represented by the CHARGE Consortium. The CHARGE Consortium [36] performed the first meta-analysis with four cohorts of 1836 cases who lived to at least 90 years and 1955 controls who passed away in the age rank of [55-80]. Replication was conducted in the Leiden Longevity Study (950 cases and 744 partners of their offspring and 680 blood bank donors) and the Danish 1905 Cohort (2,262 cases and 2007 controls). However, none of the SNPs reached genome-wide significance of $P < 5x10^{-8}$. One SNP, rs9664222 at *MINPP1*, was replicated at the second stage with a *P*-value of $6.77x10^{-7}$.

Expanding to more cohorts, the CHARGE Consortium [37] recently conducted a second meta-analysis based on 6,036 longevity cases, age ≥ 90 years, and 3,757 controls that died between 55 and 80 years from 11 cohorts of European descent. Although no associations with longevity reached the genome-wide significance threshold ($P < 5x10^{-8}$), seven loci reached the discovery threshold of significance ($P < 1x10^{-5}$) and a new genetic variant (rs9841144 at *CADM2 locus)* for longevity was found. This study further confirmed the two well-known longevity variants – *APOE* - rs2075650 and *FOXO3A* – rs10457180. In addition, using the CHARGE Consortium, Walter et al. [38] performed a meta-analysis of nine cohorts for two healthspan outcomes: all-cause mortality and survival free of major disease or death. While none of the SNPs reached a significance threshold of $P < 5x10^{-8}$, 14 independent SNPs associated with risk of death, and eight SNPs with event-free survivals were suggestive of significance at $P \leq 1x10^{-5}$. Taking a different approach from the typical GWAS, Sebastiani et al. [35] introduced the naïve Bayes classifier method and identified 281 SNPs that were not genome-wide significant, and then predicted longevity with 89% sensitivity and 89% specificity in the discovery cohort, and 58-85% sensitivity and 58-61% specificity in independent replicated samples.

To date, there are very few genetic variants that show consistent association across populations, even within European ancestry. There is no doubt that the environment plays a crucial role. Considering many genetic variants were only detected once or twice in one particular population, in addition to lack of statistical power in most GWAS, other reason could be the result of gene-gene and genotype by environment interaction complicated such associations.

3. GxE Interaction of Longevity Variants with Environmental Factors on Age-Related Diseases

The main reason that many genetic variants that show inconsistent associations with aging and longevity across populations could be the result of diverse environmental factors (exposures that individuals experience) and their interactions with genetic factors. Such environmental factors include lifestyle behaviors, dietary habits, and external or less modifiable environments, such as climate, pollution, social background, and cultural practices. Because there are limited observations on GxE interactions on longevity and aging, this chapter reviews the GxE interactions of genetic variants that are consistently and potentially associated with longevity and aging on age-related diseases, such as CVD, metabolic diseases, and cancer [39, 40]. While many of the following examples do also show GxE interactions, it must be emphasized that the associated phenotypes are not lifespan or longevity but relate indirectly by phenotypes of age-related diseases.

APOE, variants of which are consistently associated with longevity and healthy lifespan, is an important apoliproprotein that takes part in lipid metabolism [41]. Three isoforms of *APOE, E2, E3*, and *E4* coded by three haplotypes of two genetic variants, pose different risk of CVD and Alzheimer's disease to carriers of these variants [42]. *APOE E4* carriers have higher LDL-C, and higher risk of CVD and Alzheimer's disease, whereas carriers of the *APOE E2* genotype are protected [42]. However, such associations are not always true. Corella et al. [43] showed that when consuming a diet low in saturated fat, there is no significant difference in CHD risk between *E2* and *E4* carriers, but under a high saturated fat diet (> 10% of total energy), *E4* carriers had a 4-fold increased risk of CHD over *E2* carriers (OR for *E4* vs *E2*, 3.33; 95% CI, 1.61–6.90). Interestingly, when consuming a high saturated fat diet, both carriers of *APOE E2* and *APOE E4* have higher risk of MI compared to non-carriers in a Costa Rican population [44]. In addition, when consuming alcohol, *E2* male carriers have lower LDL-C, whereas *E4* male carriers have higher LDL-C [45].

CETP, a plasma cholesteryl ester transfer protein, is involved in the transfer of cholesteryl ester from high density lipoprotein (HDL) to other lipoproteins. The variants of this gene are known to be associated with both HDL concentration [46] and lifespan in Ashkenazi Jews [47]. The *CETP B2* carriers are associated with higher HDL-C and lower risk of CVD. On the other hand, the *B2* carriers of the *CETP*-TaqIB SNP (rs708272) [48] or homozygotes [49] have reduced risk of myocardial infarction than non-carriers when consuming alcohol [50].

MTHFR, methylenetetrahydro-folate reductase, plays a key role in one-carbon nutrient metabolism. Homozygotes (TT) of a well-known variant, C677T (rs1801133), have increased levels of homocysteine and higher risk of CVD. The frequency of TT seems to be reduced in the elderly [51]. However, in another study, there was no significant difference in CVD risk between genotypes [52]. The C677T genotype interacts with folate intake on CVD, and carriers of TT tend to have higher risk of CVD when folate intake is low [53]. However, there is no significant difference in CVD risk when folate status is high. In addition, C677T genotype also interacts with vitamin B12 intake influencing plasma homocysteine concentration [54].

ACE, angiotensin I converting enzyme that converts angiotensin I into a physiologically active peptide angiotensin II, is a crucial regulator of the renin-angiotensin system. An insertion/deletion of 287 bp at *ACE* is known to be associated with coronary heart disease (CHD)[16]. Carriers of the insertion (I allele) tend to have increased risk of hypertension with increased salt intake, but DD (deletion) carriers do not [55]. Furthermore, carriers of the D allele have improved gait speed in response to exercise, whereas there is no improvement among subjects with the II genotype after exercise intervention [56].

KL is a type-I membrane protein that can subdue oxidative stress and growth factor signaling when over-expressed [57, 58]. The G115A variant at *KL* is associated with healthy aging and longevity [23, 59]. Carriers of the A allele have increased risk of CAD from smoking and decreased risk from hypertension and elevated levels of HDL-cholesterol [23].

An alternative analysis is to compare the genetic findings that associate variants with aging phenotypes to known gene-environment interactions. A large catalog of GxE interactions for a collection of cardiometabolic phenotypes recently has been made available [60]. Here, we performed an analysis with the 33 unique candidate genes that previously have been assigned a role in human longevity. Eight of these 33 genes – *ACE*, *APOC3*, *APOE*, *CETP*, *IL6*, *MTHFR*, *MTTP* and *PON1* – have cardiometabolic GxE interactions. Interestingly, all eight of these genes harbor variants participating in GxEs for LDL-cholesterol. Our analysis has shown that it is much more likely to observe GxE interactions for the other primary blood lipids of HDL-cholesterol, total cholesterol and triglycerides than for LDL-C. The full implication of this result is not entirely clear, but may indicate the value in maintaining healthy serum LDL-C levels. Other findings of note include six of the eight genes are involved in GxEs for triglyceride (all excluding *ACE* and *MTTP*), four genes – *APOC3*, *APOE*, *CETP* and *MTTP* – are involved in GxEs for apolipoprotein measures in lipid particles (a rarely measured phenotype), and four genes – *APOE*, *IL6*, *MTHFR* and *PON1* – are involved in GxEs for inflammation-associated phenotypes.

On the other hand, a similar analysis of the variants identified by Sebastiani [35] as associating with human lifespan showed that not one was identical to any SNP in the cardiometabolic GxE catalog, including a large set (> 3300) of variants in high linkage disequilibrium, as mined from the 1000 Genomes Project data. This finding implies that such lifespan SNPs either do not readily engage in GxE interactions of commonly measured environmental factors, such as diet, physical activity, and alcohol intake, or have little bearing on cardiometabolic phenotypes. However, these genetic factors could have greater impact on risk of cancer, which remains to be determined.

4. Environmental-Wide Association Studies

Just as genome-wide association studies can identify genetic markers associating with disease phenotypes, an environmental-wide association study, or EWAS, will demonstrate environmental or lifestyle factors linked to population characteristics. An EWAS with its different perspective is quite likely to employ data on environmental factors that are not generally collected in genetic epidemiology studies. In such instances, EWAS can inform the selection of environmental factors on the contribution to disease [7, 61, 62], or to be used in future tests for GxE interactions (Figure 1). This winnowing of the myriad environmental factors to which cohort participants are exposed, or their exposome, can have significant, beneficial impact on the issue of multiple testing.

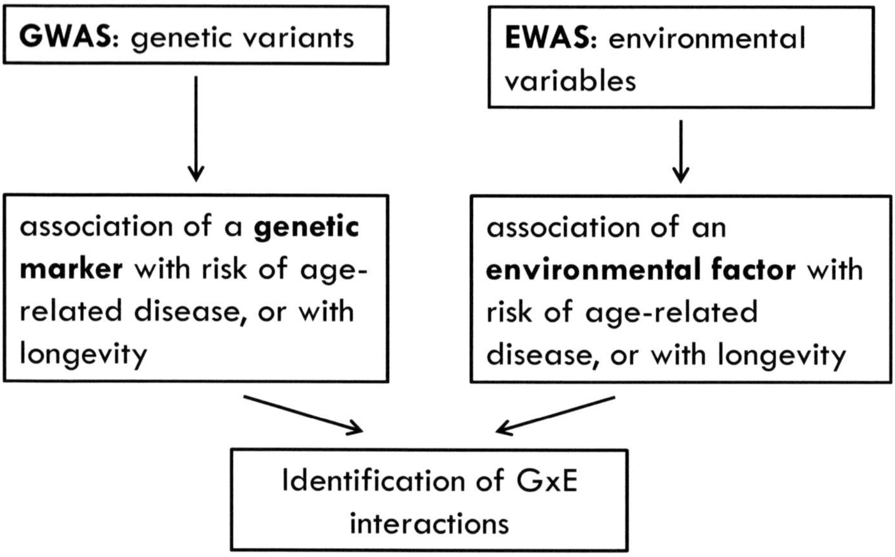

Figure 1. A genome-wide association study, GWAS, looks at a large panel of genetic variants, typically 1-20 million, across a population and seeks to associate a variant with any of a number of phenotypes for which measures are available. A variant may show association with a phenotype, while a particular allele (most variants have two alleles) associates with direction of that phenotype (i.e., increased/decreased risk, or higher/lower level of the phenotype). Further tests with the genetic markers identified as associating with a phenotype may be undertaken and may reveal associations that are modified by an environmental factor, these are termed gene-environment interactions, GxE. An environmental-wide association study, EWAS, has a similar objective of identifying factors involved in disease risk or longevity, but begins from the perspective of variation within the environment. Thus, an environmental factor, with a range of exposure values across a population, can be found to associate with disease risk or longevity. Similar to the GWAS path, environmental factors identified by EWAS can be used in tests to discover GxE interactions.

An EWAS conducted on the dataset from National Health and Nutrition Examination Survey (NHANES) with the outcome of mortality identified serum levels of cadmium and lycopene, and lifestyle factors of physical activity and smoking as mortality environmental factors [61]. Cadmium, both serum and urinary measures, and smoking correlated negatively with all-cause mortality while lycopene and physical activity showed positive correlations.

Other EWAS have been conducted with outcomes pertaining to type 2 diabetes [62, 63]. One of those EWAS on type 2 diabetes identified lower alcohol use and increased smoking in childhood and adulthood as associating with type 2 diabetes status [62]. These type 2 diabetes studies consider non-infectious exposures with the EWAS, but infectious agents are also a type of exposure. The collection of a person's exposures to infectious agents relates directly to serological, geoepidemiological and molecular evidence, and manners in which these can be considered in autoimmune diseases have been reviewed [64]. Ideas presented therein are applicable to GxE interactions affecting lifespan. Lastly, the roles of genomics, transcriptomics, proteomics and metabolomics data in assessing the impact of environmental chemical exposures on human health are explored in another review [7].

Four circulating biomarkers predicted the risk of all-cause mortality among participants from the Estonian Biobank after adjusting for conventional risk factors: ORM1 (alpha-1-acid glycoprotein), albumin, very-low-density lipoprotein (VLDL) particle size, and citrate [65]. All four biomarkers were predictive of cardiovascular mortality, as well as death from cancer and other nonvascular diseases [65]. It is noteworthy that these four Estonian plasma biomarkers each have some relationship to diet. Citrate is found in food and is a constituent of the Krebs cycle, while type and quantity of dietary fat intake affects VLDL particle size. In addition, a proteomics study conducted on individuals partaking of either an eastern or western dietary pattern identified several peptides, which could discern the dietary pattern of the individual, including albumin and ORM1 [66].

In summary, such studies as those described in brief in this section have not identified in a direct manner genetic markers of longevity / lifespan or markers supporting gene-environment interactions. However, these studies are informative as they highlight significant correlating factors. Such data can be used in the selection of environmental factors in future studies that seek to characterize GxE interactions for longevity / lifespan.

5. Assessing Variance in Lifespan by GWAS and GxE Approaches

Cardiometabolic phenotypes represent important clinical measures of disease status and, especially when considered longitudinally, of risk of progression of a given disease. Many of the diseases that are affiliated with cardiometabolic phenotypes – blood lipids, glycemic traits, obesity anthropometrics, vascular measures, inflammation and metabolic syndrome – are age-related and are well known to be sensitive to environmental or lifestyle factors. GWAS have identified a large number of genetic polymorphisms that associate with these cardiometabolic phenotypes. At the same time, genome-wide interaction studies and analyses designed to uncover the contributions of GxEs to phenotype variance are identifying still other contributors to variance in these phenotypes. The difference between the GWASs and GxE approaches is exemplified below.

CardioGxE is a recently published catalog of GxE interactions for an extensive range of cardiometabolic phenotypes, compiled from over 380 literature reports [60]. GWAS, on the other hand, are useful interrogators of the genome, and have led to the identification of numerous genetic variants associating with phenotypic variance and disease risk. Yet, GWAS-identified, main-effect associations have been described as contributing only a small proportion to total phenotype variance [67]. Our analysis has demonstrated that there is very little in common between any of four different databases providing links between genes and disease, and the GxE-based associations to relevant phenotypes cataloged by us [60]. Such lack of correspondence between catalogs of main-effect variants and GxE-based variants points to the incomplete description of contribution to phenotypic variance by those main-effect associations, and adds to the value of GxE interactions as contributors to that variance. It follows that genetic contributors by themselves fall short as diagnostic tools to assess overall disease risk. The GxE term is a necessary and important, but under-appreciated component in this assessment [60].

Some researchers are looking into the emerging field of human social genomics and its consideration of organism- and genome-wide responses in the context of social-environmental conditions as applied to, for example, pro-inflammatory gene networks but also to optimal health [68], and this is encouraging for the identification of new lifespan GxEs.

6. Common Disease Variants and Longevity

It remains to be determined if long-lived individuals (LLI) carry fewer disease risk alleles than subjects who have a normal lifespan. To address this, Beekman et al. [69] examined two case/control cohorts: (1) 723 nonagenarian siblings (mean age 94) of long-lived subjects vs 721 unrelated younger subjects (mean age 52), and (2) 979 long-lived individuals over 85 (mean age 87) vs 1,167 younger controls (mean age 41) from the Netherlands Twin cohort. Surprisingly, for tests on 30 SNPs that are known to be associated with CVD, type 2 diabetes (T2D), and cancer, both the case and control cohorts carried the same number of risk alleles (mean, 27 alleles). In addition, there was no difference in the distribution of risk alleles between elderly and younger populations. The authors argued that common disease risk alleles contribute little to longevity, or long-lived subjects have protective factors that can buffer the effect of those tested risk alleles. The latter hypothesis was supported by the follow-up observation by Mooijaart et al. [70]. That study examined glucose tolerance and 15 known T2D SNPs of offspring of the LLI in comparison with that of the offspring's spouses and other age-matched controls. While the former (LLI offspring) had a better metabolic profile and glucose tolerance than the age-matched control, the frequencies of T2D risk alleles are the same in both groups. This observation indicates that there are protective genetic factors in the offspring of LLI subjects that quench the detrimental effects of risk alleles, rather than modulating environmental factors, as offspring and their spouses live in similar environments. However, although common diseases are affected by many loci and environmental factors, in the above-mentioned studies only a few loci were examined. Thus, studies with more genetic factors that are associated with age-related diseases remain to be explored.

7. EPIGENETICS OF LONGEVITY

Environmental factors can modify phenotypic changes without alteration of the genetic code. Such heritable and reversible changes, also known as epigenetic modifications of DNA, include chromatin remodeling and DNA methylation. Numerous experiments in model systems and humans have demonstrated that throughout life – from *in utero* to death – environmental signals can impart changes to epigenetic patterns. Epigenetic changes can be used to predict chronological aging [71]. Some of these epigenetic changes affect activity of genes important in age-related diseases and longevity [72, 73]. Although epigenetics constitutes an important molecular mechanism to explain some observed GxE interactions based on intimate relationships with both genes and environmental factors, direct implications of epigenetics on longevity are poorly understood. Nonetheless, some aspects are worth exploring in detail.

In one study, an algorithm was developed using 8,000 samples from 82 different DNA methylation array datasets and over 50 non-diseased tissues and cell types to identify age-sensitive differences in DNA methylation at CpG islands [74]. This work identified 344 unique genes mapping at or near the site of the differential DNA methylation. Because DNA methylation serves as a kind of record of accumulated environmental exposures, it would be interesting to know if the methylation differences that those CpG islands harbor function in gene-environment interactions. To accomplish this, a comparison with a collection of established GxE interactions would be informative.

SNPs within the GxE interaction catalog [60] were taken from published examples and cover an array of blood lipid, glycemic, obesity anthropometric, vascular, inflammation and metabolic syndrome phenotypes. In total, 1187 significant GxEs with SNPs mapping to 189 different genes are listed. Importantly, these phenotypes are the subjects of many of the chapters in this book and are noted precursors of diseases that negatively affect lifespan. Nonetheless, none of the GxE interactions culled from the 386 published scientific reports that formed the basis of the cardiometabolic GxE catalog involved lifespan, longevity or age at death as a phenotype.

The comparison between the genes near the age-sensitive methylated CpG islands [74] and the known cataloged cardiometabolic GxEs appears to show very little in common [60]. Only two genes, *CLOCK* and *SELP*, encoding a core regulator of circadian rhythms and P-selectin platelet alpha-granule protein respectively, are noted in both lists of genes. This is too small a number of genes to be able to draw any confident conclusion on how age-related epigenetic changes can affect cardiometabolic GxEs as research groups are currently assessing such interactions. In other words, it is entirely possible that age-related changes to DNA methylation patterns will not manifest as GxE interactions. On the other hand, accumulation of more data may clarify this issue. A limitation of this comparison is the assumption that the 344 genes showing age-related differences in DNA methylation at CpG islands exert influence on cardiometabolic phenotypes.

In another study, in which subjects from the 1921 and 1936 Scottish birth cohorts were examined for genomic DNA methylation at CpG sites [75], data show that a large number of CpGs across the genome, as a result of environmental and/or genetic constraints, have stable variation in DNA methylation over the human lifetime. In addition, while DNA epigenetic variation at a number of CpGs in the population is determined by genetic factors, some sites are highly modifiable by the environment [75]. This result would imply that variation in methylation at any given CpG site can be influenced by both genetic and environmental factors, and there are heritable epigenetic patterns associated with environmental factors.

Genetic factors, on the one hand, play significant roles in regulating components of all epigenetic mechanisms, including DNA methylation, histone modification, chromatin remodeling, and microRNA function. For example, DNA methylation patterns were found to be determined by the local DNA sequences spanning approximately 1 kb [76]. It is widely observed that the phenomenon of allele specific DNA methylation is widespread across the entire human genome [77, 78]. In addition, sequence variants, such as SNPs, have been reported to affect epigenetic marks. DNA methylation was shown to be regulated by the SNPs in both *cis* (close) and *trans* (distant) positions [79-82]. It also was reported that a cancer-associated SNP was related to multiple enhancer-related histone marks, including dimethylation of histone H3 at lysine 4 (H3K4me2), acetylation of histone H3 (H3-Ac), and acetylation of histone H4 (H4-Ac) [83]. That the DNA sequence feature of an ~10 bp periodicity of the nucleosomal DNA affects DNA bending and shape indicates the potential for interactions between SNPs and chromatin structure. This was emphasized by the recent finding that the DNA sequence of the periodicity is under selective pressure from genetic evolution [84]. A gain-of-function SNP disrupting a microRNA miR-410 seed site was found to interact with dietary fatty acids to modulate blood lipids and incidence of stroke in a wide array of cohorts across the USA and Europe [85, 86]. In these studies, one allele of the SNP under study, mapping to the 3'-UTR of the lipoprotein lipase (*LPL*) gene, created a motif supporting binding of and subsequent regulation by miR-410.

On the other hand, epigenetic mechanisms have been proposed as the biomarkers of the effect of environmental factors acting on the genetic structure. Among a wide array of environmental factors, aging is the only unmodifiable factor with prominent effects on all perspectives of epigenetic mechanisms, with particular research attention in DNA methylation and histone modifications.

That aging affects DNA methylation is suggested by changes in DNA methylation patterns during each stage of development [87] and cellular differentiation[88, 89]. Through three stages of mouse development, DNA methylation patterns undergo changes from demethylation at pre-implantation stage to the bimodal methylation pattern at the implantation stage, and then to the stage- and tissue-specific methylation patterns during the stage after implantation [87]. A dramatic loss of the cellular content of 5-methylcytosine was observed during the process whereby normal diploid fibroblasts were immortalized into cell lines [88] and the treatment with DNA methylation inhibitors curtailed initial cellular growth [89]. Twin studies showed that aging was associated with not only the genomic DNA methylation content[90], but also the gene-specific DNA methylation patterns of a diverse panel of genes including the Edar associated death domain (*EDARADD*), target of myb1 (chicken)-like 1 (*TOMIL1*), neuronal pentraxin II (*NPTX2*)[91] dopamine receptor 4 gene (*DRD4*), serotonin transporter gene (*SERT/SLC6A4*), and the X-linked monoamine oxidase A gene (*MAOA*) [92]. DNA methylation patterns were also reported to be associated with multiple age-related diseases such as cancer [93-95], Alzheimer's disease (AD) [96], age-related macular degeneration (AMD) [97], and atherosclerosis [98].

The most prominent effects of aging on histone modifications relate to both acetylation and methylation. The protein levels of the acetylation of histone 4 lysines 5 and 16 (H4K5ac and H4K16ac) and histone 3 lysine 9 (H3K9ac) change gradually across different stages of the cell cycle[99], by showing a decrease in H4K5ac and H4K16ac and an increase in H3K9ac in aged cycling human fibroblasts. Consistently, their corresponding deacetylases, the sirtuin family member 1 and 6 (*SIRT1* and *SIRT6*)[100-102], were reported to have anti-aging effects [103, 104] through changes in histone acetylation [105]. A cross-sectional study with monozygotic twins found that older twins exhibit greater differences in histone acetylation than the younger twins [90]. Three histone methylation markers have been studied widely for their correlations with the aging process: the tri-methylation of histone 3 lysines 4, 9 and 27 (H3K4me3, H3K9me3, H3K27me3) and the tri-methylation of histone 4 lysine 20 (H4K20me3). However, the directions of aging on these methylation markers are inconsistent. Protein levels of H4K20me3 and H3K9me3 were down-regulated in late-passage cycling human fibroblasts [99], but up-regulated in the premature aging disorder of Hutchinson-Gilford Progeria Syndrome [106], the livers and kidneys of the elderly rats [107], and the cellular senescence induced by the *retinoblastoma* oncogene [108-110]. In contrast, H3K4me3 was reported to increase in aged diploid fibroblasts [99] but decrease in Hutchinson-Gilford Progeria Syndrome [106, 111] and senescent cells [112]. The deficiency of the methyltransferase of H3K4me3, a member of the Ash-2 complex, was found to extend the lifespan of the nematode *Caenorhabditis elegans* [113]. In terms of the H3K27me3, reports consistently have shown that aging may introduce the loss of protein levels of H3K27me3 [106, 111, 112] and its methyltransferase, EZH2 [112].

Thus, although epigenetics is recognized as an environmentally sensitive component of the GxE equation, research in this area with respect to human longevity and lifespan is sparse. Furthermore, when taken together there is much evidence to encourage research of GxE interactions in conjunction with epigenetic signals across the genome, or by inclusion of genetic variation at those specific loci where the epigenetic differences are detected.

8. Perspectives

Many potential lifespan-associated genetic variants have been reported, but only a few show consistent associations with longevity and lifespan. Complexity of longevity and low heritability (25%) implies that a large proportion of phenotypic variance in longevity can be modified by epistasis, gene by environment interactions, and environmental factors. Because there are inherent difficulties with lifespan studies in human populations, the full extent of gene-environment interactions on lifespan or longevity as phenotypes is not known. Research into genetic factors contributing to lifespan and longevity are also incompletely described.

The fact that long-lived individuals carry a similar number of common disease-associated alleles as do individuals with normal lifespan suggests that longevity could be attributable to epistatic and GxE interactions. It also suggests that the exceptionally long-lived carry other alleles that allow for better flexibility in peripheral or affiliated processes – such as chronic inflammation, and oxidative stress – that can have supporting functions in disease onset and progression. On the other hand, the complexity of longevity implies many genetic and environmental factors are involved. Thus, studies with more genetic factors that are associated with age-related diseases in this area are needed.

Longevity is correlated with healthspan, but our observation with a GxE cardiometbolic database in relation to longevity-associated genetic factors implies that genetic variants associated with extreme longevity differ from those that are associated with healthspan. The former may have a high heritability, especially in older age over 90, and thus environmental factors or GxE interactions contribute less to the longevity phenotypes. The latter are those genetic variants that are associated with age-related diseases, and thus have a large GxE contribution to lifespan (healthspan). Future studies should identify and distinguish these genetic variants.

Epigenetic changes are highly correlated with age and are heritable, although epigenetic variation is affected both by genetic and environmental factors. Identifying epigenetic changes over a lifetime and across generations, which are induced by environmental factors, is an important approach in relationship to longevity and healthspan.

Acknowledgments

This study was supported by contract 1950-51520-014-01S from the US Department of Agriculture, Agriculture Research Service. Mention of trade names or commercial products in this publication is solely for the purpose of providing specific information and does not imply recommendation or endorsement by the U.S. Department of Agriculture. The USDA is an equal opportunity provider and employer.

No conflict of interest relevant to this article was reported.

References

[1] Vaupel, JW; et al. Biodemographic trajectories of longevity. *Science* (New York, N.Y, 280, 855-860 (1998).

[2] Herskind, AM; et al. The heritability of human longevity: a population-based study of, 2872 Danish twin pairs born, 1870-1900. *Human genetics*, 97, 319-323 (1996).

[3] Skytthe, A; et al. Longevity studies in GenomEUtwin. *Twin research : the official journal of the International Society for Twin Studies*, 6, 448-454, doi:10.1375/136905203770326457 (2003).

[4] Brooks-Wilson, AR. Genetics of healthy aging and longevity. *Human genetics*, 132, 1323-1338.

[5] Wheeler, HE; Kim, SK. Genetics and genomics of human ageing. *Philosophical transactions of the Royal Society of London*, 366, 43-50.

[6] Mortensen, LH; et al. Early-life environment influencing susceptibility to cytomegalovirus infection: evidence from the Leiden Longevity Study and the Longitudinal Study of Aging Danish Twins. *Epidemiology and infection*, 140, 835-841, doi:10.1017/S0950268811001397 (2012).

[7] Mortensen, HM; Euling, SY. Integrating mechanistic and polymorphism data to characterize human genetic susceptibility for environmental chemical risk assessment in the, 21st century. *Toxicology and applied pharmacology*, 271, 395-404, doi:10.1016/j.taap.2011.01.015 (2013).

[8] Gavrilov, LA; Gavrilova, NS. New Developments in the Biodemography of Aging and Longevity. *Gerontology*, doi:10.1159/000369011 (2014).

[9] van Ommen, B; van der Greef, J; Ordovas, JM; Daniel, H. Phenotypic flexibility as key factor in the human nutrition and health relationship. *Genes & nutrition*, 9, 423, doi:10.1007/s12263-014-0423-5 (2014).

[10] Calcada, D; et al. The role of low-grade inflammation and metabolic flexibility in aging and nutritional modulation thereof: a systems biology approach. *Mechanisms of ageing and development*, 136-137, 138-147, doi:10.1016/j.mad.2014.01.004 (2014).

[11] Castillo-Fernandez, JE; Spector, TD; Bell, JT. Epigenetics of discordant monozygotic twins: implications for disease. *Genome medicine*, 6, 60, doi:10.1186/s13073-014-0060-z (2014).

[12] Tammen, SA; Friso, S; Choi, SW. Epigenetics: the link between nature and nurture. *Molecular aspects of medicine*, 34, 753-764, doi:10.1016/j.mam.2012.07.018 (2013).

[13] Christensen, K; Johnson, TE; Vaupel, JW. The quest for genetic determinants of human longevity: challenges and insights. *Nature reviews*, 7, 436-448 (2006).

[14] Manolio, TA. Study designs to enhance identification of genetic factors in healthy aging. *Nutrition reviews*, 65, S228-233 (2007).

[15] Geesaman, BJ; et al. Haplotype-based identification of a microsomal transfer protein marker associated with the human lifespan. *Proceedings of the National Academy of Sciences of the United States of America*, 100, 14115-14120, doi:10.1073/pnas.1936249100 (2003).

[16] Luft, FC. Bad genes, good people, association, linkage, longevity and the prevention of cardiovascular disease. *Clinical and experimental pharmacology & physiology*, 26, 576-579 (1999).

[17] Newman, AB; Murabito, JM. The Epidemiology of Longevity and Exceptional Survival. *Epidemiologic reviews*.

[18] Corder, EH; et al. Gene dose of apolipoprotein E type, 4 allele and the risk of Alzheimer's disease in late onset families. *Science (New York, N.Y*, 261, 921-923 (1993).

[19] Castro, E; et al. Polymorphisms at the Werner locus: II. 1074Leu/Phe, 1367Cys/Arg, longevity, and atherosclerosis. *American journal of medical genetics*, 95, 374-380 (2000).

[20] Kuningas, M; Slagboom, PE; Westendorp, RG; van Heemst, D. Impact of genetic variations in the WRN gene on age related pathologies and mortality. *Mechanisms of ageing and development*, 127, 307-313, doi:10.1016/j.mad.2005.11.005 (2006).

[21] Arking, DE; Atzmon, G; Arking, A; Barzilai, N; Dietz, HC. Association between a functional variant of the KLOTHO gene and high-density lipoprotein cholesterol, blood pressure, stroke, and longevity. *Circulation research*, 96, 412-418, doi:10.1161/01.RES.0000157171.04054.30 (2005).

[22] Koh, N; et al. Severely reduced production of klotho in human chronic renal failure kidney. *Biochemical and biophysical research communications*, 280, 1015-1020, doi:10.1006/bbrc.2000.4226 (2001).

[23] Arking, DE; et al. KLOTHO allele status and the risk of early-onset occult coronary artery disease. *American journal of human genetics*, 72, 1154-1161, doi:10.1086/375035 (2003).

[24] Lopez-Otin, C; Blasco, MA; Partridge, L; Serrano, M; Kroemer, G. The hallmarks of aging. *Cell*, 153, 1194-1217, doi:10.1016/ j.cell.2013.05.039 (2013).

[25] Willcox, BJ; et al. FOXO3A genotype is strongly associated with human longevity. *Proceedings of the National Academy of Sciences of the United States of America*, 105, 13987-13992, doi:10.1073/ pnas.0801030105 (2008).

[26] Anselmi, CV; et al. Association of the FOXO3A locus with extreme longevity in a southern Italian centenarian study. *Rejuvenation research*, 12, 95-104, doi:10.1089/rej.2008.0827 (2009).

[27] Flachsbart, F; et al. Association of FOXO3A variation with human longevity confirmed in German centenarians. *Proceedings of the National Academy of Sciences of the United States of America*, 106, 2700-2705, doi:10.1073/pnas.0809594106 (2009).

[28] Yashin, AI; et al. Genes and longevity: lessons from studies of centenarians. *The journals of gerontology*, 55, B319-328 (2000).

[29] Chen, A; Raule, N; Chomyn, A; Attardi, G. Decreased reactive oxygen species production in cells with mitochondrial haplogroups associated with longevity. *PloS one*, 7, e46473, doi:10.1371/journal.pone.0046473 (2012).

[30] Capri, M; et al. The genetics of human longevity. *Annals of the New York Academy of Sciences*, 1067, 252-263, doi:10.1196/annals.1354.033 (2006).

[31] Atzmon, G; et al. Evolution in health and medicine Sackler colloquium: Genetic variation in human telomerase is associated with telomere length in Ashkenazi centenarians. *Proceedings of the National Academy of Sciences of the United States of America*, 107 Suppl, 1, 1710-1717, doi:10.1073/pnas.0906191106 (2010).

[32] Soerensen, M; et al. Genetic variation in TERT and TERC and human leukocyte telomere length and longevity: a cross-sectional and longitudinal analysis. *Aging cell*, 11, 223-227, doi:10.1111/j.1474-9726.2011.00775.x (2012).

[33] Nebel, A; et al. A genome-wide association study confirms APOE as the major gene influencing survival in long-lived individuals. *Mechanisms of ageing and development*, 132, 324-330, doi:10.1016/ j.mad.2011.06.008 (2011).

[34] Deelen, J; et al. Genome-wide association study identifies a single major locus contributing to survival into old age; the APOE locus revisited. *Aging cell*, 10, 686-698, doi:10.1111/j.1474-9726.2011.00705.x (2011).

[35] Sebastiani, P; et al. Genetic signatures of exceptional longevity in humans. *PloS one*, 7, e29848.

[36] Newman, AB; et al. A meta-analysis of four genome-wide association studies of survival to age, 90 years or older: the Cohorts for Heart and Aging Research in Genomic Epidemiology Consortium. *The journals of gerontology*, 65, 478-487.

[37] Broer, L; et al. GWAS of Longevity in CHARGE Consortium Confirms APOE and FOXO3 Candidacy. *The journals of gerontology*, 70, 110-118.

[38] Walter, S; et al. A genome-wide association study of aging. *Neurobiology of aging*, 32, 2109 e2115-2128.

[39] Ho, V; et al. Meat-derived carcinogens, genetic susceptibility and colorectal adenoma risk. *Genes & nutrition*, 9, 430, doi:10.1007/s12263-014-0430-6 (2014).

[40] Figueiredo, JC; et al. Genome-wide diet-gene interaction analyses for risk of colorectal cancer. *PLoS genetics*, 10, e1004228, doi:10.1371/journal.pgen.1004228 (2014).

[41] Leduc, V; Jasmin-Belanger, S; Poirier, J. APOE and cholesterol homeostasis in Alzheimer's disease. *Trends in molecular medicine*, 16, 469-477, doi:10.1016/j.molmed.2010.07.008 (2010).

[42] Minihane, AM; Jofre-Monseny, L; Olano-Martin, E; Rimbach, G. ApoE genotype, cardiovascular risk and responsiveness to dietary fat manipulation. *The Proceedings of the Nutrition Society*, 66, 183-197, doi:10.1017/S0029665107005435 (2007).

[43] Corella, D; et al. Saturated fat intake and alcohol consumption modulate the association between the APOE polymorphism and risk of future coronary heart disease: a nested case-control study in the Spanish EPIC cohort. *The Journal of nutritional biochemistry*, 22, 487-494, doi:10.1016/j.jnutbio.2010.04.003 (2011).

[44] Yang, Y; Ruiz-Narvaez, E; Kraft, P; Campos, H. Effect of apolipoprotein E genotype and saturated fat intake on plasma lipids and myocardial infarction in the Central Valley of Costa Rica. *Human biology*, 79, 637-647 (2007).

[45] Corella, D; et al. Alcohol drinking determines the effect of the APOE locus on LDL-cholesterol concentrations in men: the Framingham Offspring Study. *The American journal of clinical nutrition*, 73, 736-745 (2001).

[46] Wu, Z; et al. Association of cholesteryl ester transfer protein (CETP) gene polymorphism, high density lipoprotein cholesterol and risk of coronary artery disease: a meta-analysis using a Mendelian randomization approach. *BMC medical genetics*, 15, 118, doi:10.1186/s12881-014-0118-1 (2014).

[47] Barzilai, N; Atzmon, G; Derby, CA; Bauman, JM; Lipton, RB. A genotype of exceptional longevity is associated with preservation of cognitive function. *Neurology*, 67, 2170-2175, doi:10.1212/01.wnl.0000249116.50854.65 (2006).

[48] Jensen, MK; Mukamal, KJ; Overvad, K; Rimm, EB. Alcohol consumption, TaqIB polymorphism of cholesteryl ester transfer protein, high-density lipoprotein cholesterol, and risk of coronary heart disease in men and women. *European heart journal*, 29, 104-112, doi:10.1093/eurheartj/ehm517 (2008).

[49] Fumeron, F; et al. Alcohol intake modulates the effect of a polymorphism of the cholesteryl ester transfer protein gene on plasma high density lipoprotein and the risk of myocardial infarction. *The Journal of clinical investigation*, 96, 1664-1671, doi:10.1172/JCI118207 (1995).

[50] Mehlig, K; et al. CETP TaqIB genotype modifies the association between alcohol and coronary heart disease: The INTERGENE case-control study. *Alcohol*, 48, 695-700, doi:10.1016/j.alcohol.2014.08.011 (2014).

[51] Stessman, J; et al. Candidate genes associated with ageing and life expectancy in the Jerusalem longitudinal study. *Mechanisms of ageing and development*, 126, 333-339, doi:10.1016/j.mad.2004.08.025 (2005).

[52] Kluijtmans, LA; Whitehead, AS. Reduced frequency of the thermolabile methylenetetrahydrofolate reductase genotype in the elderly. *Atherosclerosis*, 146, 395-397 (1999).

[53] Markus, HS; et al. A common polymorphism in the methylenetetrahydrofolate reductase gene, homocysteine, and ischemic cerebrovascular disease. *Stroke; a journal of cerebral circulation*, 28, 1739-1743 (1997).

[54] D'Angelo, A; et al. The role of vitamin B12 in fasting hyperhomocysteinemia and its interaction with the homozygous C677T mutation of the methylenetetrahydrofolate reductase (MTHFR) gene. A case-control study of patients with early-onset thrombotic events. *Thrombosis and haemostasis*, 83, 563-570 (2000).

[55] Zhang, L; et al. Interaction of angiotensin I-converting enzyme insertion-deletion polymorphism and daily salt intake influences hypertension in Japanese men. *Hypertens Res*, 29, 751-758 (2006).
[56] Buford, TW; et al. Genetic influence on exercise-induced changes in physical function among mobility-limited older adults. *Physiological genomics*, 46, 149-158, doi:10.1152/physiolgenomics.00169.2013 (2014).
[57] Kuro-o, M. Klotho and aging. *Biochimica et biophysica acta*, 1790, 1049-1058, doi:10.1016/j.bbagen.2009.02.005 (2009).
[58] Wang, Y; Sun, Z. Current understanding of klotho. *Ageing research reviews*, 8, 43-51, doi:10.1016/j.arr.2008.10.002 (2009).
[59] Di Bona, D; Accardi, G; Virruso, C; Candore, G; Caruso, C. Association of Klotho polymorphisms with healthy aging: a systematic review and meta-analysis. *Rejuvenation research*, 17, 212-216, doi:10.1089/ rej.2013.1523 (2014).
[60] Parnell, LD; et al. CardioGxE, a catalog of gene-environment interactions for cardiometabolic traits. *BioData mining*, 7, 21, doi:10.1186/1756-0381-7-21 (2014).
[61] Patel, CJ; Bhattacharya, J; Butte, AJ. An Environment-Wide Association Study (EWAS) on type, 2 diabetes mellitus. *PloS one*, 5, e10746, doi:10.1371/journal. pone.0010746 (2010).
[62] Hall, MA; et al. Environment-wide association study (EWAS) for type, 2 diabetes in the Marshfield Personalized Medicine Research Project Biobank. *Pacific Symposium on Biocomputing. Pacific Symposium on Biocomputing*, 200-211 (2014).
[63] Patel, CJ; Chen, R; Kodama, K; Ioannidis, JP; Butte, AJ. Systematic identification of interaction effects between genome- and environment-wide associations in type, 2 diabetes mellitus. *Human genetics*, 132, 495-508, doi:10.1007/s00439-012-1258-z (2013).
[64] Bogdanos, DP; et al. Infectome: a platform to trace infectious triggers of autoimmunity. *Autoimmunity reviews*, 12, 726-740, doi:10.1016/ j.autrev.2012.12.005 (2013).
[65] Fischer, K; et al. Biomarker profiling by nuclear magnetic resonance spectroscopy for the prediction of all-cause mortality: an observational study of, 17,345 persons. *PLoS medicine*, 11, e1001606, doi:10.1371/journal.pmed.1001606 (2014).
[66] Garcia-Bailo, B; et al. Dietary patterns and ethnicity are associated with distinct plasma proteomic groups. *The American journal of clinical nutrition*, 95, 352-361, doi:10.3945/ajcn.111.022657 (2012).
[67] Manolio, TA; et al. Finding the missing heritability of complex diseases. *Nature*, 461, 747-753, doi:10.1038/nature08494 (2009).
[68] Cole, SW. Human social genomics. *PLoS genetics*, 10, e1004601, doi:10.1371/ journal.pgen.1004601 (2014).
[69] Beekman, M; et al. Genome-wide association study (GWAS)-identified disease risk alleles do not compromise human longevity. *Proceedings of the National Academy of Sciences of the United States of America*, 107, 18046-18049.
[70] Mooijaart, SP; et al. Polymorphisms associated with type, 2 diabetes in familial longevity: The Leiden Longevity Study. *Aging*, 3, 55-62 (2011).
[71] Hannum, G; et al. Genome-wide methylation profiles reveal quantitative views of human aging rates. *Molecular cell*, 49, 359-367, doi:10.1016/ j.molcel.2012.10.016 (2013).

[72] Gravina, S; Vijg, J. Epigenetic factors in aging and longevity. *Pflugers Archiv: European journal of physiology*, 459, 247-258, doi:10.1007/s00424-009-0730-7 (2010).

[73] Ben-Avraham, D; Muzumdar, RH; Atzmon, G. Epigenetic genome-wide association methylation in aging and longevity. *Epigenomics*, 4, 503-509, doi:10.2217/ epi.12.41 (2012).

[74] Horvath, S. DNA methylation age of human tissues and cell types. *Genome biology*, 14, R115, doi:10.1186/gb-2013-14-10-r115 (2013).

[75] Shah, S; et al. Genetic and environmental exposures constrain epigenetic drift over the human life course. *Genome research*, 24, 1725-1733, doi:10.1101/gr.176933.114 (2014).

[76] Lienert, F; et al. Identification of genetic elements that autonomously determine DNA methylation states. *Nature genetics*, 43, 1091-1097, doi:10.1038/ng.946 (2011).

[77] Shoemaker, R; Deng, J; Wang, W; Zhang, K. Allele-specific methylation is prevalent and is contributed by CpG-SNPs in the human genome. *Genome research*, 20, 883-889, doi:10.1101/gr.104695.109 (2010).

[78] Schalkwyk, LC; et al. Allelic skewing of DNA methylation is widespread across the genome. *American journal of human genetics*, 86, 196-212, doi:10.1016/j.ajhg.2010.01.014 (2010).

[79] Zhang, D; et al. Genetic control of individual differences in gene-specific methylation in human brain. *American journal of human genetics*, 86, 411-419, doi:10.1016/j.ajhg.2010.02.005 (2010).

[80] Bell, CG; et al. Integrated genetic and epigenetic analysis identifies haplotype-specific methylation in the FTO type, 2 diabetes and obesity susceptibility locus. *PloS one*, 5, e14040, doi:10.1371/ journal.pone.0014040 (2010).

[81] Gertz, J; et al. Analysis of DNA methylation in a three-generation family reveals widespread genetic influence on epigenetic regulation. *PLoS genetics*, 7, e1002228, doi:10.1371/journal.pgen.1002228 (2011).

[82] Qu, W; et al. Genome-wide genetic variations are highly correlated with proximal DNA methylation patterns. *Genome research*, 22, 1419-1425, doi:10.1101/gr.140236.112 (2012).

[83] Wright, JB; Brown, SJ; Cole, MD. Upregulation of c-MYC in cis through a large chromatin loop linked to a cancer risk-associated single-nucleotide polymorphism in colorectal cancer cells. *Molecular and cellular biology*, 30, 1411-1420, doi:10.1128/MCB.01384-09 (2010).

[84] Langley, SA; Karpen, GH; Langley, CH. Nucleosomes shape DNA polymorphism and divergence. *PLoS genetics*, 10, e1004457, doi:10.1371/journal.pgen.1004457 (2014).

[85] Richardson, K; et al. Gain-of-function lipoprotein lipase variant rs13702 modulates lipid traits through disruption of a microRNA-410 seed site. *American journal of human genetics*, 92, 5-14, doi:10.1016/ j.ajhg.2012.10.020 (2013).

[86] Corella, D; et al. MicroRNA-410 regulated lipoprotein lipase variant rs13702 is associated with stroke incidence and modulated by diet in the randomized controlled PREDIMED trial. *The American journal of clinical nutrition*, 100, 719-731, doi:10.3945/ajcn.113.076992 (2014).

[87] Cedar, H; Bergman, Y. Programming of DNA methylation patterns. *Annual review of biochemistry*, 81, 97-117, doi:10.1146/annurev-biochem-052610-091920 (2012).

[88] Wilson, VL; Jones, PA. DNA methylation decreases in aging but not in immortal cells. *Science (New York, N.Y*, 220, 1055-1057 (1983).

[89] Holliday, R. Strong effects of, 5-azacytidine on the in vitro lifespan of human diploid fibroblasts. *Experimental cell research*, 166, 543-552 (1986).

[90] Fraga, MF; et al. Epigenetic differences arise during the lifetime of monozygotic twins. *Proceedings of the National Academy of Sciences of the United States of America*, 102, 10604-10609, doi:10.1073/ pnas.0500398102 (2005).

[91] Bocklandt, S; et al. Epigenetic predictor of age. *PloS one*, 6, e14821, doi:10.1371/journal.pone.0014821 (2011).

[92] Wong, CC; et al. A longitudinal study of epigenetic variation in twins. *Epigenetics : official journal of the DNA Methylation Society*, 5, 516-526 (2010).

[93] Issa, JP; et al. Methylation of the oestrogen receptor CpG island links ageing and neoplasia in human colon. *Nature genetics*, 7, 536-540, doi:10.1038/ng0894-536 (1994).

[94] Li, LC; et al. Age-dependent methylation of ESR1 gene in prostate cancer. *Biochemical and biophysical research communications*, 321, 455-461, doi:10.1016/j.bbrc. 2004.06.164 (2004).

[95] Belinsky, SA. Gene-promoter hypermethylation as a biomarker in lung cancer. *Nature reviews. Cancer*, 4, 707-717, doi:10.1038/nrc1432 (2004).

[96] De Jager, PL; et al. Alzheimer's disease: early alterations in brain DNA methylation at ANK1, BIN1, RHBDF2 and other loci. *Nature neuroscience*, 17, 1156-1163, doi:10.1038/nn.3786 (2014).

[97] Hunter, A; et al. DNA methylation is associated with altered gene expression in AMD. *Investigative ophthalmology & visual science*, 53, 2089-2105, doi:10.1167/iovs.11-8449 (2012).

[98] Dong, C; Yoon, W; Goldschmidt-Clermont, PJ. DNA methylation and atherosclerosis. *The Journal of nutrition*, 132, 2406S-2409S (2002).

[99] O'Sullivan, RJ; Kubicek, S; Schreiber, SL; Karlseder, J. Reduced histone biosynthesis and chromatin changes arising from a damage signal at telomeres. *Nature structural & molecular biology*, 17, 1218-1225, doi:10.1038/nsmb.1897 (2010).

[100] Vaquero, A; et al. Human SirT1 interacts with histone H1 and promotes formation of facultative heterochromatin. *Molecular cell*, 16, 93-105, doi:10.1016/j.molcel. 2004.08.031 (2004).

[101] Pruitt, K; et al. Inhibition of SIRT1 reactivates silenced cancer genes without loss of promoter DNA hypermethylation. *PLoS genetics*, 2, e40, doi:10.1371/journal.pgen. 0020040 (2006).

[102] Michishita, E; et al. SIRT6 is a histone H3 lysine, 9 deacetylase that modulates telomeric chromatin. *Nature*, 452, 492-496, doi:10.1038/nature06736 (2008).

[103] Mostoslavsky, R; et al. Genomic instability and aging-like phenotype in the absence of mammalian SIRT6. *Cell*, 124, 315-329, doi:10.1016/j.cell.2005.11.044 (2006).

[104] Kanfi, Y; et al. The sirtuin SIRT6 regulates lifespan in male mice. *Nature*, 483, 218-221, doi:10.1038/nature10815 (2012).

[105] Liu, Y; et al. Inhibition of SIRT6 in prostate cancer reduces cell viability and increases sensitivity to chemotherapeutics. *Protein & cell*, doi:10.1007/s13238-013-3054-5 (2013).

[106] Shumaker, DK; et al. Mutant nuclear lamin A leads to progressive alterations of epigenetic control in premature aging. *Proceedings of the National Academy of Sciences of the United States of America*, 103, 8703-8708, doi:10.1073/pnas.0602569103 (2006).

[107] Sarg, B; Koutzamani, E; Helliger, W; Rundquist, I; Lindner, HH. Postsynthetic trimethylation of histone H4 at lysine, 20 in mammalian tissues is associated with aging. *The Journal of biological chemistry*, 277, 39195-39201, doi:10.1074/jbc.M205166200 (2002).

[108] Chicas, A; et al. H3K4 demethylation by Jarid1a and Jarid1b contributes to retinoblastoma-mediated gene silencing during cellular senescence. *Proceedings of the National Academy of Sciences of the United States of America*, 109, 8971-8976, doi:10.1073/pnas.1119836109 (2012).

[109] Chandra, T; et al. Independence of repressive histone marks and chromatin compaction during senescent heterochromatic layer formation. *Molecular cell*, 47, 203-214, doi:10.1016/ j.molcel.2012.06.010 (2012).

[110] Sadaie, M; et al. Redistribution of the Lamin B1 genomic binding profile affects rearrangement of heterochromatic domains and SAHF formation during senescence. *Genes & development*, 27, 1800-1808, doi:10.1101/gad.217281.113 (2013).

[111] Shah, PP; et al. Lamin B1 depletion in senescent cells triggers large-scale changes in gene expression and the chromatin landscape. *Genes & development*, 27, 1787-1799, doi:10.1101/gad.223834.113 (2013).

[112] Bracken, AP; et al. The Polycomb group proteins bind throughout the INK4A-ARF locus and are disassociated in senescent cells. *Genes & development*, 21, 525-530, doi:10.1101/gad.415507 (2007).

[113] Greer, EL; et al. Members of the H3K4 trimethylation complex regulate lifespan in a germline-dependent manner in C. elegans. *Nature*, 466, 383-387, doi:10.1038/nature09195 (2010).

In: Gene-Environment Interactions and Human Diseases ISBN: 978-1-63483-680-7
Editor: Lu Qi

Chapter 13

GENE-ENVIRONMENT INTERACTION IN ORIGINS OF CHRONIC NON-COMMUNICABLE DISEASE

Roya Kelishadi[1,*] and Parinaz Poursafa[2]

[1]Faculty of Medicine and Child Growth and Development Research Center, Research Institute for Primordial Prevention of Non-communicable Disease, Isfahan University of Medical Sciences, Isfahan, Iran

[2]Environmental Health Engineering, Environment Research Center, Research Institute for Primordial Prevention of Non-communicable Disease, Isfahan University of Medical Sciences, Isfahan, Iran

ABSTRACT

This chapter is a broad review on epigenetic factors related to the developmental origins of health and disease from early life to adulthood. The additive effects of changes in the environment in combination with genetic susceptibility may contribute to many chronic diseases in adulthood. This chapter presents a summary of the impacts of various factors such as epigenetics and gene-environment interaction in predisposition to non-communicable disease. In addition to genetic predisposition, maternal or paternal exposure to environmental pollutants, and in turn their role in fetal growth and fetal programming of chronic disease, intrauterine growth, and post-natal trajectory of growth are presented. The complex contributions of genes, the environment, and gene-gene and gene-environment interactions to development of adult disease are discussed.

1. INTRODUCTION

Outcomes of epigenetic alterations might have a long latency period from months to decades, i.e., each person with one genome would have multiple epigenomes. Epigenetic

* Corresponding author: kelishadi@med.mui.ac.ir.

modifications might be dependent to the type of cell and tissue, as well as to gender and time. The sensitivity of the epigenetic system to environmental factors happens predominantly in the period of developmental plasticity, which is the time for critical modifications [1]. Epigenetic variation induced in early life might be the mediator of early-life programming of adult-onset diseases. The Developmental Origins of Health and Disease (DOHaD) hypothesis explains how environmental factors might impact later health and disease [2]. According to the DOHaD theory, intrauterine or postnatal adaptations to the environment might cause various morphologic, physiologic or metabolic alterations affecting health later in life. The DOHaD theory proposes an association of inappropriate intrauterine development with risk of chronic disorders as hypertension, obesity, diabetes, and atherosclerotic cardiovascular diseases in adult life. Low birth weight is considered as an indirect clinical marker of such inappropriate intrauterine growth (Figure 1).

Studying gene-environment interaction might increase our understanding of how environmental factors can induce developmental disease, and in turn it might help finding potential therapeutic and preventive measures [3]. Non-communicable diseases are the leading cause of mortality, with enormous economic burden notably in the low-and middle-income countries [4]. Primordial and primary prevention of such chronic disease might have many long-term impacts. An unfavorable intrauterine environment and intrauterine growth retardation might contribute to the early development of adult chronic diseases [5].

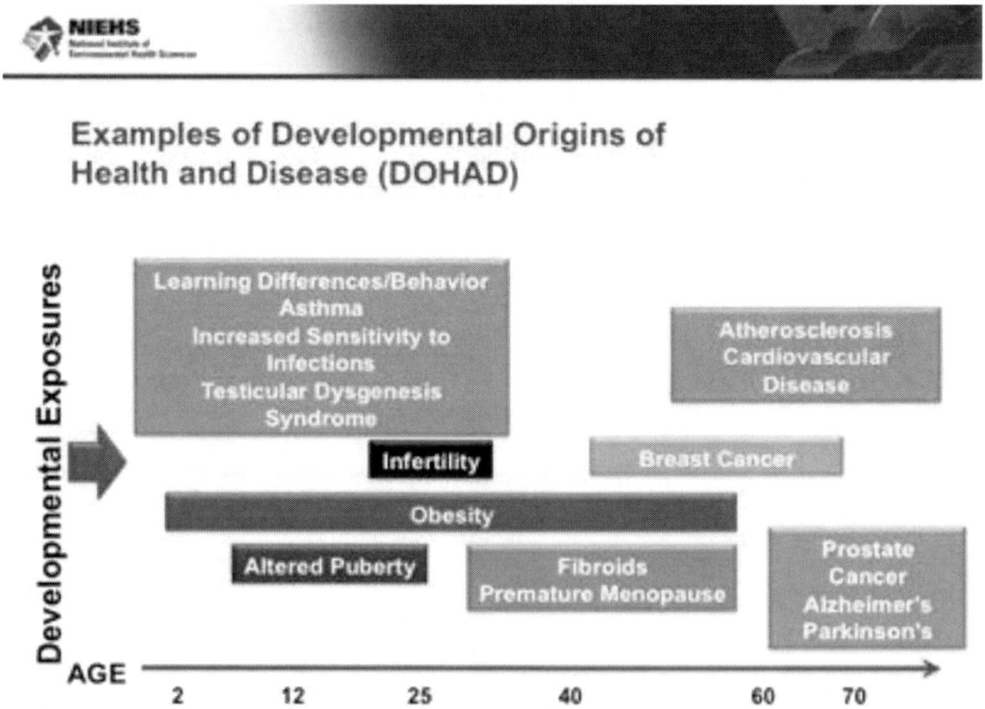

www.niehs.nih.gov/news/newsletter/2012/6/science-origins/

Figure 1. Examples of developmental origins of health and disease. http://

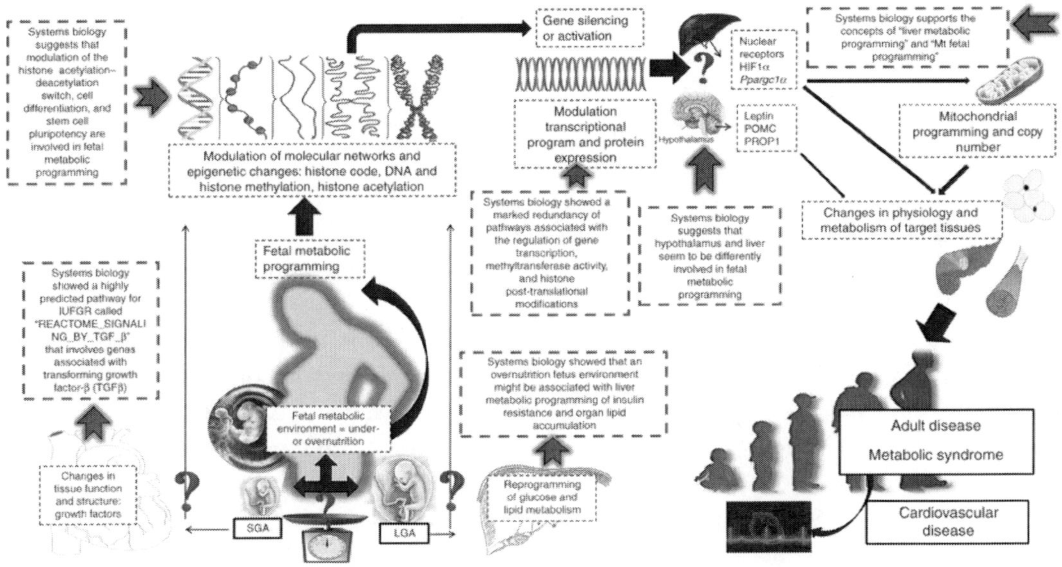

Sookoian S, Gianotti TF, Burgueño AL, Pirola CJ. Fetal metabolic programming and epigenetic modifications: a systems biology approach. Pediatr Res. 2013, 73(4 Pt),531-42.

Figure 2. The complex metabolic networks that modulate fetal metabolic programming.

It is well documented that low birth weight might increase the risk of non-communicable disease and related risk factors in later life. It should be acknowledged that rapid postnatal catch-up growth of low birth weight neonates is another important factor in this regard. Growth velocity in early postnatal life, particularly rapid "catch-up" growth during the first few months of life, as well as weight gain in childhood are associated with endothelial dysfunction and its late-onset sequels [6-8].

2. Early Life Origins of Adult Chronic Diseases

Early life risk factors track from childhood to adulthood, and have a considerable role in the etiology of several disorders including obesity, type 2 diabetes, atherosclerosis, and subsequent cardiovascular disease risk later in life (Figure 2).

In recent decades, a growing body of evidence highlighted the role of epigenetic factors in the process of chronic disease and proposed that early life risk factors may affect the development of chronic diseases in adulthood [2, 5, 9-12].

3. Thrifty Phenotype Hypothesis and Hefty Fetal Phenotype Hypothesis

The "thrifty phenotype hypothesis" proposes that by exposure to a detrimental fetal environment, the fetus would adapt its survival by preferring growth of essential organs at the expense of other organs. These organs have "reserve" capacity and are considered less important for survival under the inappropriate fetal environmental conditions (Figure 3). Such adaptation may cause various physiological and metabolic changes with adverse health consequences later in life. This phenomenon is termed as developmental plasticity or programming because the genetic program adapts to existing environmental conditions that would lead to different phenotypes. Exposure to some environmental toxins might have an impact in this regard [13, 14].

Likewise, high birth-weights also appear to have long-term health consequences [15, 16].

The "hefty fetal phenotype" is another expression proposed instead of the thrifty genotype. The "hefty fetal type" hypothesis suggests the association between high birth-weight, but not low birth-weight, and future risk of chronic diseases [17]. This is another confirmatory evidence on the role of epigenetic factors on the development of adult-onset disease from early life. It should be acknowledged that another hypothesis termed "rapid catch-up growth hypothesis" proposes that when low birth-weight is followed by rapid catch-up growth in the first years of life, it would increase the risk of chronic diseases in adulthood [7].

An inventory of candidate genes is known to be responsible for predisposition to DOHaD [18]. Several genome-wide association studies and meta-analyses have identified common genetic variants contributing to chronic diseases and their risk factors [19-22].

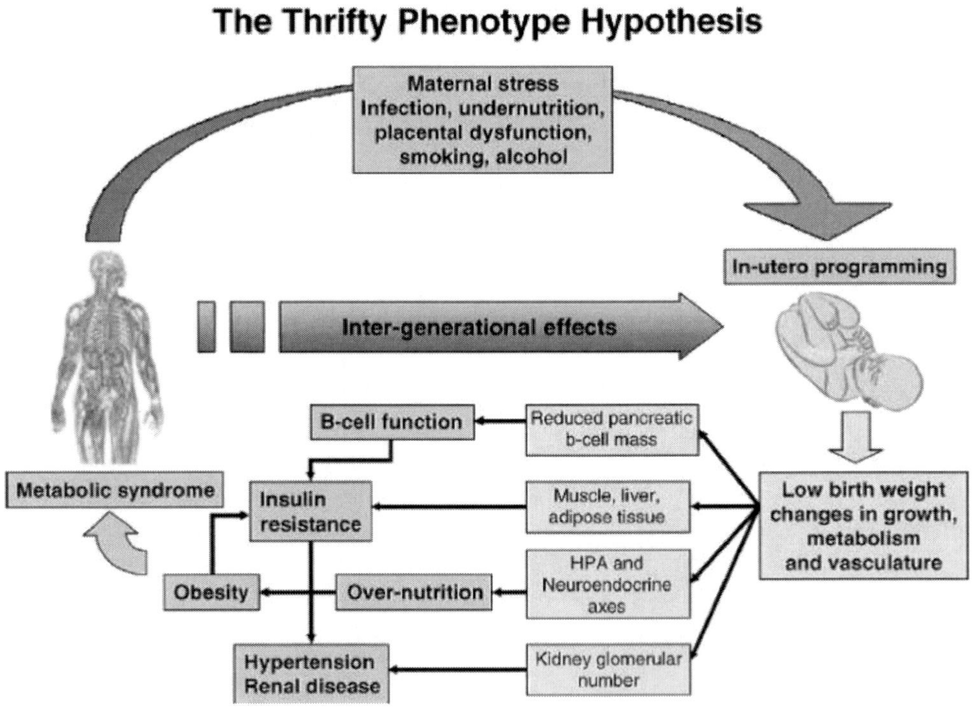

http://www.frieslandcampinainstitute.com/en/maternal-nutrition/fetal-programming-and-the-risk-of-cardiometabolic-disorders.aspx

Figure 3. Schematic representation of the Thrifty Phenotype Hypothesis.

4. Transgenerational Influence of Perinatal Factors on Programming of Chronic Diseases

Physiological and developmental steps are affected by the interaction of genetic control and environmental influence. It is suggested that changes in gene expression might have transgenerational effects [23]. Transgenerational effects may be because of mutations as well as because of factors regulating gene expression [24].

Epigenetic factors have a crucial role in this regard. For instance perinatal exposure to bisphenol A is considered as a programming factor for excess weight and its metabolic disorders later in life [25]. In addition to maternal factors, epigenetic susceptibility to environmental factors might occur during the period of sperm development. It is suggested that the mechanisms for a non-genetic transfer of paternal environmental information are based on changes in DNA methylation, histone modification, and non-coding RNAs from maturing germ cell to zygote. This would result in transgenerational epigenetic plasticity and adverse health effects in future generations [26]. Epigenetic deregulation of gene expression might be a mediator for the development of chronic disease. Epigenetic factors show individual variability at birth and are thought to be precursors of early-life programming of adult diseases. Findings on the deregulation of the early-life epigenome or the sum total of genomic epigenetic marks as DNA methylation remain to be determined [5, 27, 28].

Prenatal exposures to environmental pollutants including exposure to air pollution as well as second-hand and third-hand tobacco smoke may have long-term effects on the growth pattern of children and development of non-communicable disease in adulthood [29-33]. Such environmental factors might have direct epigenetic effects on the development of adult disease or may act indirectly by influencing fetal growth or gestational age, which in turn are associated with predisposition to adult chronic disease [34, 35].

5. Gene-Environment Interaction and Non-Communicable Diseases

Chronic non-communicable diseases are complex and have a multifactorial etiology. The effect of the interaction between genetic and environmental factors is larger than the contribution of individual risk factor [36-41]. As presented in Figure 4, diverse environmental factors have interactions with the human genetic system.

In addition to direct effects of epigenetic factors on the development of non-communicable diseases, the impact of these factors might be mediated through their effect on preterm labor and or on intrauterine growth retardation and in turn low birth weight, which are well-known risk factors for chronic diseases. In addition to prenatal period, epigenetic modifications in the metabolic pathways might be related to early childhood. Poor weight gain in the first two years of life results in subsequent stunting in childhood and impaired metabolic pathways in adulthood. Stunting might be followed the relatively early onset of prompt weight gain, i.e., “early adiposity rebound.” Potential explanations consist of genetic characteristics or, alternatively, epigenetic changes conferring an increased tendency toward rapid weight gain [12].

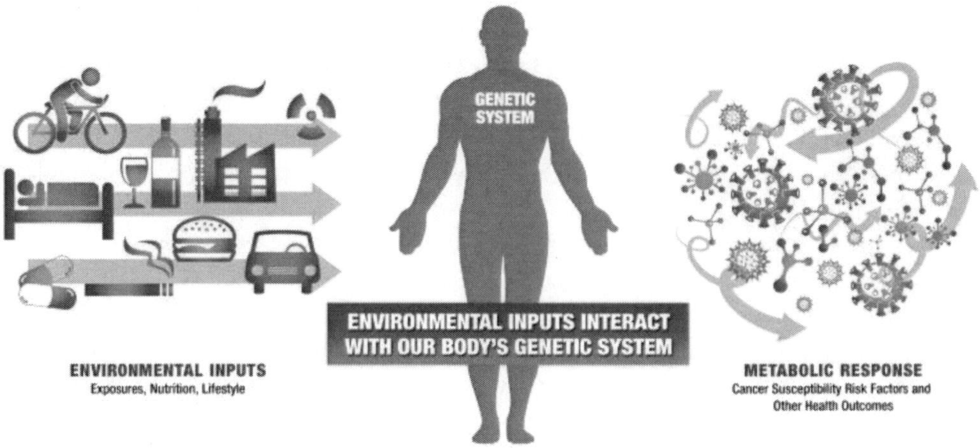

http://blog-epi.grants.cancer.gov/2013/04/15/combined-environmental-exposures/

Figure 4. Interaction of environmental inputs with human genetic system and its health consequences.

Gene-environment interaction is considered as an important factor for various non-communicable diseases; for instance exposure to air pollutants might have transgenerational metabolic consequences. Different mechanisms have been explained for such adverse cardiometabolic consequences of air-pollutants. The basic mechanisms are mainly mediated by interaction with pro-inflammatory pathways, stress oxidative, insulin resistance, vasomotor alterations, hyper-coagulability, and changes in autonomic tone [42-44].

It is of special concern that some of these changes are even found in children and adolescents. The role of gene-environment interaction on the development of some non-communicable diseases and their risk factors are presented in the following sections.

6. Gene-Environment Interaction and Diabetes Mellitus

Epigenetic rearrangements might have essential roles in insulin resistance, metabolic syndrome, pre-diabetes, and diabetes. The interaction between genetic, intra-uterine and environmental factors might lead to related metabolic disorders [45, 46].

Developmental exposures to some environmental chemicals, mainly endocrine disrupting chemicals (EDCs), would result in functional changes in gene expression and increased risk of disease, without any phenotype change detectable at birth. EDCs are found in several products including plastic dishes, bottles, metal cans, cosmetics, pesticides, and manufactured foods. They interfere with the synthesis, action and removal of natural hormones , they would act on endogenous receptors, and would affect usual control processes. A growing body of evidence suggests that the increased exposure to environmental EDCs including bisphenol A, phthalates and persistent organic pollutants, may have a crucial role in the escalating trend of diabetes at global level [47-49].

Moreover, it is documented that exposure to ambient fine particulate matter might exaggerate both insulin resistance and visceral inflammation/adiposity [50]. Particulate air pollution may also induce oxidative stress and changes of mitochondria and gene expression in brown and white adipose tissues [51]. Then, the inflammatory process in visceral adipose tissue and liver can recruit innate immune cells and might result in dysregulation of glucose and lipoprotein metabolism. It is also suggested that the effects of particulate air pollution on metabolic disorders might be mediated by triggered cell stress response in adipose tissue [52].

The other mechanism can be mediated by the impact of epigenetic factors on intrauterine growth. Low birth weight is associated with higher risk of diabetes at older ages [53, 54]. It is also documented that in those children who were born with low birth weight, serum insulin is related to rapid catch-up growth [55-57]. A systematic review reported that both low birth weight and catch-up-growth were associated with some aspects of metabolic syndrome in later life [58].

7. Gene-Environment Interaction and Atherosclerotic Cardiovascular Diseases

Interactions between genotype and exposures to environmental factors would influence cardiovascular phenotypes from conception to adulthood. Several genes are involved in the beginning, progress, and clinical symptom of atherosclerotic cardiovascular diseases, but still it remains to be determined how a person's environment would interact with these genes [2].

Several studies have confirmed the Barker hypothesis, proposed in 1989, on the association of low birth weight with future cardiovascular disease morbidity and mortality [59]. After a long latency period, adverse intrauterine environment and impaired fetal growth would contribute to the development of atherosclerotic cardiovascular diseases. Atherosclerosis begins from fetal period, and it is documented that even by preschool age, those children born as preterm and or low birth weight might have thicker carotid intima-media thickness and increased arterial stiffness [5, 60, 61].

Epigenetic factors increase the risk of low birth weight, which particularly when accompanied by further rapid catch-up growth, is a risk factor for cardiovascular disease in later life. Adaptation to prenatal and postnatal environments would determine the arrangements of interaction between different factors controlling cellular and organic functions to support the survival of prenatal and postnatal life, but posing adverse atherosclerotic cardiovascular outcomes later in life [62].

8. Gene-Environment Interaction and Obesity

Obesity has become a worldwide epidemic. Changes in lifestyle habits cannot explain the rapid escalating trend of obesity at global level. Excess weight develops from perinatal period, in addition to the abovementioned role of intrauterine growth retardation and subsequent low birth weight, maternal obesity and or excessive weight gain during pregnancy increase the risk of obesity in their offspring (Figure 5).

It is postulated that the rise in the incidence of obesity coincides with the rise in the use of industrial chemicals with possible role in the development of obesity. Substances as EDCs might be linked to the obesity epidemic [63].

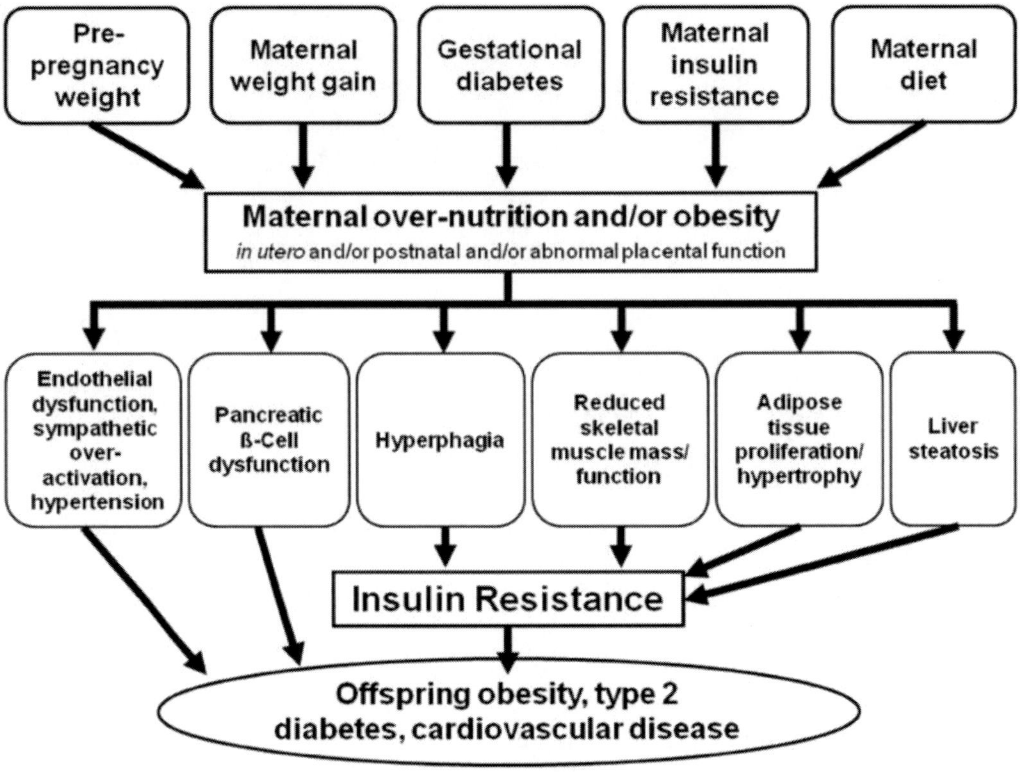

O'Dowd JF, Stocker CJ. Endocrine pancreatic development: impact of obesity and diet. Front Physiol. 2013 Jul 18, 4,170.

Figure 5. Association of maternal obesity or excess weight gain during pregnancy with obesity in offspring.

Exposure to various doses of environmental obesogen chemicals interacts with some endocrine signaling mechanisms and in turn would lead to excess weight.The epigenetic impact of environmental obesogens has several disruptive effects on fat homeostasis and weight controlling mechanisms, as well as in programming the development of obesity from early life [64].

The environmental obesogen hypothesis suggests that exposure to certain materials as EDCs during the prenatal period or early-life may predispose exposed individuals to increased fat mass and overweight. It is proposed that exposure to obesogens can modify the epigenome of stromal stem cells, promoting them to the adipocyte lineage instead of bone. Therefore, exposure to obesogens during early life might have a role on fetal programming for an adipogenic outcome [65].

The epigenetic effects of early life exposure to obesogen chemicals may be irreversible and long lasting lifelong. It is documented that exposure to low doses of EDCs during critical times of differentiation might influence the developmental programming and in turn it might result in obesity [66].

Conclusion

Epigenetics or the study of gene regulatory information without expression in DNA sequences might have transgenerational effects. Complex environmental and genetic factors, and gene-gene, and gene-environment interactions are responsible for the development of adult disease from early life. Maternal exposures to environmental contaminants, paternal factors, fetal growth, and postnatal growth trajectory have impact in this process. Understanding these factors and their interactions would provide new opportunities for investigating human health and disease and would help to reach major improvements in the diagnosis, prevention, and treatment of non-communicable disease in adulthood.

Key Facts

1. Early life risk factors track from childhood to adulthood, and have a considerable role in the etiology of chronic disorders;
2. Epigenetic variation induced in early life might be the mediator of early-life programming of adult-onset diseases;
3. Studying gene-environment interaction might increase our understanding of how environmental factors can induce developmental disease;
4. Epigenetic effects may be transgenerational;
5. Study of epigenetic and gene-environment interaction would help to reach major improvements in the diagnosis, prevention, and treatment of non-communicable disease in adulthood

References

[1] Skinner, M. K., Manikkam, M. & Guerrero-Bosagna, C. (2011). Epigenetic transgenerational actions of endocrine disruptors. *Reproductive toxicology (Elmsford, N.Y.) 31*, 337-343, doi:10.1016/ j.reprotox.2010.10.012.

[2] McCormack, S., Xia, Q. & Grant, S. F. (2012). Developmental origins of genotype-phenotype correlations in chronic diseases of old age. *Aging and disease*, *3*, 385-403.

[3] Wright, R. O. & Christiani, D. (2010). Gene-environment interaction and children's health and development. *Current opinion in pediatrics*, *22*, 197-201, doi:10.1097/ MOP.0b013e328336ebf9.

[4] Di Cesare, M., et al. (2013). Inequalities in non-communicable diseases and effective responses. *Lancet*, *381*, 585-597, doi:10.1016/ s0140-6736(12)61851-0.

[5] Sun, C., et al. (2013). Effects of early-life environment and epigenetics on cardiovascular disease risk in children: highlighting the role of twin studies. *Pediatric research*, *73*, 523-530, doi:10.1038/pr.2013.6.

[6] Saffery, R. & Novakovic, B. (2014). Epigenetics as the mediator of fetal programming of adult onset disease: what is the evidence? *Acta obstetricia et gynecologica Scandinavica*, *93*, 1090-1098, doi:10.1111/ aogs.12431.

[7] Kelishadi, R., Haghdoost, A. A., Jamshidi, F., Aliramezany, M. & Moosazadeh, M. (2015). Low birthweight or rapid catch-up growth: which is more associated with cardiovascular disease and its risk factors in later life? A systematic review and cryptanalysis. *Paediatrics and international child health*, *35*, 110-123, doi:10.1179/2046905514y.0000000136.

[8] Li, C. C., Maloney, C. A., Cropley, J. E. & Suter, C. M. (2010). Epigenetic programming by maternal nutrition: shaping future generations. *Epigenomics*, *2*, 539-549, doi:10.2217/epi.10.33.

[9] Lehnen, H., Zechner, U. & Haaf, T. (2013). Epigenetics of gestational diabetes mellitus and offspring health: the time for action is in early stages of life. *Molecular human reproduction*, *19*, 415-422, doi:10.1093/ molehr/gat020.

[10] DeBoer, M. D., et al. (2012). Early childhood growth failure and the developmental origins of adult disease: do enteric infections and malnutrition increase risk for the metabolic syndrome? *Nutrition reviews*, *70*, 642-653, doi:10.1111/j.1753-4887.2012.00543.x.

[11] Santos, M. S. & Joles, J. A. (2012). Early determinants of cardiovascular disease. *Best practice & research. Clinical endocrinology & metabolism*, *26*, 581-597, doi:10.1016/j.beem.2012.03.003.

[12] Kent, A. L. (2012). Developmental origins of health and adult disease: what should neonatologists/paediatricians be considering about the long-term health of their patients? *Journal of paediatrics and child health*, *48*, 730-734, doi:10.1111/j.1440-1754.2012.02541.x.

[13] Harville, E. W., Srinivasan, S., Chen, W. & Berenson, G. S. (2012). Is the metabolic syndrome a "small baby" syndrome?: the bogalusa heart study. *Metabolic syndrome and related disorders*, *10*, 413-421, doi:10.1089/met.2012.0031.

[14] Hayashi, Y., Ito, Y. & Nakajima, T. (2014). Effects of exposure to Di(2-ethylhexyl)phthalate during fetal period on next generation. *Nihon eiseigaku zasshi. Japanese journal of hygiene*, *69*, 86-91.

[15] Poston, L. (2012). Gestational weight gain: influences on the long-term health of the child. *Current opinion in clinical nutrition and metabolic care*, *15*, 252-257, doi:10.1097/MCO.0b013e3283527cf2.

[16] Biosca, M., et al. (2011). Central adiposity in children born small and large for gestational age. *Nutricion hospitalaria*, *26*, 971-976, doi:10.1590/s0212-16112011000500008.

[17] Dyck, R. F., Klomp, H. & Tan, L. (2001). From "thrifty genotype" to "hefty fetal phenotype": the relationship between high birthweight and diabetes in Saskatchewan Registered Indians. *Canadian journal of public health = Revue canadienne de sante publique*, *92*, 340-344.

[18] Ogawa, T. et al. (2014). Seeking genes responsible for developmental origins of health and disease from the fetal mouse liver following maternal food restriction. *Congenital anomalies*, *54*, 195-219, doi:10.1111/cga.12062.

[19] Kaskow, B. J., et al. (2014). Molecular prioritization strategies to identify functional genetic variants in the cardiovascular disease-associated expression QTL Vanin-1. *European journal of human genetics : EJHG*, *22*, 688-695, doi:10.1038/ejhg.2013.208.

[20] Kostrzewa, G., Broda, G., Konarzewska, M., Krajewki, P. & Ploski, R. (2013). Genetic polymorphism of human Y chromosome and risk factors for cardiovascular diseases: a study in WOBASZ cohort. *PloS one*, *8*, e68155, doi:10.1371/journal.pone.0068155.

[21] Muir, A. R. & Menown, I. B. (2013). Genetic biomarkers in cardiovascular disease. *Biomarkers in medicine*, *7*, 497-499, doi:10.2217/bmm.13.82.

[22] Levy, D., et al. (2000). Evidence for a gene influencing blood pressure on chromosome 17. Genome scan linkage results for longitudinal blood pressure phenotypes in subjects from the framingham heart study. *Hypertension*, *36*, 477-483.

[23] Hall, J. G. (2014). Epigenetics: What does it mean for paediatric practice? *Paediatrics & child health*, *19*, 27-30.

[24] Diamanti-Kandarakis, E., et al. (2009). Endocrine-disrupting chemicals: an Endocrine Society scientific statement. *Endocrine reviews*, *30*, 293-342, doi:10.1210/er.2009-0002.

[25] van Esterik, J. C., et al. (2014). Programming of metabolic effects in C57BL/6JxFVB mice by exposure to bisphenol A during gestation and lactation. *Toxicology*, *321*, 40-52, doi:10.1016/j.tox.2014.04.001.

[26] Soubry, A., Hoyo, C., Jirtle, R. L. & Murphy, S. K. (2014). A paternal environmental legacy: evidence for epigenetic inheritance through the male germ line. *BioEssays : news and reviews in molecular, cellular and developmental biology*, *36*, 359-371, doi:10.1002/bies.201300113.

[27] Chang, C. P. & Bruneau, B. G. (2012). Epigenetics and cardiovascular development. *Annual review of physiology*, *74*, 41-68, doi:10.1146/ annurev-physiol-020911-153242.

[28] Roth, C. L. & Sathyanarayana, S. (2012). Mechanisms affecting neuroendocrine and epigenetic regulation of body weight and onset of puberty: potential implications in the child born small for gestational age (SGA). *Reviews in endocrine & metabolic disorders*, *13*, 129-140, doi:10.1007/s11154-012-9212-x.

[29] Backes, C. H., Nelin, T., Gorr, M. W. & Wold, L. E. (2013). Early life exposure to air pollution: how bad is it? *Toxicology letters*, *216*, 47-53, doi:10.1016/j.toxlet. 2012.11.007.

[30] Kelishadi, R. & Poursafa, P. (2010). Air pollution and non-respiratory health hazards for children. *Archives of medical science: AMS*, *6*, 483-495, doi:10.5114/aoms. 2010.14458.

[31] Poursafa, P., Kelishadi, R., Haghjooy-Javanmard, S., Rafiei, L. & Keramatian, K. (2012). Synergistic effects of genetic polymorphism and air pollution on markers of endothelial dysfunction in children. *Journal of research in medical sciences : the official journal of Isfahan University of Medical Sciences*, *17*, 718-723.

[32] Hashemi, M., Afshani, M. R., Mansourian, M., Poursafa, P. & Kelishadi, R. (2012). Association of particulate air pollution and secondhand smoke on endothelium-dependent brachial artery dilation in healthy children. *Journal of research in medical sciences : the official journal of Isfahan University of Medical Sciences*, *17*, 317-321.

[33] Kelishadi, R., Poursafa, P. & Keramatian, K. (2011). Overweight, air and noise pollution: Universal risk factors for pediatric pre-hypertension. *Journal of research in medical sciences : the official journal of Isfahan University of Medical Sciences*, *16*, 1234-1250.

[34] Sinclair, K. D., Lea, R. G., Rees, W. D. & Young, L. E. (2007). The developmental origins of health and disease: current theories and epigenetic mechanisms. *Society of Reproduction and Fertility supplement*, *64*, 425-443.

[35] Kerkhof, G. F., Breukhoven, P. E., Leunissen, R. W., Willemsen, R. H. & Hokken-Koelega, A. C. (2012). Does preterm birth influence cardiovascular risk in early adulthood? *The Journal of pediatrics*, *161*, 390-396.e391, doi:10.1016/j.jpeds.2012.03.048.

[36] Panasevich, S., et al. (2013). Interaction between air pollution exposure and genes in relation to levels of inflammatory markers and risk of myocardial infarction. *BMJ open*, *3*, e003058, doi:10.1136/ bmjopen-2013-003058.

[37] Andersen, H. R., et al. (2012). Paraoxonase 1 polymorphism and prenatal pesticide exposure associated with adverse cardiovascular risk profiles at school age. *PloS one*, *7*, e36830, doi:10.1371/ journal.pone.0036830.

[38] Kerley-Hamilton, J. S., et al. (2012). Inherent and benzo[a]pyrene-induced differential aryl hydrocarbon receptor signaling greatly affects life span, atherosclerosis, cardiac gene expression, and body and heart growth in mice. *Toxicological sciences : an official journal of the Society of Toxicology*, *126*, 391-404, doi:10.1093/toxsci/kfs002.

[39] Hackett, T. L., et al. (2011). Effect of gene environment interactions on lung function and cardiovascular disease in COPD. *International journal of chronic obstructive pulmonary disease*, *6*, 277-287, doi:10.2147/ copd.s18279.

[40] Flowers, E., Froelicher, E. S. & Aouizerat, B. E. (2012). Gene-environment interactions in cardiovascular disease. *European journal of cardiovascular nursing : journal of the Working Group on Cardiovascular Nursing of the European Society of Cardiology*, *11*, 472-478, doi:10.1016/j.ejcnurse.2011.06.001.

[41] Ordovas, J. M., Robertson, R. & Cleirigh, E. N. (2011). Gene-gene and gene-environment interactions defining lipid-related traits. *Current opinion in lipidology*, *22*, 129-136, doi:10.1097/ MOL.0b013e32834477a9.

[42] Brook, R. D. & Rajagopalan, S. (2010). Particulate matter air pollution and atherosclerosis. *Current atherosclerosis reports*, *12*, 291-300, doi:10.1007/s11883-010-0122-7.

[43] Rajagopalan, S. & Brook, R. D. (2012). Air pollution and type 2 diabetes: mechanistic insights. *Diabetes*, *61*, 3037-3045, doi:10.2337/ db12-0190.

[44] Sun, Z., et al. (2013). Air-Pollution and Cardiometabolic Diseases (AIRCMD): a prospective study investigating the impact of air pollution exposure and propensity for type II diabetes. *The Science of the total environment*, *448*, 72-78, doi:10.1016/j.scitotenv.2012.10.087.

[45] Brook, R. D., Jerrett, M., Brook, J. R., Bard, R. L. & Finkelstein, M. M. (2008). The relationship between diabetes mellitus and traffic-related air pollution. *Journal of occupational and environmental medicine / American College of Occupational and Environmental Medicine*, *50*, 32-38, doi:10.1097/JOM.0b013e31815dba70.

[46] Pearson, J. F., Bachireddy, C., Shyamprasad, S., Goldfine, A. B. & Brownstein, J. S. (2010). Association between fine particulate matter and diabetes prevalence in the U.S. *Diabetes care*, *33*, 2196-2201, doi:10.2337/dc10-0698.

[47] Chevalier, N. & Fenichel, P. (2015). Endocrine disruptors: New players in the pathophysiology of type 2 diabetes? *Diabetes & metabolism*, *41*, 107-115, doi:10.1016/j.diabet.2014.09.005.

[48] Portha, B., Fournier, A., Kioon, M. D., Mezger, V. & Movassat, J. (2014). Early environmental factors, alteration of epigenetic marks and metabolic disease susceptibility. *Biochimie*, *97*, 1-15, doi:10.1016/ j.biochi.2013.10.003.
[49] Casals-Casas, C. & Desvergne, B. (2011). Endocrine disruptors: from endocrine to metabolic disruption. *Annual review of physiology*, *73*, 135-162, doi:10.1146/annurev-physiol-012110-142200.
[50] Xu, Z., et al. (2011). Ambient particulate air pollution induces oxidative stress and alterations of mitochondria and gene expression in brown and white adipose tissues. *Particle and fibre toxicology*, *8*, 20, doi:10.1186/1743-8977-8-20.
[51] Xu, X., et al. (2011). Long-term exposure to ambient fine particulate pollution induces insulin resistance and mitochondrial alteration in adipose tissue. *Toxicological sciences: an official journal of the Society of Toxicology*, *124*, 88-98, doi:10.1093/ toxsci/kfr211.
[52] Mendez, R., et al. (2013). Exposure to fine airborne particulate matter induces macrophage infiltration, unfolded protein response, and lipid deposition in white adipose tissue. *American journal of translational research*, *5*, 224-234.
[53] Euser, A. M., Dekker, F. W. & Hallan, S. I. (2010). Intrauterine growth restriction: no unifying risk factor for the metabolic syndrome in young adults. *European journal of cardiovascular prevention and rehabilitation : official journal of the European Society of Cardiology, Working Groups on Epidemiology & Prevention and Cardiac Rehabilitation and Exercise Physiology*, *17*, 314-320, doi:10.1097/ HJR.0b013e328331d2c3.
[54] Tian, J. Y., et al. (2006). Birth weight and risk of type 2 diabetes, abdominal obesity and hypertension among Chinese adults. *European journal of endocrinology / European Federation of Endocrine Societies*, *155*, 601-607, doi:10.1530/eje.1.02265.
[55] Jaquet, D., et al. (2005). Dynamic change in adiposity from fetal to postnatal life is involved in the metabolic syndrome associated with reduced fetal growth. *Diabetologia*, *48*, 849-855, doi:10.1007/s00125-005-1724-4.
[56] Mericq, V., et al. (2005). Longitudinal changes in insulin sensitivity and secretion from birth to age three years in small- and appropriate-for-gestational-age children. *Diabetologia*, *48*, 2609-2614, doi:10.1007/ s00125-005-0036-z.
[57] Reinehr, T., Kleber, M. & Toschke, A. M. (2009). Small for gestational age status is associated with metabolic syndrome in overweight children. *European journal of endocrinology / European Federation of Endocrine Societies*, *160*, 579-584, doi:10.1530/eje-08-0914.
[58] Nobili, V., Alisi, A., Panera, N. & Agostoni, C. (2008). Low birth weight and catch-up-growth associated with metabolic syndrome: a ten year systematic review. *Pediatric endocrinology reviews : PER*, *6*, 241-247.
[59] Barker, D. J. (1995). Fetal origins of coronary heart disease. *BMJ (Clinical research ed.)*, *311*, 171-174.
[60] Nakashima, Y., Fujii, H., Sumiyoshi, S., Wight, T. N. & Sueishi, K. (2007). Early human atherosclerosis: accumulation of lipid and proteoglycans in intimal thickenings followed by macrophage infiltration. *Arteriosclerosis, thrombosis, and vascular biology*, *27*, 1159-1165, doi:10.1161/atvbaha.106.134080.
[61] Skilton, M. R., et al. (2011). Fetal growth and preterm birth influence cardiovascular risk factors and arterial health in young adults: the Cardiovascular Risk in Young Finns

Study. *Arteriosclerosis, thrombosis, and vascular biology*, *31*, 2975-2981, doi:10.1161/atvbaha.111.234757.

[62] Skilton, M. R., et al. (2013). Fetal growth, omega-3 (n-3) fatty acids, and progression of subclinical atherosclerosis: preventing fetal origins of disease? The Cardiovascular Risk in Young Finns Study. *The American journal of clinical nutrition*, *97*, 58-65, doi:10.3945/ajcn.112.044198.

[63] Baillie-Hamilton, P. F. (2002). Chemical toxins: a hypothesis to explain the global obesity epidemic. *Journal of alternative and complementary medicine (New York, N.Y.)*, *8*, 185-192, doi:10.1089/ 107555302317371479.

[64] Kelishadi, R., Poursafa, P. & Jamshidi, F. (2013). Role of environmental chemicals in obesity: a systematic review on the current evidence. *Journal of environmental and public health*, 2013, 896789, doi:10.1155/2013/896789.

[65] Janesick, A. & Blumberg, B. (2011). Endocrine disrupting chemicals and the developmental programming of adipogenesis and obesity. *Birth defects research. Part C, Embryo today : reviews*, *93*, 34-50, doi:10.1002/bdrc.20197.

[66] Newbold, R. R. (2010). Impact of environmental endocrine disrupting chemicals on the development of obesity. *Hormones (Athens, Greece)*, *9*, 206-217.

In: Gene-Environment Interactions and Human Diseases
Editor: Lu Qi
ISBN: 978-1-63483-680-7

Chapter 14

GLOBAL DNA METHYLATION IN OBESITY, DIABETES AND CARDIOVASCULAR DISEASES AND THE INFLUENCE OF ENVIRONMENTAL FACTORS

***M. Garcia-Lacarte*[1], *Fermín I. Milagro*[1,2] *and Alfredo J. Martinez*[1,2]**

[1]Department of Nutrition, Food Science and Physiology and the Centre for Nutrition Research, University of Navarra, Pamplona, Spain

[2]Instituto de Salud Carlos III, CIBER Fisiología Obesidad y Nutrición (CIBERobn), Madrid, Spain

ABSTRACT

As the prevalence of obesity and related diseases such as cardiovascular disease and type 2 diabetes is increasing worldwide, different strategies and novel approaches are being investigated to manage these pathological conditions. In this context, dynamic changes in DNA as a consequence of environmental interactions have gained interest in understanding these diseases. Indeed, several studies have demonstrated that the epigenome is affected by external factors, such as diet, physical activity, stress, or exposure to chemical carcinogens. Global methylation status is commonly used as a surrogate measure of overall methylation changes. Actually, aberrations in DNA methylation are often related to disease. In this sense, global DNA methylation pattern changes, measured in *LINE-1* and *Alu* sequences, have been associated to body weight regulation and the onset of obesity, cardiovascular diseases and type 2 diabetes, and could be used as possible diagnostic and prognostic biomarkers. In relation to cardiovascular diseases, *LINE-1* is the principal sequence that trends to hyper or hypomethylation depending on the tissue analysed. In type 2 diabetes, changes in *LINE-1* methylation levels have been reported in peripheral blood mononuclear cells (PBMC). Focusing on obesity, several studies have reported changes in the methylation status of both, *LINE-1* and *Alu* sequences. For the near future, the principal challenges in nutriepigenetics are the large number of variables, intermediate markers and measurements to be considered, as well as the dynamic nature of nutrients and the limited outcome information.

Keywords: *Alu*, *LINE-1*, global methylation, CVD, obesity, insulin resistance, diabetes. epigenetic, environment

INTRODUCTION

The genome contains information related to gene structure and functions of every organism [1]. However, the epigenome is defined as the reversible hereditable alterations in gene expression that are not related with changes in DNA sequence [2]. The word epigenetics was first used by Conrad H. Waddington in the 1940s to describe events that could not be explained by conventional genetic principles [3].

Epigenetic regulation is needed for the normal function of a cell or organ. However, during disease initiation and progression, this regulation may be disturbed [4]. In the human body, all cells elicit the same genome but different epigenetic marks, supporting that genetic information is static whereas the epigenome is dynamic [5]. Furthermore, the phenotype (the observable characteristic of an organism) is determined by the epigenome [6]. Nutrients and other dietary compounds, toxins, disease states, stress, physical activity and environmental interactions, including the exposure to radiation or infectious agents, are known to affect the epigenome and alter gene expression [7-10]. Many of these studies have focused on the early developmental stage (in utero and lactation), where it is believed that the epigenetic marks are more amenable to modification due to the environmental factors. However, there is evidence that epigenetic marks can also be modified in the adult individuals, paving the way to an epigenetic-derived therapy of the disease by prescribing drugs, diet and other modifiable environmental factors.

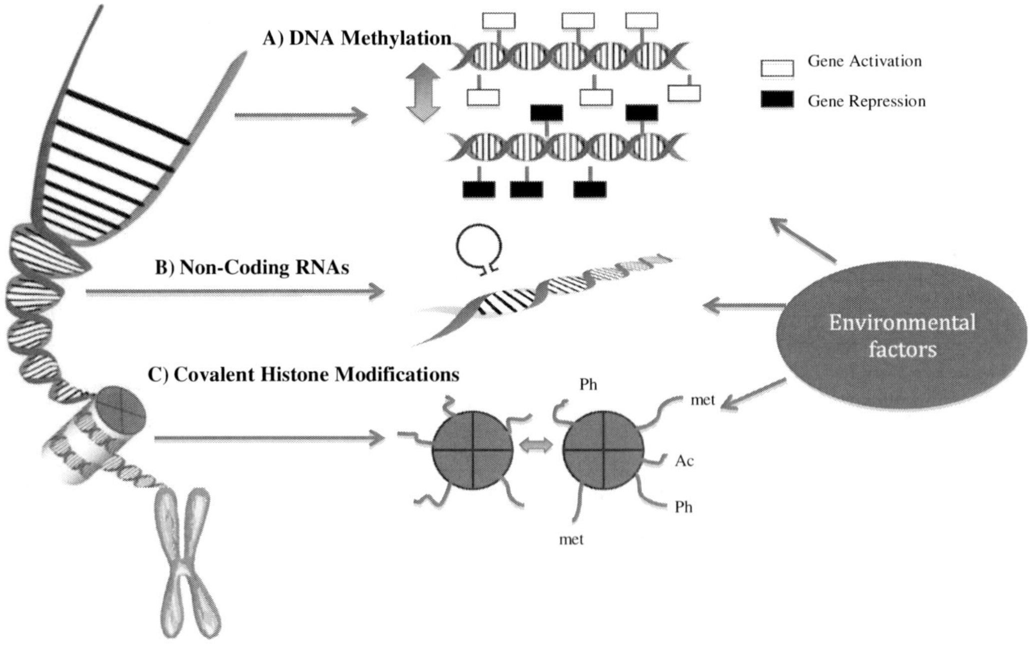

Figure 1. Main epigenetic mechanisms: A) DNA methylation, B) Non-coding RNAs, C) Covalent histone modifications.

EPIGENETIC MECHANISMS

The three main epigenetic mechanisms currently known are DNA methylation, histone covalent modification, and non-coding microRNAs [3] (Figure 1). Other existing mechanisms are genomic imprinting, non-covalent mechanisms and chromatin remodelling [10]. Indeed, every cell in the body has the same genetic information, but the way that makes cells, tissues and organs different is that distinct sets of genes are silenced or expressed [11].

DNA Methylation

DNA methylation is the best-characterized and most studied epigenetic mechanism in humans [12]. DNA methylation is essential for the development and for the survival of cells during the organism's life [13]. In mammalian cells, DNA methylation occurs at the 5' position of the cytosine residues that are followed by a guanine nucleotide (CpGs sites). CpG sites are not randomly distributed in the genome. Thus, CpG-rich regions (genomic regions with at least 200 bp, and a GC percentage that is greater than 50%), also known as CpG islands, are present in about 50% of the promoter regions of genes [14]. Hypermethylation of these CpG sites in promoters is commonly related to gene silencing, while transcriptionally active genes typically have hypomethylated promoters [15]. However, when methylation occurs at gene body, the expression is sometimes increased [16]. In summary, DNA methylation is a crucial step to control gene expression by preventing or promoting the recruitment and expression of regulatory proteins.

Changes in global (genome-wide) methylation levels are thought to be involved in the development of several diseases, particularly neoplasia and tumour growth [17, 18]. Thus, cancer-linked global genomic hypomethylation is a common characteristic in a wide variety of malignancies [19-23], and occurs not only in transcription control regions, such as promoters, but also in repetitive DNA sequences, such as heterochromatic regions and retrotransposons [19]. Hypomethylation is thought to contribute to carcinogenesis by inducing genomic instability [24]. During aging, a variety of metabolically crucial genes, such as estrogen receptors, serotonin receptor type 2A, *AQP9* and insulin-like growth factor II, become silenced due to promoter hypermethylation [25-28]. During the adipogenesis process, the expression of many genes is regulated by DNA methylation or demethylation [29]. For example, Caveolin 1 and insulin receptor are upregulated in adipocytes in comparison to preadipocytes as a result of the demethylation of their promoter regions [30]. In the adipose tissue of obese people some genes are differentially methylated when compared to lean subjects, as some authors have found [31], and may predict weight loss [32, 33].

Covalent Histone Modifications

Histone modifications influence chromatin structure, which also plays an important role in gene regulation and function. There are a variety of posttranslational modifications that can change the morphological status of histones [34], which are core proteins of the nucleosome.

These epigenetic modifications, such as phosphorylation of serine and threonine, methylation of lysine and arginine, lysine acetylation, ubiquitination, sumoylation and biotinylation, deimination, ADP ribosylation, glycosylation, and carbonylation, interact with the accessible residues of the amino-termini of H3 and H4 and the amino and carboxyl termini of H2A, H2B, and H1 histones [35], being the N-terminal histone tails the major sites for epigenetic modifications [36, 37]. These modifications may stabilize binding sites for regulatory proteins, but also may disrupt or occlude chromatin binding sites [38]. Euchromatin (decondensed and dispersed regions related to transcriptionally active genes) and heterochromatin (compact and condensed regions) are affected by covalent histone modifications [39].

Histone modifications influence chromatin structure by altering the charge distribution and DNA contacts with histone octamers [39]. For example, acetylation of histone H4K16 prevent folding of chromatin into higher-order structures [40].

Histone changes are related to disease development and histone acetylation or deacetylation are the most common histone modifications involved in disease progression. Aberrations in histones can significantly disrupt gene expression regulation, a common factor in disease, and could potentially be transmissible across generations [41].

Genome-wide analyses using Chip-Seq have shown that some histone modifications are found in diabetes and cardiovascular disease. In this context, Pirola et al., (2011) reported that hyperglycemia-mediated induction of genes and pathways was associated with acetylation of H3K9/K14 [42] in primary human vascular cells, whereas Weiguo et al., (2013) found that alterations of H3K9me 3 explained the immunological disturbance and high cardiovascular complications in uremia patients [43].

Non-Coding RNAs

MicroRNAs (miRNAs) are small non-coding RNAs of 21-25 bp length and with the ability to inhibit the translation of a gene by binding to partially complementary messenger RNAs (mRNA) [44, 45]. Mechanisms of heterochromatin silencing have been linked to miRNAs showing the importance of these transcripts in chromatin conformation [46]. miRNA knockout models have demonstrated that a single miRNA can affect mRNA and protein levels of a large number of genes, including tumor suppressors, signalling pathways associated to aging, apoptotic genes and oncogenes [47-50]. Concerning obesity, miRNAs seem to participate in crucial biological processes, such as adipocyte differentiation, metabolic integration, fat metabolism and insulin sensitivity [51, 52]. In this sense, when a high fat diet is fed to induce obesity, miR-21, is dysregulated in different tissues [53]. Furthermore, the expression of miR-21 also changes in response to the intake of polyunsaturated fatty acids [54]. There are other miRNAs whose expression is regulated by nutrients, such as the miR-200 family, which is affected by protein intake, polyunsaturated fatty acids or methyl deficiency [53], and the miR-221/222 family, which is regulated by glucose and folate [53]. Other dietary factors such as vitamins and phytoderivates, including vitamins A, D, E and B, and coenzyme Q or ubiquinone, are known to regulate microRNA expression [55].

Gene Imprinting

Another epigenetic mechanism involved in gene expression regulation is gene imprinting, which is the distinct subset of epigenetic regulation of a limited number of genes (50-80) depending on the sex of the parent that transmits each allele [4]. Without the correct imprinting control, abnormal growth occurs and some diseases may appear. Methylation of DNA occurs in the loci called imprinting control regions (ICRs). In this sense, loss of imprinting (LOI) of *IGF2* and *H19* locus has been associated with tumorigenesis. In Wilms' tumour onset, imprinted expression of *IGF2* is usually dimmed and is associated with increased methylation and reduced expression of the maternal copy of the tightly linked *H19* locus [56-58]. Parallelly, LOI shows similar patterns of these loci in patients with hepatoblastoma [59], uterine leiomyosarcomata [60], cervical carcinoma [61], renal cell carcinoma [62], rhabdomyosarcoma [63, 64], gliomas [65] and colorectal cancer [66, 67]. In relation to nutrition, several *in vitro* studies have shown that nutritional manipulation of early mouse embryos can alter allelic methylation and expression of imprinted genes [68, 69].

GLOBAL DNA METHYLATION: *LINE-1* AND *ALU* ELEMENTS

Global methylation status of easily accessible tissues (mainly blood cells) is commonly used as a surrogate measure of methylation changes at less accessible cancer target-sites [70]. At a global level, DNA methylation is usually measured by screening repeated interspersed regions, such as Long Interspersed Nucleotide Element 1 (*LINE-1*) or *Alu* sequences, a kind of Short Interspersed Nucleotide Element (*SINE*). *LINE* and *Alu* sequences represent approximately 30% of the genome [71]. The ubiquitous presence of both *LINE*/*Alu* in the genome combined with their relatively conserved sequence, make them suitable targets to measure global methylation. *LINE* elements are retrotransposons, mobile sequences that use a "copy-and-paste" mechanisms to change their location and their copy number using an RNA intermediate [72]. Retrotransposons are the most prevalent class of active mobile elements in mammalian genomes. *LINE-1* is the most spread non-long terminal repeat retrotransposon in the genome, covering 17-25% of total human genome [73,74]. *LINEs* are transcribed by RNA polymerase II and encode the protein that chaperone targets, and reverse transcribe their mRNA into a new genomic locus [75]. Bicistronic *LINE* transcripts encode two kind of proteins: ORF1 and ORF2. Interestingly, ORF1 is an RNA binding protein that is thought to act as a chaperone [75], whereas the ORF2 protein has both endonuclease and reverse transcriptase activity [76,77]. There are approximately 500,000 *LINE-1* elements in the genome, but it is unknown how many of these are of full length (6 kb). The CpG sequence targeted by pyrosequencing for *LINE-1* is in the 5′ region. However, the 5′ end of the sequence uses to be deleted (with unknown frequency) except in the more active, evolutionarily newer sequences [78].

SINEs are non-autonomous mobile elements, being the *Alu* elements the predominant *SINEs* in humans [77]. Although active at higher levels in primates, *Alu* elements are constantly inserting in modern humans, creating genetic diversity and contributing to disease through insertional mutagenesis [79]. These elements play an important role in the non-allelic homologous recombination events causing copy number variation and disease [79]. The *Alu* sequence is about 280 bases and has evolved from 7SL RNA, a component of the signal recognition particle, and forms an extensive secondary structure. This molecular organization of *Alu* is thought to favour their proximity to ongoing translation and enables "hijacking" of newly translated *LINE-1* proteins. Thus, *SINE* RNA mobility is directly dependent on *LINE-1* transcription and translation [80].

Besides, about one third of the total DNA methylation occur in these elements [81], which are typically heavily methylated (and hence maintained as heterochromatin), being their hypomethylation commonly associated with cancer development and progression [78, 82, 83].

Some studies have suggested that DNA methylation, including *LINE-1* and *Alu* methylation, may play a role in chronic diseases, specifically, obesity, cardiovascular disease and diabetes [84-87]. Furthermore, there are some evidences suggesting that environmental exposures are related to DNA hypomethylation at these repeat regions [88].

For example, as studies in monozygotic twins have shown, environmental conditions can affect the epigenome postnatally [89, 90]. In these trials, researchers quantitatively measured global and locus-specific DNA methylation of a cohort of monozygotic [89] and twin-pairs [90]. They found that, although twins are epigenetically indistinguishable during the early life, older monozygous twins exhibited remarkable differences in the overall methylation content.

Global DNA Methylation and Diseases

Aberrations in DNA methylation patterns are usually related to disease [91]. For example, imprinting diseases, such as the Angelman, Silver-Russell, Prader-Willi and Beckwith-Wiedemann syndromes, are often associated with alterations in DNA methylation [92].

In human cancers, such as colorectal, prostate, gastric or lung tumour, *LINE-1* is frequently hypomethylated, which can be related with genomic instability [93-100]. In addition, genetic mutations in epigenetic genes cause dysfunctions that lead to certain neurodevelopmental disorders [16]. Moreover, genome wide hypomethylation in *LINE-1* and regional hypermethylation occur in *Helicobacter pylori* gastritis and may contribute to the tumorigenesis of diffuse-type gastric cancers [101].

On the other hand, aberrations in imprinting have also been associated with diabetes, schizophrenia, autism and cancer, among other diseases [91]. Current knowledge in global DNA methylation and metabolic diseases, (CVD, T2D and obesity) is widening.

Global DNA Methylation and Cardiovascular Disease (CVD)

Several studies have analysed the relationship between global DNA methylation and CVD (Table 1).

Most studies have been carried out in humans, being peripheral blood mononuclear cells (PBMCs) the cell type in which global DNA methylation has been more frequently examined.

Even though there are many mechanisms to study the global methylation pattern, *LINE-1* and *Alu* sequences are the most usual techniques. Homocysteine (Hcy) is one of the most associated biomarkers related to changes in the global epigenetic status, and abnormally homocysteine plasma levels often accompany cardiovascular risk features [102]. On one hand, a decrease in global DNA methylation (hypomethylation) has been related to an increase in the Hcy levels, both in PBMC [103] and in a vascular smooth muscle cells (VSMC) cultures [104]. On the other hand, an increase in the Hcy levels has been related to hypermethylation in PBMC [86].

HDL is another biomarker related to changes in global DNA methylation. At least two studies have found associations between a decrease in the HDL levels and changes in PBMC methylation status, one showing hypermethylation [105] and the second one hypomethylation [106]. Also, an increase in methylcytosine/cytosine ratio has been related to a decrease in the HDL levels for both, Hcy and HDL [107]. However, these controversial results imply that more studies are needed to reach a conclusion in relation to the epigenetic marks.

Concerning inflammation, an association between an increase in DNA methylation levels and C-reactive protein has been reported [108]. In contrast, a rise in the vascular cell adhesion protein 1 (VCAM-1) has been associated with lower levels of methylation levels in PBMCs [109].

Table 1. Some dynamic changes in global DNA methylation related to CVD

Approach	Type of Change	Specie	Tissue/ Cells	Biomarker	Ref
LINE-1	Hypomethylation	Human	PBMC	↑LDL ↓HDL	[106]
HpaII/MspI Digestion	Hypermethylation	Human	PBMC	↑ Hcy	[86]
Global 5'-methyldioxycytidine	Hypomethylation	Human Mice	Artery	↑ pDGF-B and 15-LO mRNA	[173]
Alu + SAT2	Hypermethylation	Human	PBMC	↑BMI	[84]
HpaII/MspI Digestion	Hyper and Hypomethylation	Mice	Aortic PBMC	Atherosclerotic lesions	[174]
Alu + *LINE-1*	Hypomethylation	Human	VSMC culture	↑ Hcy	[104]
HpaII/MspI Digestion	Hypermethylation	Human	PBMC	↑ Inflammation	[108]
LINE-1	Hypomethylation	Human	PBMC	CVD risk	[85]
HpaII/MspI Digestion	Hypomethylation	Human	PBMC	↑ Hcy + ↑SAM	[103]
LINE-1	Hypomethylation	Human	PBMC	↑ VCAM-1	[109]
LINE-1	Hypermethylation	Human	PBMC	↓HDL	[105]
LINE-1	Hypomethylation	Human	VAT	↑DBP	[55]
Methylcytosine/cytosine ratio	Hypermethylation	Human	PBMC	↑HDL	[107]
Global 5'-methylcytosine	Hypomethylation	Human	PBMC	↓Triglycerides	[175]

PBMC: Peripheral blood mononuclear cells, LDL: Low density lipoprotein; HDL: High density lipoprotein; Hcy: Homocysteine; pDGF-B: Platelet-derived growth factor beta; 15-LO mRNA: 15-lipoxygenase mRNA BMI: Body mass index; VSMC: Vascular smooth muscle cell; SAM: S-adenosyl methionine; VCAM-1; Vascular cell adhesion protein 1; VAT: Visceral adipose tissue; DBP: Diastolic blood pressure.

GLOBAL DNA METHYLATION AND TYPE 2 DIABETES

Some studies suggest that DNA methylation, including *LINE-1* and *Alu* methylation, may play a role in diabetes [84-86]. Genome-wide association studies in families have shown that the genetic background also influences the risk of type 2 diabetes [110, 111]. Furthermore, hyperglycemia, aging, BMI and other risks factors for type 2 diabetes, produced differential global DNA methylation in islets of non-diabetic human in an elegant study [112].

In this sense, several studies analysing global DNA methylation in relation to the diabetic condition have been published (Table 2). Fasting glucose level is the reference used to identify type 2 diabetes, which has related changes in global DNA methylation levels [113]. However, different patterns in the methylation signatures according to the tissue examined may be found. When the visceral adipose tissue (VAT) was analysed, hyperglycemia was associated with *LINE-1* hypomethylation [55]. Meanwhile, in PBMCs, the rise in glucose levels was related with hypermethylation in *LINE*-1 [105], but also with a decrease in the methylcytosine/cytosine ratio [107].

On the other hand, the HOMA index, an indicator of insulin resistance, has been associated with hypomethylation in *LINE-1* in PBMC, especially in the presence of lower concentrations of plasma folate [114], and hypermethylation in *Alu* sequences in peripheral blood leukocytes [115].

Table 2. Dynamic changes in global DNA methylation related to Type 2 Diabetes

Approach	Type of Change	Specie	Tissue/Cells	Biomarker	Ref
LINE-1	Hypermethylation	Human	PBMC	↑Glucose	[105]
LINE-1	Hypomethylation	Human	PBMC	↑HOMA-IR ↓plasma folate	[114]
LINE-1	Hypermethylation	Human	PBMC	↑Triglycerides	[105]
Global 5'-methyldioxycytidine	Hypermethylation	Human	HepG2 Cell Culture	↑SAM ↓AdoHcy	[176]
Alu	Hypermethylation	Human	PBMC	↑ HOMA-IR	[115]
LINE-1	Hypomethylation	Human	VAT	↑Glucose	[55]
Methylcytosine/cytosine ratio	Hypomethylation	Human	PBMC	↑Glucose	[107]

PBMC: Peripheral blood mononuclear cells; HOMA-IR: HOMA Index insulin resistance; SAM: S-adenosyl methionine, AdoHcy: S-adenosyl-L-homocysteine; VAT: Visceral adipose tissue.

GLOBAL DNA METHYLATION AND OBESITY

Epigenetic dysregulation, assessed as DNA methylation changes, envisages the genetic and environmental modulation of obesity [116]. Although there are some studies that did not find a statistical association between global DNA methylation changes and obesity, other studies evidenced this association. In this sense, PBMCs from obese asthmatic children have distinctive DNA methylation patterns than those from non-obese healthy controls [117].

Michels et al., (2011) reported that high or low birth weigh, as well as premature birth, were associated with significantly lower *LINE-1* methylation [118]. Likewise, Perng et al.,

(2013) found that *LINE-1* hypomethylation was related to adiposity development in boys [119]. Two additional investigations have reported that the body mass index (BMI) increases when hypermethylation in *SAT2* [84] or *Alu* and *LINE-1* [120] occurred, although *Alu* sequences gave conflicting results in other study carried out in blood cells [121]. In the same way, another trial showed an association between the increase of waist circumference and hypomethylation of *LINE-1* [119]. A summary concerning the studies focused on global DNA methylation and obesity-related parameters, such as BMI and waist circumference, is reported (Table 3).

Furthermore, it has been found that, although both DNA methylation and gene expression differences exist between the high and low responders to a hypocaloric diet after the nutritional intervention, only epigenetic differences were found before dieting, suggesting that the epigenetic profile has the potential to discriminate between good and poor responders to caloric restriction [31]. In this context, epigenetic marks have been described as potential indicators of susceptibility or resistance to body weight gain [31,122].

Table 3. Dynamic changes in global DNA methylation related to Obesity

Approach	Type of Change	Specie	Tissue/ Cells	Biomarker	Ref
SAT2	Hypermethylation	Human	PBMC	↑BMI	[84]
LINE-1	Hypomethylation	Human	PBMC	↑Waist Circumference	[119]
Alu + *LINE-1*	Hypermethylation	Human	PBMC	↑ Hcy ↑BMI	[120]
Alu	Hypomethylation	Human	PBMC	↑BMI	[121]
Alu	Hypomethylation	Human	WB	↑BMI	[177]
HpaII/MspI Digestion	Hypermethylation	Human	Placenta and umbilical cord blood	↑BMI	[178]

PBMC: peripheral blood mononuclear cells; Hcy: Homocysteine; BMI: body mass index; WB: whole blood.

DIET AND GLOBAL DNA METHYLATION

The honeybee model, in which bees grow to be either queens or workers depending on the different type of honey they eat, suggests that there is an interaction between the diet and the phenotype that is mediated by epigenetic mechanisms [123]. In this sense, studies focused on dietary factors involved in one-carbon metabolism show the evidence of an interaction between nutrients and DNA methylation [124]. These nutrients include vitamin B12, vitamin B6, folate, methionine, biotin and choline. Surprisingly, dietary factors modify simultaneously global DNA and gene-specific methylation, but sometimes causing opposite effects. One example is folate deficiency, that causes global DNA hypomethylation, but hypermethylation of the 5' regulatory sequence of the H-cadherin gene [125]. In a model of helicobacter-associated gastric cancer in mice, folic acid increases global DNA methylation and reduces mucosal inflammation and dysplasia [126]. Some bioactive food components and nutrients, such as genistein, zinc, selenium and vitamin A, can also affect DNA methylation. In several diseases, including metabolic syndrome manifestations, effects of different dietary compounds on epigenetic mechanisms have been reported, including regulation of gene expression of methylation-related enzymes and other molecules responsible for drug absorption, metabolism, distribution, and excretion [127]. Interestingly, whereas deficiencies of some food components can induce global DNA hypomethylation, the insufficient intake of other food components induce global hypermethylation. For example, zinc deficiency and retinoic acid excess result in global DNA hypomethylation by reducing the use of methyl groups from SAM in rat liver [128, 129]. Parallelly, selenium deficiency decreases DNA methylation in Caco-2 cells and in rat liver and colon [130]. In contrast, vitamin C low levels induces DNA hypermethylation in lung cancer cells [131]. Moreover, global DNA methylation was twofold lower in monkeys fed a high fat diet compared to monkeys fed a standard diet [132].

Environmental Factors and Global DNA Methylation

Apart from the diet, other environmental and non-modificable factors, including socioeconomic status, physical exercise, gender, infections, pollutants, stress and age, have been related to changes in global DNA methylation in different models [133-137]. For instance, an epidemiological study identified a link between air pollution and global DNA methylation in blood cells [138]. Global methylation was also altered in the blood of individuals that had been exposed to benzene [139]. Locus-specific DNA methylation changes in other tissues have been associated with different environmental factors, such as chronic exposure to sunlight, asbestos and tobacco smoke, alcohol drinking and use of hair-dye [140-142]. In addition, prenatal exposure to metals has been associated with decreased global methylation in cord blood [143].

A relationship between socio-economic and lifestyle factors and global DNA methylation was reported by McGuiness et al., (2012) in an extreme socio-economic and health gradient cohort [144]. The researchers explained this result as an adaptive mechanism that allows for epigenetic plasticity in response to environmental changes. Similar results were reported by Subramanyam et al., (2013) [134] when analysing data from the Multi-Ethnic Study of Atherosclerosis. They found statistically significant race/ethnic differences in *Alu* and *LINE-1* methylation levels when comparing African-Americans and Hispanics. However, these associations were in opposite directions.

Concerning physical activity, a six months exercise intervention in 23 healthy men changed global DNA methylation levels of 39 genes related to obesity and type 2 diabetes. These shifts in DNA methylation were associated with gene expression and were validated in *in vitro* studies [133]. Parallelly, Zhang et al., (2011) found that individuals with physical activity 26-30 min/day showed significantly higher level of *LINE-1* DNA methylation when compared to those with physical activity ≤10 min/day [135].

The impact of aging on DNA methylation has been well documented [145]. For example, aging cells show a progressive loss of 5-methylcytosine content [136,146]. In addition, aging reduces global genomic DNA methylation similarly across tissue types [22]. Global DNA methylation as measured in *Alu* sequences also decreases over time [147]. Furthermore, global DNA methylation in old subjects is also correlated with frailty [148]. However, there are reports indicating that CpG islands tend to increase methylation whereas non-islands lose methylation with aging [140].

Stress is another factor that can affect the methyl epigenome [149]. Particularly, fetal growth restriction is associated with reduced DNA methylation and increased oxidative stress [150]. Unternaehrer et al., (2012) found stress-associated DNA methylation changes in *OXTR* gene after acute psychosocial stress [151]. Moreover, aberrant global DNA methylation has been associated to post-traumatic stress disorder [137].

Pathogenic bacterial infections and gut microbiota can also affect the epigenome [152] including covalent histone modifications and DNA methylation. It has been proposed that intestinal bacteria change the epigenome by producing epigenetically active metabolites, such as butyrate and acetate, or by activating cell signalling pathways that lead to epigenetic modulation of gene expression [153]. Virus infections can also modify the epigenome [154]. For instance, it is known that DNA/RNA virus interact with host cells, lacking histones in some cases [155, 156]. Moreover, the human cytomegalovirus infection is sensitive to the host cell DNA methylation state and alters global DNA methylation capacity [157]. In a recent study, Mc Erlean et al., (2014) found that the human rhinovirus infection causes different methylation changes in nasal epithelial cells [158].

Concerning metals, evidences based on in vitro and in vivo models suggest that chronic arsenic exposure induces global DNA hypomethylation, in conjunction with hypermethylation of promoter regions of tumour suppressor genes [159-162]. Some minerals, including nickel, lead, cadmium, chromium, beryllium, and arsenic, are associated with altered DNA methylation pattern [163]. Some of them, such as arsenic or cadmium, can also interfere gene expression [164-167]. In this sense, several trials have suggested that the health effects of environmental exposures, including exposure to metals, could be mediated in part by epigenetic mechanisms [168-171]. Moreover, a more recent study found a positive correlation between global DNA methylation and global hydroxymethylation in human blood samples with metals measured in urine [172].

SUMMARY AND PERSPECTIVES

There are not many studies where the global DNA methylation levels and biomarkers of metabolic diseases have been linked and conjointly analysed. Cardiovascular disease is the field in which more evidences have been found. However, as occurs with diabetes and obesity, there are some controversies about if it is the hypo or the hypermethylation which is related with these diseases. Nowadays there are no solid information if the epigenetic alterations are the cause or the effect over biomarker changes. Epigenetics in general and nutritional epigenetics in particular have the potential to make a tremendous impact in disease prevention, control, and management. In this context, it will be possible to use *LINE-1* and *Alu* methylation status as a biomarker. However, validation studies have not been completed to evaluate their potential in different diseases, ethnicities and cell types. Thus, once the human epigenome is completed and additional nutritional epigenetic studies conducted, it may be possible to achieve this goal.

The principal challenges in nutriepigenetics are the large number of variables, intermediate markers and measurements, dynamic nature of nutrients, and limited available outcomes. One strategy to address these problems could be the application of systems biology to existing databases in order to develop models to understand dietary and environmental influences on epigenetics and disease development [4]. Single pathway approaches can be expanded also to include genome-wide approaches. The combination of observational studies with experimental studies may result in risk-prediction models with implications for identifying populations at high risk of developing nutritional-related diseases, such as cardiovascular disease, diabetes, and obesity. The incorporation of genomic information in epigenetic databases may also be useful to understand the biology of the underlying disease, develop intervention targets, and ultimately improve health. Considering the evidence that excess body weight gain in early life is related to increased cardiovascular morbidity in adulthood and higher risk of type 2 diabetes, it is critical to understand the epigenetic processes that underlie childhood obesity for effective early interventions.

There is no doubt that the environment affects chromatin structure and global DNA methylation. Many studies have revealed that the exposure to dietary compounds, environmental contaminants, stress and other external factors influence DNA methylation. *LINE-1* and *Alu* sequences have shown to be reliable markers to study global methylation and its relation to obesity-related diseases. These sequences are affected by environmental and dietary factors, and changes in their methylation levels seem to be related with the development of several diseases, such as cardiovascular disease, diabetes or obesity. In these diseases, *LINE-1* and *Alu* methylation levels could be used as diagnostic/prognostic biomarkers. However, more studies with larger sample sizes and more controlled conditions are needed to produce more robust results and to clarify the kind of epigenetic changes that are the source of the disease origin, considering that epigenetic phenomena are variable in different tissues.

References

[1] Blomen, VA; Boonstra, J. Stable transmission of reversible modifications: maintenance of epigenetic information through the cell cycle. *Cellular and Molecular Life Sciences:CMLS*, 68, 27-44, doi:10.1007/s00018-010-0505-5 (2011).

[2] Ballestar, E; Esteller, M. Epigenetic gene regulation in cancer. *Advances in Genetics*, 61, 247-267, doi:10.1016/s0065-2660(07)00009-0 (2008).

[3] Korkmaz, A; et al. Epigenetic mechanisms in human physiology and diseases. *Journal of Experimental and Integrative Medicine*, 139, doi:10.5455/jeim.060611.rw.003 (2011).

[4] Mukesh, V. in *Nutrition, Epigenetic Mechanisms, and Human Disease* Ch. 1, 1-11 (CRC Press, 2010).

[5] Simmons, D. Epigenetic influence and disease. *Nature*, 1, 6 (2008).

[6] Murrell, A; Rakyan, VK; Beck, S. From genome to epigenome. *Human Molecular Genetics*, 14 Spec No, 1, R3-r10, doi:10.1093/hmg/ddi110 (2005).

[7] Martinez, JA; Cordero, P; Campion, J; Milagro, FI. Interplay of early-life nutritional programming on obesity, inflammation and epigenetic outcomes. *The Proceedings of the Nutrition Society*, 71, 276-283, doi:10.1017/s0029665112000055 (2012).

[8] Verma, M. Viral genes and methylation. *Annals of the New York Academy of Sciences*, 983, 170-180 (2003).

[9] Kumar, D; Verma, M. Methods in cancer epigenetics and epidemiology. *Methods in Molecular Biology (Clifton, N.J.)* 471, 273-288, doi:10.1007/978-1-59745-416-2_14 (2009).

[10] Milagro, FI; Mansego, ML; De Miguel, C; Martinez, JA. Dietary factors, epigenetic modifications and obesity outcomes: progresses and perspectives. *Molecular Aspects of Medicine*, 34, 782-812, doi:10.1016/j.mam.2012.06.010 (2013).

[11] Nilanjana, M; Gautam, M. in *Nutrition, Epigenetic Mechanisms, and Human Disease*, Ch. 2, 13-30 (CRC Press, 2010).

[12] Mansego, ML; Milagro, FI; Campion, J; Martinez, JA. Techniques of DNA methylation analysis with nutritional applications. *Journal of Nutrigenetics and Nutrigenomics*, 6, 83-96, doi:10.1159/000350749 (2013).

[13] Smith, ZD; Meissner, A. DNA methylation: roles in mammalian development. *Nature Reviews. Genetics*, 14, 204-220, doi:10.1038/ nrg3354 (2013).

[14] Takai, D; Jones, PA. Comprehensive analysis of CpG islands in human chromosomes, 21 and, 22. *Proceedings of the National Academy of Sciences of the United States of America*, 99, 3740-3745, doi:10.1073/pnas.052410099 (2002).

[15] Kulis, M; Esteller, M. DNA methylation and cancer. *Advances in Genetics*, 70, 27-56, doi:10.1016/b978-0-12-380866-0.60002-2 (2010).

[16] Portela, A; Esteller, M. Epigenetic modifications and human disease. *Nature Biotechnology*, 28, 1057-1068, doi:10.1038/nbt.1685 (2010).

[17] Feinberg, AP; Tycko, B. The history of cancer epigenetics. *Nature Reviews. Cancer*, 4, 143-153, doi:10.1038/nrc1279 (2004).

[18] Issa, JP. Aging, DNA methylation and cancer. *Critical Reviews in Oncology/hematology*, 32, 31-43 (1999).

[19] Ehrlich, M. DNA methylation in cancer: too much, but also too little. *Oncogene*, 21, 5400-5413, doi:10.1038/sj.onc.1205651 (2002).

[20] Feinberg, AP; Gehrke, CW; Kuo, KC; Ehrlich, M. Reduced genomic, 5-methylcytosine content in human colonic neoplasia. *Cancer Research*, 48, 1159-1161 (1988).

[21] Feinberg, AP; Vogelstein, B. Hypomethylation distinguishes genes of some human cancers from their normal counterparts. *Nature*, 301, 89-92 (1983).

[22] Hoffmann, MJ; Schulz, WA. Causes and consequences of DNA hypomethylation in human cancer. *Biochemistry and Cell Biology = Biochimie et biologie cellulaire*, 83, 296-321, doi:10.1139/o05-036 (2005).

[23] Ha, PK; Califano, JA. Promoter methylation and inactivation of tumour-suppressor genes in oral squamous-cell carcinoma. *The Lancet Oncology*, 7, 77-82, doi:10.1016/s1470-2045(05)70540-4 (2006).

[24] Eden, A; Gaudet, F; Waghmare, A; Jaenisch, R. Chromosomal Instability and Tumors Promoted by DNA Hypomethylation. *Science*, 300, 455, doi:10.1126/science.1083557 (2003).

[25] Issa, JP; et al. Methylation of the oestrogen receptor CpG island links ageing and neoplasia in human colon. *Nature Genetics*, 7, 536-540, doi:10.1038/ng0894-536 (1994).

[26] Issa, JP; Vertino, PM; Boehm, CD; Newsham, IF; Baylin, SB. Switch from monoallelic to biallelic human IGF2 promoter methylation during aging and carcinogenesis.

Proceedings of the National Academy of Sciences of the United States of America, 93, 11757-11762 (1996).

[27] Moleres, A; et al. Differential DNA methylation patterns between high and low responders to a weight loss intervention in overweight or obese adolescents: the EVASYON study. *FASEB journal : official publication of the Federation of American Societies for Experimental Biology*, 27, 2504-2512, doi:10.1096/fj.12-215566 (2013).

[28] Perez-Cornago, A; Mansego, ML; Zulet, MA; Martinez, JA. DNA hypermethylation of the serotonin receptor type-2A gene is associated with a worse response to a weight loss intervention in subjects with metabolic syndrome. *Nutrients*, 6, 2387-2403, doi:10.3390/nu6062387 (2014).

[29] José María Moreno-Navarrete, J. M. F.-R. in *Adipose Tissue Biology* Vol. 1 (ed Michael E. Symonds) Ch. 2, 17-38 (Springer, 2012).

[30] Palacios-Ortega, S; Varela-Guruceaga, M; Milagro, FI; Martínez, JA; de Miguel, C. Expression of Caveolin 1 Is Enhanced by DNA Demethylation during Adipocyte Differentiation. Status of Insulin Signaling. *PloS One* 9, e95100, doi:10.1371/journal.pone.0095100 (2014).

[31] Bouchard, L; et al. Differential epigenomic and transcriptomic responses in subcutaneous adipose tissue between low and high responders to caloric restriction. *The American Journal of Clinical Nutrition*, 91, 309-320, doi:10.3945/ajcn.2009.28085 (2010).

[32] Lomba, A; et al. Weight gain induced by an isocaloric pair-fed high fat diet: a nutriepigenetic study on FASN and NDUFB6 gene promoters. *Molecular Genetics and Metabolism*, 101, 273-278, doi:10.1016/ j.ymgme.2010.07.017 (2010).

[33] Cordero, P; et al. Leptin and TNF-alpha promoter methylation levels measured by MSP could predict the response to a low-calorie diet. *Journal of physiology and biochemistry*, 67, 463-470, doi:10.1007/ s13105-011-0084-4 (2011).

[34] Tollefsbol, TO. in *Epigenetics in Human Disease* (ed Trygve O. Tollefsbol) 1-6 (Academic Press, 2012).

[35] Margueron, R; Trojer, P; Reinberg, D. The key to development: interpreting the histone code? *Current Opinion in Genetics & Development*, 15, 163-176, doi:10.1016/j.gde.2005.01.005 (2005).

[36] Cheung, P; Lau, P. Epigenetic regulation by histone methylation and histone variants. *Molecular Endocrinology (Baltimore, Md.)* 19, 563-573, doi:10.1210/me.2004-0496 (2005).

[37] Strahl, BD; Allis, CD. The language of covalent histone modifications. *Nature*, 403, 41-45, doi:10.1038/47412 (2000).

[38] Campion, J; Milagro, F; Martinez, JA. Epigenetics and obesity. *Progress in Molecular Biology and Translational Science*, 94, 291-347, doi:10.1016/S1877-1173(10)94011-810.1016/B978-0-12-375003-7.00011-X (2010).

[39] Smolle, M; Workman, JL. Transcription-associated histone modifications and cryptic transcription. *Biochimica et Biophysica Acta (BBA) - Gene Regulatory Mechanisms*, 1829, 84-97, doi:10.1016/ j.bbagrm.2012.08.008 (2013).

[40] Shogren-Knaak, M; et al. Histone H4-K16 acetylation controls chromatin structure and protein interactions. *Science*, 311, 844-847, doi:10.1126/science.1124000 (2006).

[41] Cavalli, G; Paro, R. The Drosophila Fab-7 chromosomal element conveys epigenetic inheritance during mitosis and meiosis. *Cell*, 93, 505-518 (1998).

[42] Pirola, L; et al. Genome-wide analysis distinguishes hyperglycemia regulated epigenetic signatures of primary vascular cells. *Genome Research*, 21, 1601-1615, doi:10.1101/gr.116095.110 (2011).

[43] Sui, W; et al. Genome-wide analysis of histone H3 lysine9 trimethylation by ChIP-seq in peripheral blood mononuclear cells of uremia patients. *Hemodialysis international. International Symposium on Home Hemodialysis*, 17, 493-501, doi:10.1111/hdi.12051 (2013).

[44] Ku, G; McManus, MT. Behind the scenes of a small RNA gene-silencing pathway. *Human Gene Therapy*, 19, 17-26, doi:10.1089/ hum.2007.1226 (2008).

[45] Chen, Y; Gelfond, JA; McManus, LM; Shireman, PK. Reproducibility of quantitative RT-PCR array in miRNA expression profiling and comparison with microarray analysis. *BMC genomics*, 10, 407, doi:10.1186/1471-2164-10-407 (2009).

[46] van Wolfswinkel, JC; Ketting, RF. The role of small non-coding RNAs in genome stability and chromatin organization. *Journal of Cell Science*, 123, 1825-1839, doi:10.1242/jcs.061713 (2010).

[47] Calin, GA; et al. A MicroRNA signature associated with prognosis and progression in chronic lymphocytic leukemia. *The New England Journal of Medicine*, 353, 1793-1801, doi:10.1056/NEJMoa050995 (2005).

[48] Davalos, V; Esteller, M. MicroRNAs and cancer epigenetics: a macrorevolution. *Current Opinion in Oncology*, 22, 35-45, doi:10.1097/CCO.0b013e328333dcbb (2010).

[49] Othman, N; Nagoor, NH. The Role of microRNAs in the Regulation of Apoptosis in Lung Cancer and Its Application in Cancer Treatment. *BioMed Research International*, 2014, 318030, doi:10.1155/2014/318030 (2014).

[50] Bronze-da-Rocha, E. MicroRNAs Expression Profiles in Cardiovascular Diseases. *BioMed research international*, 2014, 985408, doi:10.1155/2014/985408 (2014).

[51] McGregor, RA; Choi, MS. microRNAs in the regulation of adipogenesis and obesity. *Current Molecular Medicine*, 11, 304-316 (2011).

[52] Milagro, FI; et al. High-throughput sequencing of microRNAs in peripheral blood mononuclear cells: identification of potential weight loss biomarkers. *PloS One*, 8, e54319, doi:10.1371/journal.pone.0054319 (2013).

[53] Palmer, JD; et al. MicroRNA expression altered by diet: Can food be medicinal? *Ageing Research Reviews*, 17c, 16-24, doi:10.1016/ j.arr.2014.04.005 (2014).

[54] Shah, MS; et al. Integrated microRNA and mRNA expression profiling in a rat colon carcinogenesis model: effect of a chemo-protective diet. *Physiological Genomics*, 43, 640-654, doi:10.1152/physiolgenomics. 00213.2010 (2011).

[55] Turcot, V; et al. LINE-1 methylation in visceral adipose tissue of severely obese individuals is associated with metabolic syndrome status and related phenotypes. *Clinical Epigenetics*, 4, 10, doi:10.1186/1868-7083-4-10 (2012).

[56] Moulton, T; et al. Epigenetic lesions at the H19 locus in Wilms' tumour patients. *Nature Genetics*, 7, 440-447, doi:10.1038/ng0794-440 (1994).

[57] Steenman, MJ; et al. Loss of imprinting of IGF2 is linked to reduced expression and abnormal methylation of H19 in Wilms' tumour. *Nature Genetics*, 7, 433-439, doi:10.1038/ng0794-433 (1994).

[58] Taniguchi, T; Sullivan, MJ; Ogawa, O; Reeve, AE. Epigenetic changes encompassing the IGF2/H19 locus associated with relaxation of IGF2 imprinting and silencing of H19

in Wilms tumor. *Proceedings of the National Academy of Sciences of the United States of America*, 92, 2159-2163 (1995).

[59] Rainier, S; Dobry, CJ; Feinberg, AP. Loss of imprinting in hepatoblastoma. *Cancer Research*, 55, 1836-1838 (1995).

[60] Vu, TH; Yballe, C; Boonyanit, S; Hoffman, AR. Insulin-like growth factor II in uterine smooth-muscle tumors: maintenance of genomic imprinting in leiomyomata and loss of imprinting in leiomyosarcomata. *The Journal of Clinical Endocrinology and Metabolism*, 80, 1670-1676, doi:10.1210/jcem.80.5.7745016 (1995).

[61] Douc-Rasy, S; et al. High incidence of loss of heterozygosity and abnormal imprinting of H19 and IGF2 genes in invasive cervical carcinomas. Uncoupling of H19 and IGF2 expression and biallelic hypomethylation of H19. *Oncogene*, 12, 423-430 (1996).

[62] Oda, H; Kume, H; Shimizu, Y; Inoue, T; Ishikawa, T. Loss of imprinting of igf2 in renal-cell carcinomas. *International Journal of Cancer. Journal International du Cancer*, 75, 343-346 (1998).

[63] Pedone, PV; et al. Mono- and bi-allelic expression of insulin-like growth factor II gene in human muscle tumors. *Human Molecular Genetics*, 3, 1117-1121 (1994).

[64] Zhan, S; Shapiro, DN; Helman, LJ. Activation of an imprinted allele of the insulin-like growth factor II gene implicated in rhabdomyosarcoma. *The Journal of Clinical Investigation*, 94, 445-448, doi:10.1172/ jci117344 (1994).

[65] Uyeno, S; et al. IGF2 but not H19 shows loss of imprinting in human glioma. *Cancer Research*, 56, 5356-5359 (1996).

[66] Cui, H; Horon, IL; Ohlsson, R; Hamilton, SR; Feinberg, A. P. Loss of imprinting in normal tissue of colorectal cancer patients with microsatellite instability. *Nature Medicine*, 4, 1276-1280, doi:10.1038/ 3260 (1998).

[67] Takano, Y; Shiota, G; Kawasaki, H. Analysis of genomic imprinting of insulin-like growth factor, 2 in colorectal cancer. *Oncology*, 59, 210-216, doi:12163 (2000).

[68] Doherty, AS; Mann, MR; Tremblay, KD; Bartolomei, MS; Schultz, R. M. Differential effects of culture on imprinted H19 expression in the preimplantation mouse embryo. *Biology of Reproduction*, 62, 1526-1535 (2000).

[69] Khosla, S; Dean, W; Brown, D; Reik, W; Feil, R. Culture of preimplantation mouse embryos affects fetal development and the expression of imprinted genes. *Biology of Reproduction*, 64, 918-926 (2001).

[70] Ashbury, JE; et al. Biomarkers measured in buccal and blood leukocyte DNA as proxies for colon tissue global methylation. *International Journal of Molecular Epidemiology and Genetics*, 5, 120-124 (2014).

[71] Miki, Y. Retrotransposal integration of mobile genetic elements in human diseases. *Journal of Human Genetics*, 43, 77-84, doi:10.1007/ s100380050045 (1998).

[72] NL, C. in *Mobile DNA II* (ASM, Washington, D.C; 2002.).

[73] Belancio, VP; Roy-Engel, AM; Pochampally, RR; Deininger, P. Somatic expression of LINE-1 elements in human tissues. *Nucleic Acids Research*, 38, 3909-3922, doi:10.1093/nar/gkq132 (2010).

[74] Kazazian, HH; Jr. Mobile elements: drivers of genome evolution. *Science*, 303, 1626-1632, doi:10.1126/science.1089670 (2004).

[75] Ostertag, EM; Kazazian, HH; Jr. Biology of mammalian L1 retrotransposons. *Annual Review of Genetics*, 35, 501-538, doi:10.1146/ annurev.genet.35.102401.091032 (2001).

[76] Feng, Q; Moran, JV; Kazazian, HH; Jr; Boeke, JD. Human L1 retrotransposon encodes a conserved endonuclease required for retrotransposition. *Cell*, 87, 905-916 (1996).

[77] Mathias, SL; Scott, AF; Kazazian, HH; Jr., Boeke, JD; Gabriel, A. Reverse transcriptase encoded by a human transposable element. *Science*, 254, 1808-1810 (1991).

[78] Nelson, HH; Marsit, CJ; Kelsey, KT. Global methylation in exposure biology and translational medical science. *Environmental Health Perspectives*, 119, 1528-1533, doi:10.1289/ehp.1103423 (2011).

[79] Deininger, P. Alu elements: know the SINEs. *Genome Biology*, 12, 236, doi:10.1186/gb-2011-12-12-236 (2011).

[80] Powell, WT; LaSalle, JM. in *Epigenetic Regulation in the Nervous System* (eds J. David Sweatt, Michael J. Meaney, Eric J. Nestler, & Schahram Akbarian) 267-279 (Academic Press, 2013).

[81] Yang, AS; et al. A simple method for estimating global DNA methylation using bisulfite PCR of repetitive DNA elements. *Nucleic Acids Research*, 32, e38, doi:10.1093/nar/gnh032 (2004).

[82] Zhu, ZZ; et al. Repetitive element hypomethylation in blood leukocyte DNA and cancer incidence, prevalence, and mortality in elderly individuals: the Normative Aging Study. *Cancer Causes & Control : CCC*, 22, 437-447, doi:10.1007/s10552-010-9715-2 (2011).

[83] Sunami, E; de Maat, M; Vu, A; Turner, RR; Hoon, DS. LINE-1 hypomethylation during primary colon cancer progression. *PloS One*, 6, e18884, doi:10.1371/journal.pone.0018884 (2011).

[84] Kim, M; et al. DNA methylation as a biomarker for cardiovascular disease risk. *PloS One*, 5, e9692, doi:10.1371/journal.pone.0009692 (2010).

[85] Baccarelli, A; et al. Ischemic heart disease and stroke in relation to blood DNA methylation. *Epidemiology*, 21, 819-828, doi:10.1097/ EDE.0b013e3181f20457 (2010).

[86] Sharma, P; et al. Detection of altered global DNA methylation in coronary artery disease patients. *DNA and Cell Biology*, 27, 357-365, doi:10.1089/dna.2007.0694 (2008).

[87] Wei, L; et al. LINE-1 hypomethylation is associated with the risk of coronary heart disease in Chinese population. *Arquivos Brasileiros de Cardiologia*, 102, 481-488 (2014).

[88] Rusiecki, JA; et al. Global DNA hypomethylation is associated with high serum-persistent organic pollutants in Greenlandic Inuit. *Environmental Health Perspectives*, 116, 1547-1552, doi:10.1289/ ehp.11338 (2008).

[89] Fraga, MF; et al. Epigenetic differences arise during the lifetime of monozygotic twins. *Proceedings of the National Academy of Sciences of the United States of America*, 102, 10604-10609, doi:10.1073/ pnas.0500398102 (2005).

[90] Wong, CC; et al. A longitudinal study of epigenetic variation in twins. *Epigenetics : Official Journal of the DNA Methylation Society*, 5, 516-526 (2010).

[91] Tollefsbol, TO. in *Epigenetic in Human DIsease* (ed Trygve O. Tollefsbol) Ch. 1, 1-6 (2012).

[92] Falls, JG; Pulford, DJ; Wylie, AA; Jirtle, RL. Genomic imprinting: implications for human disease. *The American Journal of Pathology*, 154, 635-647, doi:10.1016/s0002-9440(10)65309-6 (1999).

[93] Daskalos, A; et al. Hypomethylation of retrotransposable elements correlates with genomic instability in non-small cell lung cancer. *International Journal of Cancer. Journal international du cancer*, 124, 81-87, doi:10.1002/ijc.23849 (2009).
[94] Ogino, S; et al. LINE-1 hypomethylation is inversely associated with microsatellite instability and CpG island methylator phenotype in colorectal cancer. *International Journal of Cancer. Journal international du cancer*, 122, 2767-2773, doi:10.1002/ijc.23470 (2008).
[95] Park, SY; Yoo, EJ; Cho, NY; Kim, N; Kang, GH. Comparison of CpG island hypermethylation and repetitive DNA hypomethylation in premalignant stages of gastric cancer, stratified for Helicobacter pylori infection. *The Journal of Pathology*, 219, 410-416, doi:10.1002/ path.2596 (2009).
[96] Saito, K; et al. Long interspersed nuclear element, 1 hypomethylation is a marker of poor prognosis in stage IA non-small cell lung cancer. *Clinical cancer research : an official journal of the American Association for Cancer Research*, 16, 2418-2426, doi:10.1158/1078-0432.ccr-09-2819 (2010).
[97] Yegnasubramanian, S; et al. DNA hypomethylation arises later in prostate cancer progression than CpG island hypermethylation and contributes to metastatic tumor heterogeneity. *Cancer Research*, 68, 8954-8967, doi:10.1158/0008-5472.can-07-6088 (2008).
[98] Ogino, S; et al. A cohort study of tumoral LINE-1 hypomethylation and prognosis in colon cancer. *Journal of the National Cancer Institute*, 100, 1734-1738, doi:10.1093/jnci/djn359 (2008).
[99] Estecio, MR; et al. LINE-1 hypomethylation in cancer is highly variable and inversely correlated with microsatellite instability. *PloS One*, 2, e399, doi:10.1371/journal.pone.0000399 (2007).
[100] Richards, KL; et al. Genome-wide hypomethylation in head and neck cancer is more pronounced in HPV-negative tumors and is associated with genomic instability. *PloS One*, 4, e4941, doi:10.1371/journal.pone. 0004941 (2009).
[101] Yamamoto, E; et al. LINE-1 hypomethylation is associated with increased CpG island methylation in Helicobacter pylori-related enlarged-fold gastritis. *Cancer Epidemiology, Biomarkers & Prevention : a publication of the American Association for Cancer Research, cosponsored by the American Society of Preventive Oncology*, 17, 2555-2564, doi:10.1158/1055-9965.epi-08-0112 (2008).
[102] Cordero, P; Milagro, FI; Campion, J; Martinez, JA. Maternal methyl donors supplementation during lactation prevents the hyperhomocysteinemia induced by a high-fat-sucrose intake by dams. *International Journal of Molecular Sciences*, 14, 24422-24437, doi:10.3390/ ijms141224422 (2013).
[103] Castro, R; et al. Increased homocysteine and S-adenosylhomocysteine concentrations and DNA hypomethylation in vascular disease. *Clinical Chemistry*, 49, 1292-1296 (2003).
[104] Yideng, J; et al. Homocysteine-mediated expression of SAHH, DNMTs, MBD2, and DNA hypomethylation potential pathogenic mechanism in VSMCs. *DNA and Cell Biology*, 26, 603-611, doi:10.1089/ dna.2007.0584 (2007).
[105] Pearce, MS; et al. Global LINE-1 DNA methylation is associated with blood glycaemic and lipid profiles. *International Journal of Epidemiology*, 41, 210-217, doi:10.1093/ije/dys020 (2012).

[106] Cash, HL; et al. Cardiovascular disease risk factors and DNA methylation at the LINE-1 repeat region in peripheral blood from Samoan Islanders. *Epigenetics : Official Journal of the DNA Methylation Society*, 6, 1257-1264, doi:10.4161/epi.6.10.17728 (2011).

[107] Luttmer, R; et al. Metabolic syndrome components are associated with DNA hypomethylation. *Obesity Research & Clinical Practice*, 7, e106-e115, doi:10.1016/j.orcp.2012.06.001 (2013).

[108] Stenvinkel, P; et al. Impact of inflammation on epigenetic DNA methylation - a novel risk factor for cardiovascular disease? *Journal of Internal Medicine*, 261, 488-499, doi:10.1111/j.1365-2796.2007.01777.x (2007).

[109] Baccarelli, A; et al. Repetitive element DNA methylation and circulating endothelial and inflammation markers in the VA normative aging study. *Epigenetics : Official Journal of the DNA Methylation Society*, 5 (2010).

[110] Kobberling, J; Tillil, H. Genetic and nutritional factors in the etiology and pathogenesis of diabetes mellitus. *World Review of Nutrition and Dietetics*, 63, 102-115 (1990).

[111] McCarthy, MI. Genomics, type, 2 diabetes, and obesity. *The New England Journal of Medicine*, 363, 2339-2350, doi:10.1056/ NEJMra0906948 (2010).

[112] Dayeh, T; et al. Genome-wide DNA methylation analysis of human pancreatic islets from type, 2 diabetic and non-diabetic donors identifies candidate genes that influence insulin secretion. *PLoS Genetics*, 10, e1004160, doi:10.1371/journal.pgen.1004160 (2014).

[113] Hidalgo, B; et al. Epigenome-wide association study of fasting measures of glucose, insulin, and HOMA-IR in the Genetics of Lipid Lowering Drugs and Diet Network study. *Diabetes*, 63, 801-807, doi:10.2337/db13-1100 (2014).

[114] Piyathilake, CJ; Badiga, S; Alvarez, RD; Partridge, EE; Johanning, GL. A lower degree of PBMC L1 methylation is associated with excess body weight and higher HOMA-IR in the presence of lower concentrations of plasma folate. *PloS One*, 8, e54544, doi:10.1371/journal.pone.0054544 (2013).

[115] Zhao, J; Goldberg, J; Bremner, JD; Vaccarino, V. Global DNA methylation is associated with insulin resistance: a monozygotic twin study. *Diabetes*, 61, 542-546, doi:10.2337/db11-1048 (2012).

[116] Sharma, S; et al. The impact of self-identified race on epidemiologic studies of gene expression. *Genetic Epidemiology*, 35, 93-101, doi:10.1002/gepi.20560 (2011).

[117] Rastogi, D; Suzuki, M; Greally, JM. Differential epigenome-wide DNA methylation patterns in childhood obesity-associated asthma. *Scientific Reports*, 3, 2164, doi:10.1038/srep02164 (2013).

[118] Michels, KB; Harris, HR; Barault, L. Birthweight, maternal weight trajectories and global DNA methylation of LINE-1 repetitive elements. *PloS One*, 6, e25254, doi:10.1371/journal.pone.0025254 (2011).

[119] Perng, W; et al. A prospective study of LINE-1DNA methylation and development of adiposity in school-age children. *PloS One*, 8, e62587, doi:10.1371/journal. pone.0062587 (2013).

[120] Perng, W; et al. Dietary intake, plasma homocysteine, and repetitive element DNA methylation in the Multi-Ethnic Study of Atherosclerosis (MESA). *Nutrition, Metabolism, and Cardiovascular Dseases : NMCD*, doi:10.1016/j.numecd.2013.11.011 (2013).

[121] Na, YK; Hong, HS; Lee, DH; Lee, WK; Kim, DS. Effect of body mass index on global DNA methylation in healthy Korean women. *Molecules and Cells*, 37, 467-472, doi:10.14348/molcells.2014.0073 (2014).
[122] Campion, J; Milagro, FI; Goyenechea, E; Martinez, JA. TNF-alpha promoter methylation as a predictive biomarker for weight-loss response. *Obesity*, 17, 1293-1297, doi:10.1038/oby.2008.679 (2009).
[123] Kucharski, R; Maleszka, J; Foret, S; Maleszka, R. Nutritional control of reproductive status in honeybees via DNA methylation. *Science*, 319, 1827-1830, doi:10.1126/science.1153069 (2008).
[124] Shafi, A. A. A. a. G. in *DNA Methylation - From Genomics to Technology* (ed Tatiana Tatarinova and Owain Kerton) (2012).
[125] Jhaveri, MS; Wagner, C; Trepel, JB. Impact of extracellular folate levels on global gene expression. *Molecular Pharmacology*, 60, 1288-1295 (2001).
[126] Gonda, TA; et al. Folic acid increases global DNA methylation and reduces inflammation to prevent Helicobacter-associated gastric cancer in mice. *Gastroenterology*, 142, 824-833.e827, doi:10.1053/ j.gastro.2011.12.058 (2012).
[127] Remely, M; et al. Therapeutic perspectives of epigenetically active nutrients. *British Jjournal of Pharmacology*, doi:10.1111/bph.12854 (2014).
[128] Wallwork, JC; Duerre, JA. Effect of zinc deficiency on methionine metabolism, methylation reactions and protein synthesis in isolated perfused rat liver. *The Journal of Nutrition*, 115, 252-262 (1985).
[129] Rowling, MJ; McMullen, MH; Schalinske, KL. Vitamin A and its derivatives induce hepatic glycine N-methyltransferase and hypomethylation of DNA in rats. *The Journal of Nutrition*, 132, 365-369 (2002).
[130] Davis, CD; Uthus, EO; Finley, JW. Dietary selenium and arsenic affect DNA methylation in vitro in Caco-2 cells and in vivo in rat liver and colon. *The Journal of Nutrition*, 130, 2903-2909 (2000).
[131] Halliwell, B. Vitamin C and genomic stability. *Mutation Research*, 475, 29-35 (2001).
[132] Pheiffer, C; Dias, S; Muller, C; Louw, J. Decreased global DNA methylation in the white blood cells of high fat diet fed vervet monkeys (Chlorocebus aethiops). *Journal of Physiology and Biochemistry*, 70, 725-733, doi:10.1007/s13105-014-0341-4 (2014).
[133] Ronn, T; et al. A six months exercise intervention influences the genome-wide DNA methylation pattern in human adipose tissue. *PLoS Genetics*, 9, e1003572, doi:10.1371/journal.pgen.1003572 (2013).
[134] Zhang, FF; et al. Significant differences in global genomic DNA methylation by gender and race/ethnicity in peripheral blood. *Epigenetics: Official Journal of the DNA Methylation Society*, 6, 623-629 (2011).
[135] Zhang, FF; et al. Physical activity and global genomic DNA methylation in a cancer-free population. *Epigenetics: Official Journal of the DNA Methylation Society*, 6, 293-299 (2011).
[136] Fuke, C; et al. Age related changes in 5-methylcytosine content in human peripheral leukocytes and placentas: an HPLC-based study. *Annals of Human Genetics*, 68, 196-204, doi:10.1046/j.1529-8817.2004.00081.x (2004).
[137] Smith, AK; et al. Differential immune system DNA methylation and cytokine regulation in post-traumatic stress disorder. *American Journal of Medical Genetics.*

Part B, Neuropsychiatric genetics : the Official Publication of the International Society of Psychiatric Genetics, 156b, 700-708, doi:10.1002/ajmg.b.31212 (2011).

[138] Baccarelli, A; et al. Rapid DNA methylation changes after exposure to traffic particles. *American Journal of Respiratory and Critical Care Medicine*, 179, 572-578, doi:10.1164/rccm.200807-1097OC (2009).

[139] Bollati, V; et al. Changes in DNA methylation patterns in subjects exposed to low-dose benzene. *Cancer Research*, 67, 876-880, doi:10.1158/0008-5472.can-06-2995 (2007).

[140] Christensen, BC; et al. Aging and environmental exposures alter tissue-specific DNA methylation dependent upon CpG island context. *PLoS Genetics*, 5, e1000602, doi:10.1371/journal.pgen.1000602 (2009).

[141] Langevin, SM; et al. The influence of aging, environmental exposures and local sequence features on the variation of DNA methylation in blood. *Epigenetics: Official Journal of the DNA Methylation Society*, 6, 908-919 (2011).

[142] Gronniger, E; et al. Aging and chronic sun exposure cause distinct epigenetic changes in human skin. *PLoS Genetics*, 6, e1000971, doi:10.1371/journal.pgen.1000971 (2010).

[143] Pilsner, JR; et al. Influence of prenatal lead exposure on genomic methylation of cord blood DNA. *Environmental Health Perspectives*, 117, 1466-1471, doi:10.1289/ehp.0800497 (2009).

[144] McGuinness, D; et al. Socio-economic status is associated with epigenetic differences in the pSoBid cohort. *International Journal of Epidemiology*, 41, 151-160, doi:10.1093/ije/dyr215 (2012).

[145] Issa, J. P. Age-related epigenetic changes and the immune system. *Clinical Immunology (Orlando, Fla.)* 109, 103-108 (2003).

[146] Thompson, RF; et al. Tissue-specific dysregulation of DNA methylation in aging. *Aging Cell*, 9, 506-518, doi:10.1111/j.1474-9726.2010.00577.x (2010).

[147] Bollati, V; et al. Decline in genomic DNA methylation through aging in a cohort of elderly subjects. *Mechanisms of Ageing and Development*, 130, 234-239, doi:10.1016/j.mad.2008.12.003 (2009).

[148] Bellizzi, D; et al. Global DNA methylation in old subjects is correlated with frailty. *Age (Dordrecht, Netherlands)* 34, 169-179, doi:10.1007/ s11357-011-9216-6 (2012).

[149] Fang, L; et al. Environmental-stress-induced Chromatin Regulation and its Heritability. *Journal of Carcinogenesis & Mutagenesis*, 5, doi:10.4172/2157-2518.1000156 (2014).

[150] Ergaz, Z; et al. Placental oxidative stress and decreased global DNA methylation are corrected by copper in the Cohen diabetic rat. *Toxicology and Applied Pharmacology*, 276, 220-230, doi:10.1016/j.taap.2014.02.017 (2014).

[151] Unternaehrer, E; et al. Dynamic changes in DNA methylation of stress-associated genes (OXTR, BDNF) after acute psychosocial stress. *Translational Psychiatry*, 2, e150, doi:10.1038/tp.2012.77 (2012).

[152] Bierne, H; Hamon, M; Cossart, P. Epigenetics and bacterial infections. *Cold Spring Harbor Perspectives in Medicine*, 2, a010272, doi:10.1101/cshperspect.a010272 (2012).

[153] Hesson, LB. Gut microbiota and obesity-related gastrointestinal cancer: a focus on epigenetics. *Translational Gastrointestinal Cancer*, 2, 204-210 (2013).

[154] Paschos, K; Allday, MJ. Epigenetic reprogramming of host genes in viral and microbial pathogenesis. *Trends in Microbiology*, 18, 439-447, doi:10.1016/j.tim.2010.07.003 (2010).

[155] Murphy, JC; Fischle, W; Verdin, E; Sinclair, JH. Control of cytomegalovirus lytic gene expression by histone acetylation. *The EMBO Journal*, 21, 1112-1120, doi:10.1093/emboj/21.5.1112 (2002).

[156] Nitzsche, A; Paulus, C; Nevels, M. Temporal dynamics of cytomegalovirus chromatin assembly in productively infected human cells. *Journal of Virology*, 82, 11167-11180, doi:10.1128/jvi.01218-08 (2008).

[157] Esteki-Zadeh, A; et al. Human cytomegalovirus infection is sensitive to the host cell DNA methylation state and alters global DNA methylation capacity. *Epigenetics : Official Journal of the DNA Methylation Society*, 7, 585-593, doi:10.4161/epi.20075 (2012).

[158] McErlean, P; et al. Human rhinovirus infection causes different DNA methylation changes in nasal epithelial cells from healthy and asthmatic subjects. *BMC Medical genomics*, 7, 37, doi:10.1186/1755-8794-7-37 (2014).

[159] Benbrahim-Tallaa, L; et al. Molecular events associated with arsenic-induced malignant transformation of human prostatic epithelial cells: aberrant genomic DNA methylation and K-ras oncogene activation. *Toxicology and Applied Pharmacology*, 206, 288-298, doi:10.1016/ j.taap.2004.11.017 (2005).

[160] Chen, H; et al. Chronic inorganic arsenic exposure induces hepatic global and individual gene hypomethylation: implications for arsenic hepatocarcinogenesis. *Carcinogenesis*, 25, 1779-1786, doi:10.1093/ carcin/bgh161 (2004).

[161] Reichard, JF; Schnekenburger, M; Puga, A. Long term low-dose arsenic exposure induces loss of DNA methylation. *Biochem Biophys Res Commun*, 352, 188-192, doi:10.1016/j.bbrc.2006.11.001 (2007).

[162] Niedzwiecki, MM; et al. A dose-response study of arsenic exposure and global methylation of peripheral blood mononuclear cell DNA in Bangladeshi adults. *Environmental Health Perspectives*, 121, 1306-1312, doi:10.1289/ehp.1206421 (2013).

[163] Kile, ML; et al. A panel study of occupational exposure to fine particulate matter and changes in DNA methylation over a single workday and years worked in boilermaker welders. *Environmental Health : A Global Access Science Source*, 12, 47, doi:10.1186/1476-069x-12-47 (2013).

[164] Andrew, AS; et al. Drinking-water arsenic exposure modulates gene expression in human lymphocytes from a U.S. population. *Environmental Health Perspectives*, 116, 524-531, doi:10.1289/ ehp.10861 (2008).

[165] Bourdonnay, E; et al. Global effects of inorganic arsenic on gene expression profile in human macrophages. *Molecular Immunology*, 46, 649-656, doi:10.1016/j.molimm. 2008.08.268 (2009).

[166] Castillo, P; Ibanez, F; Guajardo, A; Llanos, MN; Ronco, AM. Impact of cadmium exposure during pregnancy on hepatic glucocorticoid receptor methylation and expression in rat fetus. *PloS one*, 7, e44139, doi:10.1371/journal.pone.0044139 (2012).

[167] Cheng, TF; Choudhuri, S; Muldoon-Jacobs, K. Epigenetic targets of some toxicologically relevant metals: a review of the literature. *Journal of Applied Toxicology : JAT*, 32, 643-653, doi:10.1002/jat.2717 (2012).

[168] Arita, A; Costa, M. Epigenetics in metal carcinogenesis: nickel, arsenic, chromium and cadmium. *Metallomics: Integrated Biometal Science*, 1, 222-228, doi:10.1039/ b903049b (2009).

[169] Ren, X; et al. An emerging role for epigenetic dysregulation in arsenic toxicity and carcinogenesis. *Environmental Health Perspectives*, 119, 11-19, doi:10.1289/ehp.1002114 (2011).

[170] Reichard, JF; Puga, A. Effects of arsenic exposure on DNA methylation and epigenetic gene regulation. *Epigenomics*, 2, 87-104, doi:10.2217/ epi.09.45 (2010).

[171] Smeester, L; et al. Epigenetic changes in individuals with arsenicosis. *Chemical Research in Toxicology*, 24, 165-167, doi:10.1021/tx1004419 (2011).

[172] Tellez-Plaza, M; et al. Association of global DNA methylation and global DNA hydroxymethylation with metals and other exposures in human blood DNA samples. *Environmental Health Perspectives*, 122, 946-954, doi:10.1289/ehp.1306674 (2014).

[173] Hiltunen, MO; et al. DNA hypomethylation and methyltransferase expression in atherosclerotic lesions. *Vascular Medicine*, 7, 5-11, doi:10.1191/1358863x02vm418oa (2002).

[174] Lund, G; et al. DNA methylation polymorphisms precede any histological sign of atherosclerosis in mice lacking apolipoprotein E. *The Journal of Biological Chemistry*, 279, 29147-29154, doi:10.1074/ jbc.M403618200 (2004).

[175] Smolarek, I; et al. Global DNA methylation changes in blood of patients with essential hypertension. *Medical Science Monitor : International Medical Journal Of Experimental And Clinical Research*, 16, Cr149-155 (2010).

[176] Chiang, EP; Wang, YC; Chen, WW; Tang, FY. Effects of insulin and glucose on cellular metabolic fluxes in homocysteine transsulfuration, remethylation, S-adenosylmethionine synthesis, and global deoxyribonucleic acid methylation. *The Journal of Clinical Endocrinology and Metabolism*, 94, 1017-1025, doi:10.1210/jc.2008-2038 (2009).

[177] Jintaridth, P; Tungtrongchitr, R; Preutthipan, S; Mutirangura, A. Hypomethylation of Alu elements in post-menopausal women with osteoporosis. *PloS One*, 8, e70386, doi:10.1371/journal.pone.0070386 (2013).

[178] Nomura, Y; et al. Global methylation in the placenta and umbilical cord blood from pregnancies with maternal gestational diabetes, preeclampsia, and obesity. *Reproductive Sciences (Thousand Oaks, Calif.)* 21, 131-137, doi:10.1177/1933719113492206 (2014).

In: Gene-Environment Interactions and Human Diseases
Editor: Lu Qi
ISBN: 978-1-63483-680-7

Chapter 15

MICROBIOME AND HUMAN HEALTH

Naima G. Cortes-Perez[1] and Luis G. Bermúdez-Humarán[2,3,*]
[1]INRA, UR496 Immuno-Allergie Alimentaire,
CEA/IBiTeC-S/SPI, CEA de Saclay, Gif-sur-Yvette, France
[2]INRA, UMR1319 Micalis, Jouy-en-Josas, France
[3]AgroParisTech, UMR Micalis, Jouy-en-Josas, France

ABSTRACT

The Microbiome, defined as the collection of genomes of the microbiota present in the human body, has recently emerged as an important factor in human physiology and disease. As in all ecosystems, the microbiota (to whole microbial population present within the human body) reaches a dynamic equilibrium in the human body, which can be altered by environmental factors and external stimuli. Today, new technologies (such as metagenomics) are beginning to reveal important aspects of host-microbiota interactions. Thus, microbiota has been shown to interact with the host in several ways in health and disease, including modulating the inflammatory host response to the gut, synthesizing small molecules and proteins that are taken up by the host, changing the amount of available energy in the diet and more recently has been shown to influence also brain development. As such, several studies have examined the role of the microbiome in several diseases including inflammatory bowel diseases (IBD), obesity, diabetes, allergy and even autism. The aim of this chapter is to summarize the link between the human microbiota and health, highlighting the contribution of metagenomics in the advance of this field.

INTRODUCTION

The term Microbiome refers to the genetic material of the catalog of microbial taxa associated with humans [1]. Being frequently confused with the term microbiota, the microbiome was first defined by Joshua Lederberg in 2001[2] and is sometimes referred to as

[*] Corresponding author: E-mail: luis.bermudez@jouy.inra.fr. Phone: +33-134-65-24-63, Fax: +33-134-65-24-62.

our second genome [3]. The term microbiota (previously named microflora) refers to the microbial population present within the human body [4], including bacteria, viruses, archea, protozoans and fungi. The human gut microbiota has been matter of study for decades; however, the high diversity of microorganisms in the gut and the lack of culture techniques in order to identify these microorganisms have largely limited their study. Only within the past 10 years, with the advent of new techniques such as next-generation sequencing methods and array-based microbial identification, has allowed the microorganism diversity in the gut become evident. It is estimated that the human gut hold approximately 10 [14] cells (around 10-fold greater than the rest of the human body combined) [5]. This microorganism diversity contains 3.3 million microbial genes which represent around 150-fold more than the human genome, demonstrating a co-evolutionary relationship.

However, due to the abundance of microhabitats in the human body and the large number of interactions between the different species with the host and the external environment, microbiota can also be conceptualized as a dynamic ecological community [6]. Indeed, each microbial community within the human body has its own structure, depending on the exact environment where it is localized [7]. Many essential body processes require the presence of diverse microorganisms, as they provide the host with essential nutrients, metabolize indigestible compounds, and defend against colonization by opportunistic pathogens, as well as possessing immunomodulatory properties [8].

As with all ecosystems, a balance exists between the human body and the microbiota. However, this dynamic equilibrium can be altered at any time by environmental factors and external interferences, such as the use of antibiotics [9]. These alterations frequently result in microbial imbalances on or inside the body, a phenomenon also called dysbiosis. Thus, in some ecosystems, such as the gut, a high biodiversity is associated with a healthy status, while low biodiversity is more linked to pathological conditions [10]. In this context, there is an increasing interest in the use of microorganisms (*i.e* probiotics) to treat dysbioses.

To achieve the study of microbiota biodiversity, traditional methods were based for many years on culture dependent techniques. However, although the use of these techniques provided a large and interesting set of data, they also resulted in an erroneous view of the human microbiota composition in certain cases. Certainly, many microorganisms need special growth conditions, such as the extremely oxygen sensitive (EOS) bacteria, that makes their culturing and even their detection difficult [11]; whereas others have never been grown in culture and may require special, as yet unknown, growth conditions preventing their identification by culture dependent methods [7]. Recently, several culture independent techniques have been developed allowing for a qualitative and quantitative means of identification and are mostly based on PCR and DNA hybridization techniques. These simple methods have completely changed the notion of the human microbiota, opening the door to new and more complete fields such as metagenomics [12]. Metagenomics is a relatively new field of study (first introduced in 1998) [13] of microbial genomes within diverse environmental samples, which is of increasing importance in microbiology. The introduction of this ecological perception of microbiology is the key to achieving real knowledge about the influence of the microbiota in human health and disease.

Microbes reside on numerous sites of the body including the skin, the conjunctiva, the oral cavity, the respiratory tract, the vagina and in the gastrointestinal (GI) tract. The composition and diversity of these populations is characteristic of the location. This chapter will specifically focus on the composition of the gut microbiota and the implications for human health.

GENOMIC APPROACHES TO STUDY THE HUMAN MICROBIOTA

In the last years a great effort has been made to the study of the human microbiota which is reflected in large-scale projects such as the NIH funded Human Microbiome Project (HMP) (http://commonfund.nih.gov/hmp; HMPC 2012) and the European Commission's Metagenomics of the Human Intestinal Tract, MetaHIT (http://www.metahit.eu/). The HMP launched in 2008 with a funding of $157 million aims to characterize the microbial populations that are found at numerous sites on the human body, including skin, oral cavity and nasal, urogenital and GI tract, and to explore the role of these microbes in human health and disease. The goal of the MetaHit project is to establish associations between the genes of the human intestinal microbiota and health and disease with particular focus on two disorders of increasing importance in Europe: IBD and obesity.

Although the term metagenomics was originally used for the shotgun characterization of total DNA, it is also presently being applied for studies of marker genes such as the 16S rRNA [1]. Both shotgun characterization of total DNA and marker genes are used mainly to analyze community structure of the human microbiome. All these powerful tools provide information on community diversity and structure [14]. Thus, several micro-ecological processes have been defined using metagenomics such as microbiota establishment [15], effect of diet on gut microbiome [16], microbiota changes in IBD [23] and obesity [17, 18] among others.

THE MICROBIOME AND HUMAN HEALTH

Human microbiota is a collection of dynamic ecological communities; thus the perturbation of one of these communities has a direct impact on health and well-being of the host [19, 20]. Since microbiota (microbiome by extension) is sensitive to a myriad of stimuli such as diet, stress, disease, use of antibiotics and pharmacological intervention, it is clear that its composition varies at different stages of life. The physiological changes of the aging intestine are likely to have profound and noticeable effects on the microbial composition. As a result of recent technological advances, the GI tract has been well characterized and is to date the most probably studied human micro-ecosystems, representing high complexity communities [21]. Most of the microorganisms colonizing humans are bacteria present in the GI tract at a density of approximately 10 [14, 22] The intestinal microbiota also contains a minority population of eukaryotic microorganisms (fungi, yeasts), viruses, and archae [7]. While microbiota provides numerous nutritional benefits to the host, including vitamins [23] and short chain-fatty acids (SCFAs) [24], its presence is also vitally important for the development and functionality of the intestinal immune system [25]. Indeed, animals devoid of intestinal microbial stimulation exhibit large defects in the organization and activity of immune structures in the gut, and proper activity can be restored via microbial stimulation [25]. Individually, members of the microbiome (e.g., commensal bacteria) can also have profound effects on host mucosal homeostasis, and specific microbes have been demonstrated to promote inflammatory [26, 27] or anti-inflammatory [28, 29] responses in the gut. Hence, cross-talk between the microbiome and the intestinal immune system is critical in the maintenance of mucosal homeostasis.

One of the key points when considering the role of the gut microbiota in human health and disease is its composition across the lifespan [30, 31]. In this context, the development of molecular ecology (in particular the 16S rDNA-based approaches) has dramatically changed our vision of the gut microbiome [17]: Microbial colonization begins at birth with the exposure of the newborn to the environment. The gut microbiota is relatively simple during the newborn's first year of life and is under continuous evolution throughout the host's lifetime in particular the first three years of life during which time a stable microbiome is established [30]. The dominant genus during the first year of life is Bifidobacterium, a member of the phylum Actinobacteria. During the first 2 years of life, the microbial community grows more complex, while at the same time obtaining a more stable composition. Although the composition of the microbiota differs along the GI tract, with a significant inter-individual variability (which can probably be attributed to both host genetics and environmental factors such as health status, diet, ethnic background and geographical location), [32] Firmicutes and Bacteroidetes become the dominant phyla representing 90% of the human microbiome) [32].

In spite of this relative heterogeneity it is possible to distinguish three main clusters in the gut microbiome named Enterotypes. Each of these three enterotypes (which are determined by species composition) is identifiable by the variation in the levels of one of three genera *Bacteroides, Prevotella* and *Ruminococcus* [33]. Their abundance and proportions vary between individuals and is associated with long-term dietary habits [34]. However, the enterotype is quite a complicated concept because of the necessity to consider the intestinal microbiome according to different factors that could eventually impact its composition (e.g.,. aging, geographical origin, nutritional needs and habits, physiological variations and the impact of westernization, etc.) [35]. Thus, the enterotype concept is one possible way to simplify the microbiome complexity.

The gut microbiota represents a real functional barrier allowing the inhibition of pathogen growth and colonization. Metagenomic sequencing has enabled systematic and unbiased characterization of microbial populations including the viral spectrum and it will enable the development of therapeutic strategies and/or vaccines in the near future [36]. Indeed, the use of this approach also allows the identification of new potential pathogens. Also, a modification of the GI tract's microbial barrier could be an influencing factor in diseases such as IBD (one of the most studied diseases in this field). From this perspective, recent advances in "omics" approaches (i.e., genomics, transcriptomics, proteomics and metabolomics) have opened the door for further investigation of the structure and function of the gut microbiome without the need to cultivate, identifying some promising approaches for future therapeutic and diagnostic applications [37]. Finally, although dysbiosis has been mainly linked to the GI tract, it can take place on any exposed surface or mucus membrane such as the vagina, the skin, or the respiratory system. Indeed, this kind of variation in the microbial population can strongly impact human health.

Inflammatory Bowel Diseases (IBD) and Microbiome

IBD comprises Crohn's disease (CD) and ulcerative colitis (UC), two disorders of the GI tract (which differ mostly by the region of the gut where inflammation progresses and the depth of inflammatory damage), characterized by an abnormal activation of the immune system associated with the gut, resulting in a chronic inflammation of the digestive system. The major symptoms of IBD include diarrhea, fever, abdominal pain, GI bleeding, weight loss, malnutrition and exhaustion, and variable periods of activity and recurrence. Although there are many diseases connected to the gut microbiota and their imbalance such as obesity, celiac disease, irritable bowel syndrome (IBS), antibiotic-associated diarrhea and colorectal cancer, among others, the most evident impact of microbiota in human health is provided by studies on IBD. Indeed, IBD have long been considered to reflect interactions between gut microbiota and the host. For instance, IBD susceptibility is associated with host polymorphisms in bacterial sensor genes such as NOD2[38,39] and TLR4 [40]. Moreover, early infancy exposure to antibiotics has been associated with a significantly increased risk for CD, suggesting that gut microbiota perturbations are important for disease risk [41]. Interestingly, microbiota diversity is significantly diminished in CD, suggesting a decreased gut microbiome resilience that could affect immune interactions. In this context, the beneficial effects in the physiopathology of CD patients of the commensal bacterium *Faecalibacterium prausnitzii* [8, 29, 42] on one side and the undesired effects of the opportunistic pathogen, adherent and invasive *Escherichia coli* (AIEC) [43] on the other, are two well-illustrated examples of this gut microbiota perturbation. Indeed, different studies based on metagenomics methods have reported a reduction of *F. prausnitzii* levels in CD patients and in other patients with different intestinal disorders [11]. In contrast, it was recently shown by DNA pyrosequencing that in TLR5-deficient mice, AIEC colonization might induce lasting changes in the microbiota [44]. On the other hand, specific bacteria of the Enterobacteriaceae family may act together with a disordered microbiota to increase the risk of UC [45]. For example, between twins that are discordant for UC, those affected had significantly reduced bacterial diversity, but increased proportions of Actinobacteria and Proteobacteria [46]. Patients with CD have over-representation of *Enterococcus faecium* and of several Proteobacteria compared with controls [47].

Irritable Bowel Syndrome (IBS) and Microbiome

IBS is a common functional gut disorder with an estimated prevalence of up to 20% of the population and has a negative social and economic impact on patients [48, 49]. Symptoms include abdominal pain, altered stool consistency and frequency, bloating and distension. Even if our current understanding of the pathogenesis of IBS is incomplete, in the last years the gut microbiota has emerged as an important factor that may contribute to its pathophysiology [50, 51]. Some of the proposed beneficial effects of the gut microbiota include maintenance of intestinal homoeostasis, maintenance of peristalsis, intestinal mucosal integrity, and protection against pathogens [52, 53, 54]. For example, animal studies have shown that alterations in the gut microbiota can lead to changes in GI functions such as an altered intestinal motility and visceral hypersensitivity (two disorders often observed in IBS patients) [52, 53, 54]. In addition, epidemiological observations linking bacterial gastroenteritis and small intestinal bacterial overgrowth to IBS, microbiology studies describing quantitative and qualitative alterations in the gut microbiota in patients with IBS, and clinical studies demonstrating the efficacy of antimicrobial and probiotic treatments in patients with IBS clearly demonstrate the role of an altered intestinal microbiota in the pathogenesis of IBS [55]. The onset of IBS symptoms after a gastroenteritis episode or the occurrence of an enteric infection represents one of the strongest points of evidence highlighting the importance of gut microbiota in IBS [56]. Chronic low-grade inflammation is a common feature in many IBS patients [31]. It has also been demonstrated that subgroups of IBS patients may have an altered microbiota composition compared to healthy individuals (based on the analysis of fecal microbiota) [57]. Both an increase and a in the diversity of the microbiota have been reported in IBS patients decrease [58, 59]. From this perspective, the abnormal variation likely reflects a loss of homeostasis, in which the bacterial community is unable to maintain its normal composition. Finally, recent studies have demonstrated the important role of the microbiome-brain-gut communication in the etiology of IBS and altered central nervous system (CNS) control of visceral pain and inflammatory responses are now considered as integral pathophysiological features of this syndrome [60].

Brain-Gut Axis and Microbiome

It is now well-recognized that the gut and the brain are closely related, and that this interaction plays an important part not only in GI function but also in certain feeling states (such as depression) and in intuitive decision making. Recent neurobiological insights into this gut-brain crosstalk have revealed a complex, bidirectional communication system that not only ensures the proper maintenance of GI homeostasis and digestion but is probable to have multiple effects on affect, motivation and higher cognitive functions. Although this axis is regulated at neural, hormonal, and immunological levels, alterations in brain-gut interactions are associated with intestinal inflammation, chronic abdominal pain syndromes, and eating disorders [61]. For example, the high co-morbidity between stress-related psychiatric symptoms such as anxiety with GI disorders including IBD and IBS is further evidence of the importance of this axis [62]. Currently, researchers now recognize the microbiome itself as an active and highly-influential contributing factor in brain-gut axis bidirectional communication network [63, 64]. Although the vast amount of knowledge regarding signaling along this axis is primarily confined to neuronal communication between the enteric nervous system (ENS) and the CNS, the exact role the microbiota plays has yet to be elucidated [61]. Research from both human and animal studies has shown that emotional stimuli (e.g., stress) can negatively impact on the gut microbiota. For instance, maternal separation (model represents one of the best characterized animal models in relation to the long term effects of stress in early life on microbiota), restraint conditions, crowding and heat stress, among others, have all been shown to alter the composition of gut microbiota (for Review see ref. [65]). In addition, a recent study have determined the role of gut microbiota in cerebral metabolism, where Germ-Free mice were found to have an altered metabolic profile to their conventionally colonized counterparts with up to 10 of these metabolites thought to be specifically involved in brain function [66].

METABOLIC DISORDERS (OBESITY/DIABETES) AND MICROBIOME

Obesity and its associated disorders, such as type 2 diabetes mellitus, are a major public health problem, mainly caused by a combination of overeating and sedentary lifestyle, with contributions from host genetics, environment, and adipose tissue inflammation. Moreover, in the last years an increase number of studies have also demonstrated that an altered composition and diversity of gut microbiota could play an important role in the development of metabolic disorders such as obesity and diabetes [67]. From this perspective, recent studies of the distal gut microbiome have proposed that changes in gut microbiota are related to obesity, the metabolic syndrome, and Western diet. Indeed, changes in the composition of the intestinal microbiota have been reported in obese humans and mice with a relative decrease in the proportion of Bacteroidetes in obese *versus* lean individuals [22, 68, 69] suggesting a potential link between microbial diversity and obesity. Therefore, although composition of the gut microbiota varies between individuals, a ‘‘core gut microbiome’’ has been identified that confers greater risk of obesity [69]. In addition, preclinical studies have shown that a 7 week exposure in mice to low doses of the antibiotics induced taxonomic changes in the microbiome characterized by reductions in the Bacteroidetes/Firmicutes ratio, which resembles the microbiota profile in obese humans, and was paralleled by increased body fat [70]. Further support for a causal link between microbiota and obesity is provided by GF mice studies where these animals have reduced total body fat relative to their conventionally reared counterparts, and are resistant to diet-induced obesity [71, 72]. Interestingly, colonization of GF mice with commensal bacteria harvested from conventionally-reared mice produces a 60% increase in body weight within 2 weeks despite decreases in food consumption [73]. Probiotic treatment has also been demonstrated to have beneficial anti-obesity effects in both mouse and human studies [74, 75]. A single dose of *Lactobacillus plantarum* in GF mice maintained on high-fat/high-sugar diet resulted in the up-regulation of genes involved in carbohydrate transport and metabolism [75]. Similarly, the presence of the probiotic strain of *Bifidobacterium longum* elicits an expansion in the diversity of polysaccharides targeted for degradation by a prominent component of the adult human gut microbiota, *Bacteroides thetaiotaomicron*, when simultaneously colonized in GF mice [76]. Furthermore, a link between the metabolic disease, diabetes, and bacterial populations in the gut has been established. Specifically, the proportions of Firmicutes and class Clostridia were significantly reduced in the diabetic group compared to the control group [77]. More recently a correlation between gut microbiome profiles and type 2 diabetes was demonstrated in a Chinese population to the extent that composition of the gut microbiota could predict the disorder in a second cohort [78]. Later, a study in Danish non-obese and obese study participants performed demonstrated that subjects harboring a gut microbiota that is characterized by a low diversity of species are more often affected by obesity and low-grade inflammation than individuals whose microbiome shows a high microbial richness [79]. Moreover, dietary practices might be a predictor for the microbial complexity of the gut associated microbiota. In a study with overweight French subjects, the subgroup with a low bacterial richness of the gut microbiota had an increased risk for metabolic deviation compared to those overweight persons bearing a complex gut mircobiota. In addition, these individuals were more susceptible to dietary intervention by increasing the microbial diversity in the gut [80]. Taken together, these data suggest that high microbial richness of the gut microbiome may be an indicator for metabolic homeostasis and a protecting factor against metabolic deviation and disease.

Allergy and Microbiome

Given the high prevalence over the last years, allergies are considered a modern life epidemic. A new version of the hygiene hypothesis proposes a close association between microbiota and the development of allergic diseases [81]. Indeed, human epidemiological studies show that gut microbiota is strongly altered in Western countries [82]. Recently, a study in Sweden has shown that the reduction of microbiota diversity in the newborn children might favor the development of allergic disease later in their life at school age [83]. In addition, data from a recent research suggest that intestinal homeostasis result of an orchestrated and fine coordination between the intestinal immune system and the bacterial colonization phases, particularly at perinatal period [84]. It has been show that different mediators such as complex carbohydrates, antimicrobial factors, food antigens and antibodies can be found in breast milk thus during lactation period they could shape both the intestinal physiology and the immune systems [85, 86]. It now accepted that diets components (e.g., presence of whole proteins) are essentials for the appropriate development and function of gut- lymphoid structures [87]. Indeed, others studies suggest a synergistic effect between the microbiota and the diet, for example in the digestion of certain diet constituents such as proteins and in the production of metabolites such as fatty acids, probably implicated in the immune modulation [88]. Taken together, these data suggest that the microbiota, but also diet constituents contribute to post-natal education and consequently to gut associated diseases such as allergies food; however litter is known about the mechanisms involved. In this context, recent research work of Nagler and colleagues open light toward the better understanding of role played by microbiota in the development of oral tolerance and allergic sensitization; using a mice model of selective colonization with Clostridia spp., they show this intestinal bacterium use an IL-22-dependent mechanism to upset al.lergen permeability and thereby reduce serum allergen concentrations. This work highlights the potential use of Clostridia spp. (and probably other intestinal bacteria) on the treatment of food allergies [89]. However, further studies are needed to analyze the fingerprint of the gut microbiota on the immune system and their connection with the development of food allergies.

Summary/Perspectives

In conclusion, it is well recognized now the need to study the human microbiome as a whole ecosystem to better understands the relation between microbiota and host health. Thus, new powerful approaches are required to globally analyze these ecosystems, with metagenomic approaches being key for further analysis of the human microbiome. In the future, this kind of strategies (*ie.* metagenomics) would identify biomarkers of well being that correspond to a general and more balanced microbiota. In addition, metagenomics can also provide us with a better understanding of the relationship between us and our microbiome and the role of this interaction with health.

REFERENCES

[1] Ursell, LK; Metcalf, JL; Parfrey, LW; Knight, R. Defining the human microbiome. *Nutr Rev*, 2012, 70 Suppl 1, S38-44.

[2] Lederberg, JM. A. Ome sweet omics, a genealogical treasury of words. *The Scientist*, 2001, 15, 8.

[3] Bruls, T; Weissenbach, J. The human metagenome, our other genome? *Hum Mol Genet*, 2011, 20(R2), R142-148.

[4] Bakhtiar, SM; LeBlanc, JG; Salvucci, E; Ali, A; Martin, R; Langella, P; et al. Implications of the human microbiome in inflammatory bowel diseases. *FEMS Microbiol Lett*, 2013, 342(1), 10-17.

[5] Turnbaugh, PJ; Ley, RE; Hamady, M; Fraser-Liggett, CM; Knight, R; Gordon, JI. The human microbiome project. *Nature*, 2007, 449(7164), 804-810.

[6] Foxman, B; Goldberg, D; Murdock, C; Xi, C; Gilsdorf, JR. Conceptualizing human microbiota: from multicelled organ to ecological community. *Interdiscip Perspect Infect Dis*, 2008, 613979.

[7] Weinstock, GM. Genomic approaches to studying the human microbiota. *Nature*, 2012, 489(7415), 250-256.

[8] Martin, R; Miquel, S; Ulmer, J; Kechaou, N; Langella, P; Bermudez-Humaran, LG. Role of commensal and probiotic bacteria in human health: a focus on inflammatory bowel disease. *Microb Cell Fact*, 2013, 12, 71.

[9] Dethlefsen, L; Huse, S; Sogin, ML; Relman, DA. The pervasive effects of an antibiotic on the human gut microbiota, as revealed by deep 16S rRNA sequencing. *PLoS Biol*, 2008, 6(11), e280.

[10] Manichanh, C; Rigottier-Gois, L; Bonnaud, E; Gloux, K; Pelletier, E; Frangeul, L; et al. Reduced diversity of faecal microbiota in Crohn's disease revealed by a metagenomic approach. *Gut*, 2006, 55(2), 205-211.

[11] Miquel, S; Martin, R; Rossi, O; Bermudez-Humaran, L; Chatel, J; Sokol, H; et al. Faecalibacterium prausnitzii and human intestinal health. *Current opinion in microbiology*, 2013.

[12] Dimon, MT; Wood, HM; Rabbitts, PH; Arron, ST. IMSA: integrated metagenomic sequence analysis for identification of exogenous reads in a host genomic background. *PloS one*, 2013, 8(5), e64546.

[13] Handelsman, J; Rondon, MR; Brady, SF; Clardy, J; Goodman, RM. Molecular biological access to the chemistry of unknown soil microbes: a new frontier for natural products. *Chem Biol*, 1998, 5(10), R245-249.

[14] Gans, J; Wolinsky, M; Dunbar, J. Computational improvements reveal great bacterial diversity and high metal toxicity in soil. *Science*, 2005, 309(5739), 1387-1390.

[15] Koenig, JE; Spor, A; Scalfone, N; Fricker, AD; Stombaugh, J; Knight, R; et al. Succession of microbial consortia in the developing infant gut microbiome. *Proceedings of the National Academy of Sciences of the United States of America*, 2011, 108 Suppl 1, 4578-4585.

[16] De Filippo, C; Cavalieri, D; Di Paola, M; Ramazzotti, M; Poullet, JB; Massart, S; et al. Impact of diet in shaping gut microbiota revealed by a comparative study in children

from Europe and rural Africa. *Proceedings of the National Academy of Sciences of the United States of America*, 2010, 107(33), 14691-14696.
[17] Qin, J; Li, R; Raes, J; Arumugam, M; Burgdorf, KS; Manichanh, C; et al. A human gut microbial gene catalogue established by metagenomic sequencing. *Nature*, 2010, 464(7285), 59-65.
[18] Delzenne, NM; Neyrinck, AM; Backhed, F; Cani, PD. Targeting gut microbiota in obesity, effects of prebiotics and probiotics. *Nat Rev Endocrinol*, 2011, 7(11), 639-646.
[19] Antonio, MA; Rabe, LK; Hillier, SL. Colonization of the rectum by Lactobacillus species and decreased risk of bacterial vaginosis. *The Journal of infectious diseases*, 2005, 192(3), 394-398.
[20] Reid, G; Beuerman, D; Heinemann, C; Bruce, AW. Probiotic Lactobacillus dose required to restore and maintain a normal vaginal flora. *FEMS immunology and medical microbiology*, 2001, 32(1), 37-41.
[21] Eloe-Fadrosh, EA; Rasko, DA. The human microbiome: from symbiosis to pathogenesis. *Annu Rev Med*, 2013, 64, 145-163.
[22] Ley, RE; Turnbaugh, PJ; Klein, S; Gordon, JI. Microbial ecology: human gut microbes associated with obesity. *Nature*, 2006, 444(7122), 1022-1023.
[23] Kau, AL; Ahern, PP; Griffin, NW; Goodman, AL; Gordon, JI. Human nutrition; the gut microbiome and the immune system. *Nature*, 2011, 474(7351), 327-336.
[24] Topping, DL; Clifton, PM. Short-chain fatty acids and human colonic function, roles of resistant starch and nonstarch polysaccharides. *Physiol Rev*, 2001, 81(3), 1031-1064.
[25] Hooper, LV. Bacterial contributions to mammalian gut development. *Trends Microbiol*, 2004, 12(3), 129-134.
[26] Ivanov, II; Atarashi, K; Manel, N; Brodie, EL; Shima, T; Karaoz, U; et al. Induction of intestinal Th17 cells by segmented filamentous bacteria. *Cell*, 2009, 139(3), 485-498.
[27] Ivanov, II; Frutos Rde, L; Manel, N; Yoshinaga, K; Rifkin, DB; Sartor, RB; et al. Specific microbiota direct the differentiation of IL-17-producing T-helper cells in the mucosa of the small intestine. *Cell Host Microbe*, 2008, 4(4), 337-349.
[28] Mazmanian, SK; Round, JL; Kasper, DL. A microbial symbiosis factor prevents intestinal inflammatory disease. *Nature*, 2008, 453(7195), 620-625.
[29] Sokol, H; Pigneur, B; Watterlot, L; Lakhdari, O; Bermudez-Humaran, LG; Gratadoux, JJ; et al. Faecalibacterium prausnitzii is an anti-inflammatory commensal bacterium identified by gut microbiota analysis of Crohn disease patients. *Proceedings of the National Academy of Sciences of the United States of America*, 2008, 105(43), 16731-16736.
[30] O'Toole, PW; Claesson, MJ. Gut microbiota: Changes throughout the lifespan from infancy to elderly. *International Dairy Journal*, 2010, 20(4), 281-291.
[31] Moloney, RD; Desbonnet, L; Clarke, G; Dinan, TG; Cryan, JF. The microbiome: stress, health and disease. *Mammalian Genome*, 2014, 25(1-2), 49-74.
[32] Eckburg, PB; Bik, EM; Bernstein, CN; Purdom, E; Dethlefsen, L; Sargent, M; et al. Diversity of the human intestinal microbial flora. *Science*, 2005, 308(5728), 1635-1638.
[33] Arumugam, M; Raes, J; Pelletier, E; Le Paslier, D; Yamada, T; Mende, DR; et al. Enterotypes of the human gut microbiome. *Nature*, 2011, 473(7346), 174-180.

[34] Wu, GD; Chen, J; Hoffmann, C; Bittinger, K; Chen, YY; Keilbaugh, SA; et al. Linking long-term dietary patterns with gut microbial enterotypes. *Science*, 2011, 334(6052), 105-108.

[35] Yatsunenko, T; Rey, FE; Manary, MJ; Trehan, I; Dominguez-Bello, MG; Contreras, M; et al. Human gut microbiome viewed across age and geography. *Nature*, 2012, 486(7402), 222-227.

[36] Finkbeiner, SR; Allred, AF; Tarr, PI; Klein, EJ; Kirkwood, CD; Wang, D. Metagenomic analysis of human diarrhea, viral detection and discovery. *PLoS Pathog*, 2008, 4(2), e1000011.

[37] Lamendella, R; VerBerkmoes, N; Jansson, JK. 'Omics' of the mammalian gut--new insights into function. *Curr Opin Biotechnol*, 23(3), 491-500.

[38] Ogura, Y; Bonen, DK; Inohara, N; Nicolae, DL; Chen, FF; Ramos, R; et al. A frameshift mutation in NOD2 associated with susceptibility to Crohn's disease. *Nature*, 2001, 411(6837), 603-606.

[39] Hugot, JP; Chamaillard, M; Zouali, H; Lesage, S; Cezard, JP; Belaiche, J; et al. Association of NOD2 leucine-rich repeat variants with susceptibility to Crohn's disease. *Nature*, 2001, 411(6837), 599-603.

[40] Franchimont, D; Vermeire, S; El Housni, H; Pierik, M; Van Steen, K; Gustot, T; et al. Deficient host-bacteria interactions in inflammatory bowel disease? The toll-like receptor (TLR)-4 Asp299gly polymorphism is associated with Crohn's disease and ulcerative colitis. *Gut*, 2004, 53(7), 987-992.

[41] Hviid, A; Svanstrom, H; Frisch, M. Antibiotic use and inflammatory bowel diseases in childhood. *Gut*, 2011, 60(1), 49-54.

[42] Miquel, S; Martin, R; Rossi, O; Bermudez-Humaran, LG; Chatel, JM; Sokol, H; et al. Faecalibacterium prausnitzii and human intestinal health. *Current opinion in microbiology*, 2013, 16(3), 255-261.

[43] Miquel, S; Peyretaillade, E; Claret, L; de Vallee, A; Dossat, C; Vacherie, B; et al. Complete genome sequence of Crohn's disease-associated adherent-invasive E. coli strain LF82. *PloS one*, 2010, 17,15(19). .

[44] Chassaing, B; Koren, O; Carvalho, FA; Ley, RE; Gewirtz, AT. AIEC pathobiont instigates chronic colitis in susceptible hosts by altering microbiota composition. *Gut*, 2013, Jul 29.

[45] Garrett, WS; Gallini, CA; Yatsunenko, T; Michaud, M; DuBois, A; Delaney, ML; et al. Enterobacteriaceae act in concert with the gut microbiota to induce spontaneous and maternally transmitted colitis. *Cell Host Microbe*, 2010, 8(3), 292-300.

[46] Lepage, P; Hasler, R; Spehlmann, ME; Rehman, A; Zvirbliene, A; Begun, A; et al. Twin Study Indicates Loss of Interaction Between Microbiota and Mucosa of Patients With Ulcerative Colitis. *Gastroenterology*, 2011, 141(1), 227-236.

[47] Mondot, S; Kang, S; Furet, JP; Aguirre de Carcer, D; McSweeney, C; Morrison, M; et al. Highlighting new phylogenetic specificities of Crohn's disease microbiota. *Inflamm Bowel Dis*, 2011, 17(1), 185-192.

[48] Ford, AC; Talley, NJ. Irritable bowel syndrome. *BMJ*, 2012, 345, e5836.

[49] Dupont, HL. Review article: evidence for the role of gut microbiota in irritable bowel syndrome and its potential influence on therapeutic targets. *Aliment Pharmacol Ther*, 2014, 39(10), 1033-1042.

[50] Ohman, L; Simren, M. Intestinal microbiota and its role in irritable bowel syndrome (IBS). *Curr Gastroenterol Rep*, 2013, 15(5), 323.

[51] Ohman, L; Simren, M. New insights into the pathogenesis and pathophysiology of irritable bowel syndrome. *Digestive and Liver Disease*, 2007, 39(3), 201-215.

[52] Bercik, P; Wang, L; Verdu, EF; Mao, YK; Blennerhassett, P; Khan, WI; et al. Visceral hyperalgesia and intestinal dysmotility in a mouse model of postinfective gut dysfunction. *Gastroenterology*, 2004, 127(1), 179-187.

[53] Lee, BJ; Bak, YT. Irritable Bowel Syndrome, Gut Microbiota and Probiotics. *Journal of Neurogastroenterology and Motility*, 2011, 17(3), 252-266.

[54] Collins, S; Verdu, E; Denou, E; Bercik, P. The Role of Pathogenic Microbes and Commensal Bacteria in Irritable Bowel Syndrome. *Digestive Diseases*, 2009, 27, 85-89.

[55] Ringel, Y; Maharshak, N. Intestinal microbiota and immune function in the pathogenesis of irritable bowel syndrome. *American Journal of Physiology-Gastrointestinal and Liver Physiology*, 2013, 305(8), G529-G541.

[56] Spiller, R; Garsed, K. Postinfectious Irritable Bowel Syndrome. *Gastroenterology*, 2009, 136(6), 1979-1988.

[57] Salonen, A; de Vos, WM; Palva, A. Gastrointestinal microbiota in irritable bowel syndrome, present state and perspectives. *Microbiology* 2010, 156(Pt 11), 3205-3215.

[58] Codling, C; O'Mahony, L; Shanahan, F; Quigley, EM; Marchesi, JR. A molecular analysis of fecal and mucosal bacterial communities in irritable bowel syndrome. *Dig Dis Sci*, 2010, 55(2), 392-397.

[59] Claesson, MJ; Jeffery, IB; Conde, S; Power, SE; O'Connor, EM; Cusack, S; et al. Gut microbiota composition correlates with diet and health in the elderly. *Nature*, 2012, 488(7410), 178-184.

[60] Collins, SM; Bercik, P. The Relationship Between Intestinal Microbiota and the Central Nervous System in Normal Gastrointestinal Function and Disease. *Gastroenterology*, 2009, 136(6), 2003-2014.

[61] Mayer, EA. Gut feelings: the emerging biology of gut-brain communication. *Nat Rev Neurosci*, 2011, 12(8), 453-466.

[62] Kennedy, PJ; Clarke, G; Quigley, EM; Groeger, JA; Dinan, TG; Cryan, JF. Gut memories: towards a cognitive neurobiology of irritable bowel syndrome. *Neurosci Biobehav Rev*, 2012, 36(1), 310-340.

[63] Bercik, P. The microbiota-gut-brain axis: learning from intestinal bacteria? *Gut*, 2011, 60(3), 288-289.

[64] Collins, SM; Surette, M; Bercik, P. The interplay between the intestinal microbiota and the brain. *Nature Reviews Microbiology*, 2012, 10(11), 735-742.

[65] Collins, SM; Bercik, P. Gut microbiota: Intestinal bacteria influence brain activity in healthy humans. *Nat Rev Gastroenterol Hepatol*, 2013, 10(6), 326-327.

[66] Matsumoto, M; Kibe, R; Ooga, T; Aiba, Y; Sawaki, E; Koga, Y; et al. Cerebral low-molecular metabolites influenced by intestinal microbiota: a pilot study. *Front Syst Neurosci*, 2013, 7, 9.

[67] Parekh, PJ; Arusi, E; Vinik, AI; Johnson, DA. The role and influence of gut microbiota in pathogenesis and management of obesity and metabolic syndrome. *Front Endocrinol (Lausanne)*, 2014, 5, 47.

[68] Ley, RE; Backhed, F; Turnbaugh, P; Lozupone, CA; Knight, RD; Gordon, JI. Obesity alters gut microbial ecology. *Proc Natl Acad Sci U S A*, 2005, 102(31), 11070-11075.

[69] Turnbaugh, PJ; Hamady, M; Yatsunenko, T; Cantarel, BL; Duncan, A; Ley, RE; et al. A core gut microbiome in obese and lean twins. *Nature*, 2009, 457(7228), 480-484.

[70] Cho, I; Yamanishi, S; Cox, L; Methe, BA; Zavadil, J; Li, K; et al. Antibiotics in early life alter the murine colonic microbiome and adiposity. *Nature*, 2012, 488(7413), 621-626.

[71] Backhed, F; Manchester, JK; Semenkovich, CF; Gordon, JI. Mechanisms underlying the resistance to diet-induced obesity in germ-free mice. *Proc Natl Acad Sci U S A*, 2007, 104(3), 979-984.

[72] Rabot, S; Membrez, M; Bruneau, A; Gerard, P; Harach, T; Moser, M; et al. Germ-free C57BL/6J mice are resistant to high-fat-diet-induced insulin resistance and have altered cholesterol metabolism. *Faseb Journal*, 2010, 24(12), 4948-+.

[73] Backhed, F; Ding, H; Wang, T; Hooper, LV; Koh, GY; Nagy, A; et al. The gut microbiota as an environmental factor that regulates fat storage. *Proceedings of the National Academy of Sciences of the United States of America*, 2004, 101(44), 15718-15723.

[74] Arora, T; Singh, S; Sharma, RK. Probiotics: Interaction with gut microbiome and antiobesity potential. *Nutrition*, 2013, 29(4), 591-596.

[75] Marco, ML; Peters, THF; Bongers, RS; Molenaar, D; van Hemert, S; Sonnenburg, JL; et al. Lifestyle of Lactobacillus plantarum in the mouse caecum. *Environmental Microbiology*, 2009, 11(10), 2747-2757.

[76] Sonnenburg, JL; Chen, CTL; Gordon, JI. Genomic and metabolic studies of the impact of probiotics on a model gut symbiont and host. *Plos Biology*, 2006, 4(12), 2213-2226.

[77] Larsen, N; Vogensen, FK; van den Berg, FWJ; Nielsen, DS; Andreasen, AS; Pedersen, BK; et al. Gut Microbiota in Human Adults with Type 2 Diabetes Differs from Non-Diabetic Adults. *Plos One*, 2010, 5(2).

[78] Qin, JJ; Li, YR; Cai, ZM; Li, SH; Zhu, JF; Zhang, F; et al. A metagenome-wide association study of gut microbiota in type 2 diabetes. *Nature*, 2012, 490(7418), 55-60.

[79] Le Chatelier, E; Nielsen, T; Qin, JJ; Prifti, E; Hildebrand, F; Falony, G; et al. Richness of human gut microbiome correlates with metabolic markers. *Nature*, 2013, 500(7464), 541-+.

[80] Cotillard, A; Kennedy, SP; Kong, LC; Prifti, E; Pons, N; Le Chatelier, E; et al. Dietary intervention impact on gut microbial gene richness. *Nature*, 2013, 500(7464), 585-+.

[81] Molloy, J; Allen, K; Collier, F; Tang, MLK; Ward, AC; Vuillermin, P. The Potential Link between Gut Microbiota and IgE-Mediated Food Allergy in Early Life. *Int J Env Res Pub He*, 2013, 10(12), 7235-7256.

[82] Feehley, T; Stefka, AT; Cao, S; Nagler, CR. Microbial regulation of allergic responses to food. *Seminars in immunopathology*, 2012, 34(5), 671-688.

[83] Abrahamsson, TR; Jakobsson, HE; Andersson, AF; Bjorksten, B; Engstrand, L; Jenmalm, MC. Low gut microbiota diversity in early infancy precedes asthma at school age. *Clin Exp Allergy*, 2014, 44(6), 842-850.

[84] Blottiere, HM; de Vos, WM; Ehrlich, SD; Dore, J. Human intestinal metagenomics, state of the art and future. *Current opinion in microbiology*, 2013, 16(3), 232-239.

[85] Hansen, CH; Nielsen, DS; Kverka, M; Zakostelska, Z; Klimesova, K; Hudcovic, T; et al. Patterns of early gut colonization shape future immune responses of the host. *PloS one*, 2012, 7(3), e34043.

[86] Bernard, H; Ah-Leung, S; Drumare, MF; Feraudet-Tarisse, C; Verhasselt, V; Wal, JM; et al. Peanut allergens are rapidly transferred in human breast milk and can prevent sensitization in mice. *Allergy*, 2014, 69(7), 888-897.

[87] Faria, AM; Gomes-Santos, AC; Goncalves, JL; Moreira, TG; Medeiros, SR; Dourado, LP; et al. Food components and the immune system: from tonic agents to allergens. *Frontiers in immunology*, 2013, 4, 102.

[88] Arpaia, N. Keeping peace with the microbiome: acetate dampens inflammatory cytokine production in intestinal epithelial cells. *Immunology and cell biology*, 2014, 92(7), 561-562.

[89] Stefka, AT; Feehley, T; Tripathi, P; Qiu, J; McCoy, K; Mazmanian, SK; et al. Commensal bacteria protect against food allergen sensitization. *Proceedings of the National Academy of Sciences of the United States of America*, 2014, 111(36), 13145-13150.

In: Gene-Environment Interactions and Human Diseases
Editor: Lu Qi
ISBN: 978-1-63483-680-7

Chapter 16

THE ROLE OF GENOMICS IN PERSONALISED HEALTH MANAGEMENT

Denis C. Bauer
CSIRO, Sydney, NSW, Australia

ABSTRACT

High-throughput data collection and unprecedented computational capabilities have the potential to revolutionize health management. Specifically, genome information can be used to diagnose rare genetic diseases, modulate treatment, and devise risk-stratification to prevent or delay onset of common diseases. The potential to provide truly personalized healthcare is currently hampered by insufficient understanding of genomic influences on disease and the fact that many diseases are modulated by environmental factors such as diet, lifestyle, and microbiome. Large-scale, longitudinal research programs need to study the intrinsic properties between both genetics and environment to unravel their risk-contribution. During this discovery process, frameworks need to be established to counteract unrealistic expectations and promote good scientific practice to interpret sources of uncertainty and inform decision-making for clinical management and personal health.

INTRODUCTION

Will Technological Advances Lead to a Runaway Health Management?

Unprecedented computational capabilities and high-throughput data collection methods promise a new era of personalised, evidence-based healthcare, utilising individual genomic profiles to tailor health management [1]. Despite reported successes, amongst others, in genome-informed treatment [2], pre-natal diagnoses [3], and research relating lifestyle-choices and cancer risk [4], there are barriers to the use of genetic tests by healthcare practitioners [5-7]. Foremost amongst these is unfamiliarity with genetic tests and their interpretation. For instance, a recent survey identified the clinician's uncertainty about the available genetic tests as well as lack of training and guidelines as the largest influences on the decision to include genetic testing into patient care when assessing predisposition to diseases and/or drug response [8].

While the medical community is hesitant to base decisions on these still experimental methodologies, genetic testing is available to patients from stakeholders with commercial interest through direct-to-consumer (DTC) testing. As forecasted by Knoppers [9], the growing yet insufficiently regulated commercial market has let these stakeholders to overstate the potential benefits of genetic testing without adequate heed to the ethical and social implications. As a result, policy makers have been put under substantial pressure to extend public screening programs before the scientific merit of such tests has been adequately demonstrated [10]. This prompted the European Academies Science Advisory Council (EASAC) to submit a report to the European Union advising regulation of DTC genetic testing to exclude diagnostic, presymptotic, and prenatal testing. It recommended that policy makers consider the full implications of introducing genomic sequencing in screening programs [11]. In the US, the Food and Drug Administration (FDA) has stopped marketing campaigns featuring genetic tests until claims of their clinical utility are verified and cleared by the FDA (Document Number: GEN1300666).

However, scientific rigor and political regulations widen the gap between what is technologically possible and the current medical practice [12, 13]. In light of this, this chapter reviews the state-of-the-art in personalised health care, investigates the continuation to personalised whole-of-life health approaches, discusses technical issues that may prevent future realisation of these ideas and outlines how life science in conjunction with technological advances can improve clinical practice and enable people to "self manage in the face of physical, emotional and social challenges" [14].

GENOMIC PROFILES CAN IMPROVE DIAGNOSES AND RISK PREDICTION

Diagnosis for Rare Diseases through Sequencing

Traditionally, a patient's sample is tested for known genomic variants (mutations) in a specific gene (or set of genes). These genetic tests for highly penetrant "pathogenic" variants have been used in clinical practice for several decades. They include preimplantation and prenatal diagnoses, paediatric and adult-onset genetic conditions, and predictive testing for highly penetrant adult-onset conditions and treatment of cancer [15]. Genetic testing has also been applied in screening programs of the general population (i.e., newborn screening) and of high-risk populations (e.g., Tay Sachs screening in the Jewish community).

However, these traditional test for specific genetic aberrations are able to provide a clear diagnosis in only an estimated 42% of cases attending a genetic clinic (8% could be diagnosed to have no genetic disorder) [15]. For the remaining 50% who could not be diagnosed with the first test, the cost per diagnosis as the result of sequential traditional testing has been estimated conservatively at US $25,000 [15]. A clinically and economically beneficial alternative for patients with an unknown high penetrance genetic disorder (Figure 1, first panel) is whole genome/exome sequencing as it can identify germline variants from across the genome simultaneously (Figure 1, second panel). Genomic sequencing can lead to a successful diagnosis in up to 50% of cases where traditional genetic testing failed [15].

Disease Risk Prediction through Sequencing

The largest fraction of the disease burden stems from common conditions such as diabetes, heart disease and stroke. Here, genomic information is used towards disease risk prediction for early detection and prevention (Figure 1, third panel). Unlike highly penetrant inherited diseases, common conditions are complex multi-gene disorders arising from the interplay of multiple genetic variants. For example, hereditary breast and ovarian cancer, caused by dominantly inherited mutations in the *BRCA1* and *BRCA2* genes accounts for only 5% of breast cancer cases [16]. Each variant individually explains only a small proportion of the risk, but collectively and in concert with both identifiable and non-identifiable environmental and stochastic factors (as discussed in the next section), define the total risk. This is known as the "liability threshold model" [17]. Therefore, while genome sequencing provides simultaneously all variants in the genome, quantifying which ones contribute towards the disease remains an area of active research.

To provide informative testing for multiple variants for common complex disorders, large-scale studies are needed to assess the strength of the association between a specific combination of variants and the disease, as well as to determine the predictive value of the profile. However, making predictions from genomic profiles about an individual's complex-disease risk to inform outcome or treatment will not be possible on currently feasible sample sizes [18, 19]. This is due to the "curse of dimensionality" (the $p>>n$ problem), which states that prediction accuracy requires a training set to have a sufficiently large number of samples (individuals, n) to learn the contribution from each of the measured features (e.g., Single Nucleotide Polymorphisms (SNPs), p).

Accordingly, more studies are needed to better understand the multifactorial contribution to disease of individual variants, interacting variants and exposure risk factors before we can start to provide advice based on risk profiles. Efforts are currently underway with the initiation of the "Human Phenome Project" [20] and the "Evaluation of Genomic Applications in Practice and Prevention" (EGAPP) [21].

Ethical Flipside of Identifying Genome Wide Variants

Automatically obtaining genome-wide information may reveal diseases or risks unrelated to the original clinical indication for which the test was performed. This introduces ethical dilemmas regarding the reporting of these "incidental" findings. Recommendations that known causative variants for 37 adult-onset, treatable diseases should always be reported in clinical testing have proven highly contentious [22]. Although preliminary data indicate that over 93% of patients choose to receive secondary incidental findings from genomic sequencing [23], further genetic counselling studies are required in this area to inform academic debate prior to widespread adoption of genomic sequencing.

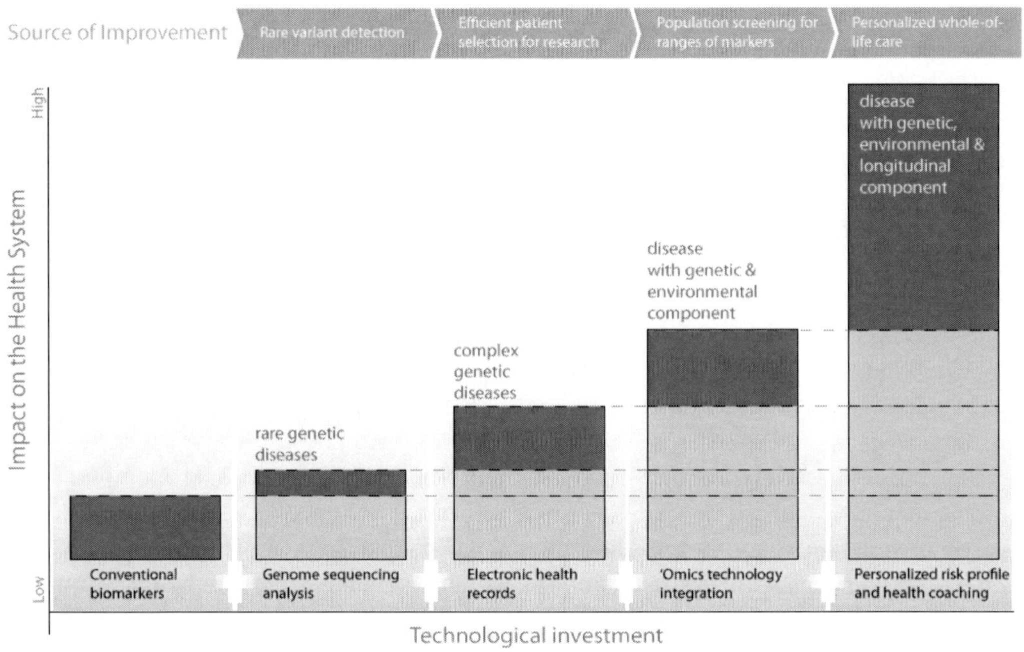

Figure 1. Technological advancement towards whole-of-life care. Patients with rare genetic diseases are first to benefit from genomic sequencing based-testing as affected genes and pathways can be identified (second panel). Utilizing electronic health records will improve diagnosis and treatment of complex genetic diseases, such as cancer, as informative patient groupings can be achieved more efficiently (third panel). Complex diseases with a clear environmental component, such as diabetes, will benefit from the integration of other 'omics based assays e.g., DNA methylation (fourth panel), however, personalised preventative and intervention strategies (rightmost panel) are required to manage the health of this high-burden multi-causal group optimally.

The potential of this technology is only beginning to be realised with the majority of genomic variants as yet clinically uncharacterized, and therefore of unknown significance [24]. Collaborative efforts across genomic laboratories, such as the Human Variome project [20] and ClinVar [25], play an important role in documenting genetic variants and their clinical associations. As these repositories of genotype-phenotype associations continue to expand, the diagnostic yield from genome sequencing will continue to grow.

STRATIFYING BY GENOMIC PROFILE IMPROVES TREATMENT AND DISEASE AVERSION

Environment Influences Disease Risk

Other than a multitude of variants collectively influencing the disease risk, there is strong evidence that environmental factors may play a role in certain diseases. This is especially the case for complex diseases, as these are not 100% determined by genetic profiles alone, as primarily determined through GWAS, requiring even more dimensions to be factored into the risk calculation [26]. This phenomenon, dubbed as "missing heritability" [27], has been debated to be the result of various potential mechanisms, including allelic architecture, rare variants, epistasis, parent of origin effects, and epigenetics (see reviews [28] and [29]). The latter is exemplified by twin studies showing variable disease penetrance in genetically identical individuals. The combination of environment and lifestyle choices has been shown to contribute at comparable levels as genetic loci and has been proposed to explain a large fraction of the observed variance that is not explained by GWAS hits (Figure 1, fourth panel) [30]. A recent Environment-Wide Association Study (EWAS) on type 2 diabetes mellitus (T2D) discovered significant association with a wide range of exposure factors such as the pesticide-derivative heptachlor epoxide, polychlorinated biphenyls and the vitamins E and A [31]. Ashley et al. [32] extended this work to study adverse drug reactions. Despite difficulty in ascertaining causality, the potential for novel exposure factors of large effect associated with T2D justify the use of EWAS given that environmental factors can be found with effect sizes comparable to the best loci found by GWAS [33]. EWAS will become even more important once the capacity to define and quantify the complete exposome (i.e., the combination of physical, chemical, nutritional, life-style, and psychosocial exposure factors) becomes feasible [34-36].

Stratified Prevention by Disease Risk

In the mean time, however, applying genetic testing as a screening rather than a diagnostic tool can identify people at high risk and focus preventative strategies, known as risk stratification. For example, a model can be trained to predict a genotype risk score for type 1 diabetes (T1D). If the threshold is set so that it returns a positive result for 18% of the population, as T1D has a prevalence of 0.004, this means that for every 50 who receive preventative intervention one T1D case can be prevented, an improvement over the traditional diagnostic test of one for every 250. In this way a weak diagnostic test can be repurposed as a useful risk stratification tool [37].

Stratifying Patients for Treatment

Even more complex are the interaction of somatic mutation and the environment as relevant in cancer treatment. Current cancer treatments are largely determined by the organ affected; yet, genomic profiling shows enormous heterogeneity even among tumours with shared tissue origins [18]. As cancer is ultimately driven by genetic alterations, genomic tumour profiling is increasingly recognised as a valuable additional tool to inform treatment. A recent comprehensive survey of cancer genomes demonstrated a correlation between a patient's tumour genome and the clinical features of disease, drug response, and patient outcomes [18]. The collaborative efforts from The Cancer Genome Atlas (TCGA) has subsequently identified 291 driver genes acting on 3,205 tumours from 12 different cancer types [38]. This paves the way to driver-gene-stratified tumour-diagnosis and treatment advice in the future.

Nutrigenomics and Whole of Life Health

Genome Profile-Driven Modulations of Lifestyle

Unlike mortality caused by injuries, genetic disorders or infectious diseases, premature deaths caused by hazardous life-style choices are missed prevention opportunities [39]. Modifiable risk factors such as smoking and poor diet were found to be responsible for 33% of all US deaths in 2000 [40]. Danaei et al. [41], who extended the analysis to include other dietary and metabolic risk factors, argue that influencing behaviour to mitigate dietary, lifestyle, and metabolic risk factors can avert substantially more deaths than further improving the health care system. Based on the notion that improved lifestyle can reduce morbidity and significantly extend the productive lifespan of individuals, investment in patient education and long-term engagement-strategies is likely to have a higher return on investment and save more lives than extending the exclusive investment in traditional health provision.

Therefore adapting food and nutritional intake to individual needs, depending on the individual's life stage, life style, and life situation is desirable for disease prevention (Figure 1, fifth panel) [42]. While lifestyle interventions are effective for preventing type 2 diabetes (T2DM) at any level of genetic risk, the benefit increases for individuals at high genetic susceptibility [43]. Hence disease risk may be reduced by changing in health behaviour in response to better understanding of all aspects that have an influence, such as an individual's genome, diet and the human gut microbiota, which encompasses a complex ecosystem in the intestine with a profound impact on the host metabolism [42]. An attack on the entirety of this problem is likely to run directly into the "curse of dimensionality." It is hence not surprising that only modest progress has been made in this field [44]. Nevertheless, personalised nutrition has proven to be important and successful in the case of inherited or acquired metabolic disorders in which the molecular mechanisms are well established, such as in phenylketonuria, lactose intolerance, and coeliac disease [45-47].

Paucity of Knowledge in How Diet Influences Overall Health

Establishing robust dietary recommendations to prolong pre-symptomatic states is specifically complicated as emerging properties from this complex system can have unpredictable negative effects. For example, an increased intake of riboflavin may be recommended to lower blood pressure in those homozygous for the Methylenetetrahydrofolate reductase *(MTHFR)* C677T polymorphism [48], yet increasing riboflavin in a low folate physiological background can increase genomic instability and thus cancer risk [49]. Only long-term large-scale scientific studies can investigate which risk is higher and what intervention should be undertaken depending on genotype, lifestyle or disease state (e.g., cancer survivor, smoker). In the interim, in the absence of reliable mathematical predictive models of nutrient-gene interactive effects of personalised intervention outcomes, it remains possible to utilise nutriome/nutrient array *in vitro* test systems to determine the optimal nutrient combinations to deliver a desired health outcome at the cellular level (e.g., genome stability, mitochondrial function) for an individual utilising their own cells (e.g., lymphocytes, fibroblasts, stem cells) [49].

Therefore, to tailor dietary advice around a person's genetic profile, well-designed studies interrogating specific hypotheses are necessary, as planned by the "Micronutrients Genomics Project" [50] and "Personalized Nutrition Project" (http://www.personalnutrition.org/).

Challenges in Data Analysis and Community Concerns Hamper Clinical Adoption

Data Analysis and Computational Challenges

The clinical use of whole exome or genome sequencing requires validated, robust and reproducible methods for sequence generation and analysis. The complexity of high throughput sequence data does not easily fit within existing laboratory standards for quality management, reference material development and independent measures of test performance. Additionally, rapid changes and developments in analytical components including chemistry, instrumentation and bioinformatics protocols present challenges in maintaining best practice whilst simultaneously complying with more rigid regulatory and quality management system requirements and standards [51]. This is especially challenging for commercial software products. Furthermore, the computational requirements necessary to analyse genomes at a scale relevant for the health-system command a large investment in compute hardware and IT personnel, which constitutes a barrier to entry for small laboratories and is difficult to maintain at peak times for larger institutes.

Recent projects, like NGSANE [52], aim to cater for this resource-hungry, fast pace yet sensitive environment of personalized medicine. NGSANE governs data analysis thereby reducing the overhead between projects yet maintains bespoke project-specific bioinformatics scripting. Furthermore, it offers an interface to commercial cloud computing frameworks, like Amazon Web Services (AWS), providing an economical alternative to in-house compute clusters. Finally, it provides an interactive Web2.0-based presentation layer for quality inspection and intuitive tools to interrogate the resulting data. Other projects, like SeqAn [53], take the need for reproducible robust bioinformatics analysis to the core algorithms used by the software tools. However, it will take time to establish best practice of data analysis and deliver production grade software frameworks that are viable in a clinical setting.

Privacy Concerns and Information Abuse

Public concern about the misuse of genetic information has been longstanding, particularly relating to the potential for "genetic discrimination." This is commonly defined as a person being treated differently on the basis of their actual or assumed genetic makeup due to their potential to develop a specific condition. While some countries have legislation specifically prohibiting genetic discrimination, for example, Belgium, France, Germany and Sweden [54] or the USA, these laws typically do not cover life insurance [55]. Applicants for life or income insurance are expected to disclose participation in studies that return genetic or genomic information to participants, impacting on the affordability of life insurance products as well as the number of consenting participants in scientific studies. As it is to be expected that the majority of people have some deleterious variants [24], predicting concrete disease risk therefore also presents a challenge for insurance underwriters. A practical approach therefore is to hold off ramification to insurance policies until evidence of the disease is present. However, individuals will have been flagged and may be under closer surveillance. Furthermore, this extends to family members as the risk allele may be shared. A trade-off therefore arises between the need to protect individual privacy and the benefits that can be realised through international data sharing to ensure high quality interpretation about the significance of an individual's genomic profile. Practices and measures that permit confidential data sharing and use while protecting confidentiality are needed by pathology services, researchers, policy analysts, and others (see for example [56] and [57]).

Investing in a Technological Health Care Era Enables Better Study Design and Empowers Patients to Better Self-Manage

A New Era of Fact-Based Study Design

Electronic records are a key enabler in the cost-effective development of evidence for practice. For instance, a number of recent initiatives (eMERGE, BioVu) aim to exploit hospital electronic health records (EHRs) for the recruitment of patient cohorts and for undertaking Electronic Health Driven Genome Research (EDGR). The rationale for EDGR is the dramatic efficiencies that can be attained by recruiting patients and mining phenotypic data by searching already existing EHRs. A number of EDGR studies undertaken to date illustrate the spectacular success of this approach [58]. For example, analysis of EHRs coupled with microarray genotyping data from 1317 cases of hypothyroidism and 5053 controls was able to identify 4 novel genetic variants associated with this disease [59]. Such studies can be undertaken at a fraction (~10%) of the cost of conventionally-recruited targeted GWAS.

EDGR also enables Phenome Wide Association Studies (PheWAS). PheWAS is effectively a reverse GWAS to associate genomic regions with patient phenotypes, a generalization, which is outside the scope of traditional, disease focused GWAS. One recent example illustrating the utility of this approach mined EHRs from 1,290 cases of rheumatoid arthritis with 1,236 controls to identify a previously unknown association of autoantibodies with autoimmune risk alleles and clinical diagnoses [60]. Another application of EDGR is to detect drug-gene interactions (pharmacogenomics). For example, to test if concurrent use of tamoxifen and antidepressants (that function by inhibiting Cytochrome P450 2D6) is associated with increase breast cancer recurrence, EHRs of 1,962 patients using tamoxifen were analysed [61]. The investigators concluded that despite the biological rationale, there was no increased breast cancer recurrence in concomitant use of Cytochrome P450 2D6 inhibitors such as tamoxifen.

Larger Cohorts Allow for Less Stringent Information Collection

Allowing people to volunteer for study participation and provide phenotypic information themselves results in an even larger cohort of individuals to be incorporated in a study. While this has the obvious downside of less trustworthy information reporting, the larger number of samples is likely to compensate the noise. Services such as Patients Like Me (PLM), where patients themselves provide information and ideas to one other to manage their diseases [62]. PLM also provides the opportunity for researchers to recruit patients to specific trials, including GWAS. However, while PLM is a specific social network, general services such as Facebook and Twitter can also be utilised to provide health information to consumers. A recent trial used Facebook for a web based diet survey and recruited a representative cohort for minimal cost [63]. Meanwhile Twitter feeds can be monitored for signals indicative of disease outbreaks, the progression of the flu season, or reports of adverse drug events [64, 65]. While much of the information available to individuals will come from medical doctors and health services, social networking shows great potential to enable individuals to self manage and receive peer support from the community without the intervention of overly stretched health services [74].

Automated Data Collection Marries Large Cohorts with Highly Accurate Data

There is an increasing trend away from personal computers to mobile devices and the use of applications (Apps) in particular, as the vehicle by which to interact with information from the World Wide Web. Apps provide the opportunity to capture biological data [66] and enable service providers to reach patients, and *vice versa*, in a ubiquitous manner. Mobile applications for health typically deliver location-based information and interactive tools to record progress and reward health behaviours. Personalised, adaptive technologies that support individual preferences or knowledge and that respond to individual behaviours can be expected to be more powerful than generic one size fits all applications.

An emerging trend in the health technology space is the popularity of personal sensing through mobile phones and standalone commercial products. These generic, cheap, often wearable, devices, that include FitBit, Jawbone and Nike Fuelband, have the ability to monitor location, movement, heart rate, blood pressure, oxygen levels via skin coloration, temperature and more throughout the day to provide opportunity for building personalized models of an individual's health or health needs and detect abnormalities that are predictive of disease. As research has demonstrated that simple physiological measurements, like heart rate, can be independent predictors of overall health outcome and general mortality [67], there are enormous opportunities to exploit user profiles in targeting information, services and interactions with individuals to further personalise existing services and increase their impact. These devices allow for accurate information capture while still be easily manageable for large cohorts, as done, for example, in the "Personalized Nutrition Project" where a continuous glucometer is used to automatically collect glucose levels over a week in response to specific diets.

Mobile and telehealth services such as rehabilitation and chronic disease care can also greatly benefit from the use of mobile sensors. Current research is trialling applications with specific patients, such as those with a chronic disease or those undergoing secondary rehabilitation [68, 69]. These approaches pave the way for the health service industry to provide whole-of-life care to patients. Increasingly, this approach will move to being preventative in nature [70]. For example, a recent randomised controlled trial showed that secondary cardiac rehabilitation services offered via a mobile phone rather than clinic showed a 80% increase in uptake and adherence [71].

SUMMARY

Current successes of utilizing genomic information in health management are mostly limited to rare genetic diseases. However, larger, better curated study cohorts could enable polygenic predictors of some complex genetic traits on a population-level as rapidly as within next five years [13]. With a better understanding of the genomic variability and a library of clinically annotated variants, extending risk prediction to include environmental factors will further improve early detection, diagnosis and intervention in complex diseases. The largest impact on the health system, however, will come from addressing common multi-causal diseases. Whole-of-life approaches based on risk-stratification form the basis for ongoing health monitoring and automated personalized health advice. To achieve this, three priority areas need to be addressed

(1) Genotype-Phenotype Annotation for Diagnostic Applications

Hospitals and international research groups need to share genomes annotated with clinical information, which will provide the basis for identifying causative variants for rare diseases and improve risk-predictors for common disorders. This information helps in developing a personal and precision medicine system that is primarily geared to health optimisation rather than crisis management.

(2) Improvements in Data Interoperability to Future-Proof Health Research

Collecting and providing data from genomic profiling and environmental risk factors needs to be done in a way which is machine readable, uses common language (ontology) and shows relationships between data items. This will ensure future use of this expensive resource in studies with cohort sizes that can cope with the added complexity from multivariate environmental risk factors.

(3) Clinical Trails for Longitudinal Health Monitoring of Individuals

New studies utilizing personal sensing devices are needed to build baseline models that enable the development of early detection warning systems, which will form the foundation of interventions to prevent disease and delivery of whole-of-life health care. Careful consideration will need to be given to ensuring the effective delivery personalised risk prediction and stratification of disease screening and prevention based on this risk.

KEY FACTS

1) Technical advances in high-throughput data collection, such as genome sequencing, as well as computational capabilities promise personalised, evidence-based healthcare.
2) Current knowledge and technical capabilities limit the successful application of genome sequencing to rare genetic diseases and some cancer treatments.
3) Common complex diseases such as diabetes are influenced by multiple genomic loci as well as environmental factors
4) Much larger studies are needed to better understand how genomic and environmental factors contribute towards common diseases.
5) Electronic Health Records can aid in curating cohorts for whole-of-life risk-predictors
6) Risk-stratified intervention is a realistic strategy for targeting common diseases, while a better understanding of genomic and environmental factors on disease is built.
7) Personal sensing devices will enable the development of early detection warning systems, which will govern the maintenance of health.

GLOSSARY

Genome wide association study (GWAS): A study designed to determine the association between a specific phenotype and the genotype of common genetic variants.

Environment wide association studies (EWAS): A study designed to determine the association between a specific phenotype and environmental factors.

Phenome Wide Association Studies (PheWAS): A study designed to determine the association between specific genomic regions and the observed phenotypes.

Exposome: Combination of physical, chemical, nutritional, life-style, and psychosocial exposure factors.

Genomic profile: Set of genetic variations observed in an individual

Risk stratification: Subdividing population into groups based on genetic profiles

Genetic diagnosis: Determining the genotype of a known disease marker

Genetic screening: Obtaining the genomic profile of an individual

Receiver operator curve (ROC): Visualization of sensitivity and specificity of a prediction method.

Prevalence: Proportion of a population found to have a specific disease condition

Personalised nutrition: Adapting food and nutritional intake depending on the individual's genomic profile, life stage, life style to prevent disease or delay its onset.

Hospital electronic health records (EHRs): Medical data and treatment history in a machine-readable format shared with all relevant healthcare providers of the patient.

Electronic Health Driven Genome Research (EDGR): Utilizing EHRs for the cost-effective recruitment of a precise and well-annotated patient cohort.

Genetic discrimination: Different treatment based on a person's actual or assumed genetic makeup

REFERENCES

[1] Bauer, DC; et al. Genomics and personalised whole-of-life healthcare. *Trends in molecular medicine*, doi:10.1016/j.molmed.2014.04.001 (2014).

[2] Bainbridge, MN; et al. Whole-genome sequencing for optimized patient management. *Sci Transl Med*, 3, 87re83, doi:10.1126/ scitranslmed.3002243 (2011).

[3] Talkowski, ME; et al. Clinical diagnosis by whole-genome sequencing of a prenatal sample. *N Engl J Med*, 367, 2226-2232, doi:10.1056/ NEJMoa1208594 (2012).

[4] Pellatt, AJ; Wolff, RK; Lundgreen, A; Cawthon, R; Slattery, ML. Genetic and lifestyle influence on telomere length and subsequent risk of colon cancer in a case control study. *Int J Mol Epidemiol Genet*, 3, 184-194 (2012).

[5] Andermann, A; Blancquaert, I; Beauchamp, S; Costea, I. Guiding policy decisions for genetic screening: developing a systematic and transparent approach. *Public Health Genomics*, 14, 9-16 (2011).

[6] Bloss, CS; Wineinger, NE; Darst, BF; Schork, NJ; Topol, EJ. Impact of direct-to-consumer genomic testing at long term follow-up. *Journal of medical genetics*, 50, 393-400, doi:10.1136/jmedgenet-2012-101207 (2013).

[7] Chute, CG; Kohane, IS. Genomic medicine, health information technology, and patient care. *JAMA*, 309, 1467-1468 (2013).

[8] Najafzadeh, M; et al. Barriers to integrating personalized medicine into clinical practice: a best-worst scaling choice experiment. *Genet Med*, 14, 520-526 (2012).

[9] Knoppers, BM; Hirtle, M; Glass, KC. Policy forum: genetic technologies. Commercialization of genetic research and public policy. *Science*, 286, 2277-2278 (1999).

[10] Andermann, A; Blancquaert, I; Déry, V. Genetic screening: a conceptual framework for programmes and policy-making. *J Health Serv Res Policy*, 15, 90-97 (2010).

[11] Fears, R; ter Meulen, V. The perspective from EASAC and FEAM on direct-to-consumer genetic testing for health-related purposes. *Eur J Hum Genet*, 21, 703-707, doi:10.1038/ejhg.2012.238 (2013).

[12] Green, RC; Farahany, NA. Regulation: The FDA is overcautious on consumer genomics. *Nature*, 505, 286-287, doi:doi:10.1038/505286a (2014).

[13] Chatterjee, N; et al. Projecting the performance of risk prediction based on polygenic analyses of genome-wide association studies. *Nat Genet*, 45, 400-405 (2013).

[14] Huber, M; et al. How should we define health? *BMJ*, 343, d4163 (2011).

[15] Shashi, V; et al. The utility of the traditional medical genetics diagnostic evaluation in the context of next-generation sequencing for undiagnosed genetic disorders. *Genet Med*, 16, 176-182, doi:10.1038/gim.2013.99 (2014).

[16] Campeau, PM; Foulkes, WD; Tischkowitz, MD. Hereditary breast cancer: new genetic developments, new therapeutic avenues. *Hum Genet*, 124, 31-42 (2008).

[17] So, HC; Kwan, JSH; Cherny, SS; Sham, PC. Risk prediction of complex diseases from family history and known susceptibility loci, with applications for cancer screening. *Am J Hum Genet*, 88, 548-565 (2011).

[18] Biankin, AV; Hudson, TJ. Somatic variation and cancer: therapies lost in the mix. *Hum Genet*, 130, 79-91 (2011).

[19] Visscher, PM; Brown, MA; McCarthy, MI; Yang, J. Five years of GWAS discovery. *Am J Hum Genet*, 90, 7-24 (2012).

[20] Oetting, WS; et al. Getting ready for the Human Phenome Project: the, 2012 forum of the Human Variome Project. *Hum Mutat*, 34, 661-666 (2013).

[21] The EGAPP initiative: lessons learned. *Genet Med*, 16, 217-224, doi:10.1038/gim.2013.110 (2014).

[22] Rehm, HL; et al. ACMG clinical laboratory standards for next-generation sequencing. *Genet Med*, 15, 733-747, doi:10.1038/gim.2013.92 (2013).

[23] Shahmirzadi, L; et al. Patient decisions for disclosure of secondary findings among the first 200 individuals undergoing clinical diagnostic exome sequencing. *Genet Med*, doi:10.1038/gim.2013.153 (2013).

[24] Abecasis, GR; et al. A map of human genome variation from population-scale sequencing. *Nature*, 467, 1061-1073, doi:10.1038/nature09534 (2010).

[25] Landrum, MJ; et al. ClinVar: public archive of relationships among sequence variation and human phenotype. *Nucleic acids research*, 42, D980-985, doi:10.1093/nar/gkt1113 (2014).

[26] Thomas, D. Gene--environment-wide association studies: emerging approaches. *Nat Rev Genet*, 11, 259-272, doi:10.1038/nrg2764 (2010).

[27] Eichler, EE; et al. Missing heritability and strategies for finding the underlying causes of complex disease. *Nat Rev Genet*, 11, 446-450 (2010).

[28] Manolio, TA; et al. Finding the missing heritability of complex diseases. *Nature*, 461, 747-753, doi:10.1038/nature08494 (2009).

[29] Yang, J; et al. Common SNPs explain a large proportion of the heritability for human height. *Nat Genet*, 42, 565-569, doi:10.1038/ng.608 (2010).

[30] Furrow, RE; Christiansen, FB; Feldman, MW. Environment-sensitive epigenetics and the heritability of complex diseases. *Genetics*, 189, 1377-1387 (2011).

[31] Patel, CJ; Bhattacharya, J; Butte, AJ. An Environment Wide Association Study (EWAS) on type 2 diabetes mellitus. *PLoS One*, 5, e10746, doi:10.1371/journal.pone.0010746 (2010).

[32] Ashley, EA; et al. Clinical assessment incorporating a personal genome. *Lancet*, 375, 1525-1535, doi:10.1016/S0140-6736(10)60452-7 (2010).

[33] Patel, CJ; et al. Whole genome sequencing in support of wellness and health maintenance. *Genome medicine*, 5, 58, doi:10.1186/gm462 (2013).

[34] Rappaport, SM; Smith, MT. Epidemiology. Environment and disease risks. *Science*, 330, 460-461, doi:10.1126/science.1192603 (2010).

[35] Wild, CP. The exposome: from concept to utility. *International journal of epidemiology*, 41, 24-32, doi:10.1093/ije/dyr236 (2012).

[36] Fenech, M. A public health genomics approach to improving well-being based on DNA damage prevention using nutritional, life-style, environmental and psycho-social strategies. *Journal of the Australasian College of Nutritional and Environmental Medicine*, 32, 9-16 (2013).

[37] Polychronakos, C; Li, Q. Understanding type 1 diabetes through genetics: advances and prospects. *Nat Rev Genet*, 12, 781-792 (2011).

[38] Tamborero, D; et al. Comprehensive identification of mutational cancer driver genes across, 12 tumor types. *Scientific reports*, 3, 2650, doi:10.1038/srep02650 (2013).

[39] McGinnis, JM; Foege, WH. Actual causes of death in the United States. *JAMA*, 270, 2207-2212 (1993).

[40] Mokdad, AH; Marks, JS; Stroup, DF; Gerberding, JL. Actual causes of death in the United States, 2000. *JAMA*, 291, 1238-1245 (2004).

[41] Danaei, G; et al. The preventable causes of death in the United States: comparative risk assessment of dietary, lifestyle, and metabolic risk factors. *PLoS Med*, 6, e1000058-e1000058 (2009).

[42] Kussmann, M; Van Bladeren, PJ. The Extended Nutrigenomics - Understanding the Interplay between the Genomes of Food, Gut Microbes, and Human Host. *Front Genet*, 2, 21-21 (2011).

[43] Hivert, MF; Vassy, JL; Meigs, JB. Susceptibility to type, 2 diabetes mellitus-from genes to prevention. *Nature reviews. Endocrinology*, 10, 198-205, doi:10.1038/nrendo. 2014.11 (2014).

[44] Minihane, AM. The genetic contribution to disease risk and variability in response to diet: where is the hidden heritability? *Proc Nutr Soc*, 72, 40-47 (2013).

[45] Fenech, M; et al. Nutrigenetics and nutrigenomics: viewpoints on the current status and applications in nutrition research and practice. *J Nutrigenet Nutrigenomics*, 4, 69-89 (2011).

[46] Ordovas, JM; Corella, D. Nutritional genomics. *Annu Rev Genomics Hum Genet*, 5, 71-118 (2004).

[47] Kaput, J; et al. The case for strategic international alliances to harness nutritional genomics for public and personal health. *Br J Nutr*, 94, 623-632 (2005).

[48] Wilson, CP; et al. Blood pressure in treated hypertensive individuals with the MTHFR, 677TT genotype is responsive to intervention with riboflavin: findings of a targeted randomized trial. *Hypertension*, 61, 1302-1308, doi:10.1161/HYPERTENSIONAHA. 111.01047 (2013).

[49] Fenech, MF. Nutriomes and nutrient arrays - the key to personalised nutrition for DNA damage prevention and cancer growth control. *Genome Integr*, 1, 11-11 (2010).

[50] van Ommen, B; et al. The Micronutrient Genomics Project: a community-driven knowledge base for micronutrient research. *Genes Nutr*, 5, 285-296 (2010).

[51] Gargis, AS; et al. Assuring the quality of next-generation sequencing in clinical laboratory practice. *Nat Biotechnol*, 30, 1033-1036 (2012).

[52] Buske, FA; French, HJ; Smith, MA; Clark, SJ; Bauer, DC. NGSANE: a lightweight production informatics framework for high-throughput data analysis. *Bioinformatics*, 30, 1471-1472, doi:10.1093/ bioinformatics/ btu036 (2014).

[53] Doring, A; Weese, D; Rausch, T; Reinert, K. SeqAn an efficient, generic C++ library for sequence analysis. *BMC bioinformatics*, 9, 11, doi:10.1186/1471-2105-9-11 (2008).

[54] Van Hoyweghen, I; Horstman, K. European practices of genetic information and insurance: lessons for the Genetic Information Nondiscrimination Act. *JAMA*, 300, 326-327, doi:10.1001/jama.2008.62 (2008).

[55] Joly, Y; Ngueng Feze, I; Simard, J. Genetic discrimination and life insurance: a systematic review of the evidence. *BMC medicine*, 11, 25, doi:10.1186/1741-7015-11-25 (2013).

[56] Malin, BA; Emam, KE; O'Keefe, CM. Biomedical data privacy: problems, perspectives, and recent advances. *J Am Med Inform Assoc*, 20, 2-6 (2013).

[57] O'Keefe, CM; Connolly, CJ. Privacy and the use of health data for research. *Med J Aust*, 193, 537-541 (2010).

[58] Kohane, IS. Using electronic health records to drive discovery in disease genomics. *Nat Rev Genet*, 12, 417-428 (2011).

[59] Denny, JC; et al. Variants near FOXE1 are associated with hypothyroidism and other thyroid conditions: using electronic medical records for genome- and phenome-wide studies. *Am J Hum Genet*, 89, 529-542 (2011).

[60] Liao, KP; et al. Associations of autoantibodies, autoimmune risk alleles, and clinical diagnoses from the electronic medical records in rheumatoid arthritis cases and non-rheumatoid arthritis controls. *Arthritis Rheum*, 65, 571-581 (2013).

[61] Dezentjé, VO; et al. Effect of concomitant CYP2D6 inhibitor use and tamoxifen adherence on breast cancer recurrence in early-stage breast cancer. *J Clin Oncol*, 28, 2423-2429 (2010).

[62] Frost, J; Okun, S; Vaughan, T; Heywood, J; Wicks, P. Patient-reported outcomes as a source of evidence in off-label prescribing: analysis of data from PatientsLikeMe. *J Med Internet Res*, 13, e6-e6 (2011).

[63] Lohse, B. Facebook is an effective strategy to recruit low-income women to online nutrition education. *J Nutr Educ Behav*, 45, 69-76 (2013).

[64] Lyon, A; Nunn, M; Grossel, G; Burgman, M. Comparison of web-based biosecurity intelligence systems: BioCaster, EpiSPIDER and HealthMap. *Transbound Emerg Dis*, 59, 223-232 (2012).

[65] Chew, C; Eysenbach, G. Pandemics in the age of Twitter: content analysis of Tweets during the 2009 H1N1 outbreak. *PLoS One*, 5, e14118-e14118 (2010).

[66] Mosa, ASM; Yoo, I; Sheets, L. A systematic review of healthcare applications for smartphones. *BMC Med Inform Decis Mak*, 12, 67-67 (2012).

[67] Seccareccia, F; et al. Heart rate as a predictor of mortality: the MATISS project. *Am J Public Health*, 91, 1258-1263 (2001).

[68] Smith, JC; Schatz, BR. Feasibility of mobile phone-based management of chronic illness. *AMIA Annu Symp Proc*, 2010, 757-761 (2010).

[69] Ryan, D; et al. Clinical and cost effectiveness of mobile phone supported self monitoring of asthma: multicentre randomised controlled trial. *BMJ*, 344, e1756-e1756 (2012).

[70] Miron-Shatz, T; Ratzan, SC. The potential of an online and mobile health scorecard for preventing chronic disease. *J Health Commun*, 16 Suppl, 2, 175-190 (2011).

[71] Varnfield, M; et al. Uptake of a technology-assisted home-care cardiac rehabilitation program. *Med J Aust*, 194, S15-S19 (2011).

[72] Johnson, AD; Leslie, R; O'Donnell, CJ. Temporal trends in results availability from genome-wide association studies. *PLoS Genet*, 7, e1002269, doi:10.1371/journal.pgen.1002269 (2011).

[73] Gymrek, M; McGuire, AL; Golan, D; Halperin, E; Erlich, Y. Identifying personal genomes by surname inference. *Science*, 339, 321-324 (2013).

[74] Swan, M. Crowdsourced health research studies: an important emerging complement to clinical trials in the public health research ecosystem. *J Med Internet Res*, 14, e46-e46 (2012).

EDITOR CONTACT INFORMATION

Dr. Lu Qi, MD, PhD, FAHA, FACN, FTOS
HCA Regents Distinguished Chair and Professor
Director, Tulane University Obesity Research Center
Department of Epidemiology
Tulane University, New Orleans, LA, US
School of Public Health and Tropical Medicine
Phone: 504.988.7259
Fax: 504.988.3548; 504.988.1568
E-mail: Lqi1@tulane.edu

and

Adjunct Professor
Department of Nutrition
Harvard School of Public Health, Boston, MA, US
E-mail: luqi@hsph.harvard.edu

INDEX

A

B

C

D

E

F

G

H

I

L

M

N

O

P

R

S

T

V

W

α